AF646943

TRANSGENIC ANIMALS IN AGRICULTURE

Transgenic Animals in Agriculture

Edited by

J.D. Murray, G.B. Anderson, A.M. Oberbauer

Department of Animal Science
University of California
Davis, California, USA

and

M.M. McGloughlin

Biotechnology Program
University of California
Davis, California, USA

CABI *Publishing*

New India Publishing Agency

New Delhi - 110088, (INDIA)

First Indian Print: 2005

New India Publishing Agency
101, Vikas Surya Plaza, CU Block, L.S.C. Mkt.,
Pitam Pura, New Delhi - 110 088, (INDIA)
Phone: 011-27341717, Fax: 011-27341616
E-mail: spjain_niph@rediffmail.com
Web: www.bookfactoryindia.com

ISBN: 81-89422-07-3

© CAB INTERNATIONAL 1999. All rights reserved. No part of this publication may be reproduced in any form or by any means, electronically, mechanically, by photocopying, recording or otherwise, without the prior permission of the copyright owners.For Sale Only In India, Pakistan, Bangladesh, Nepal and Sri Lanka

Printed at :
Salasar Imaging Systems
Lawrence Road, New Delhi

Contents

List of Contributors **vii**

Preface **xi**

1. Transgenic Farm Animals 1
C.A. Pinkert and J.D. Murray

2. Development of Genetic Tools for Transgenic Animals 19
P.B. Hackett, Z. Izsvak, Z. Ivics and L. Caldovic

3. One Gene is Not Enough: Transgene Detection, Expression and Control 37
K.D. Wells and R.J. Wall

4. Embryonic Stem Cells in Agricultural Species 57
G.B. Anderson

5. Nuclear Transfer in the Production of Transgenic Farm Animals 67
I. Wilmut, E. Schnieke, J. McWhir, A.J. Kind, A. Colman and K.H.S. Campbell

6. Embryonic Stem Cell Chimeras and Somatic Cell Nuclear Transplantation for Production of Transgenic Cattle 79
J.M. Robl, J.B. Cibelli, P.G. Golueke, J.J. Kane, C. Blackwell, J. Jerry, É.S. Dickenson, F.A. Ponce de Leon and S.L. Stice

7. Status of Sperm-mediated Delivery Methods for Gene Transfer 87
E.J. Squires

8. Understanding the Origin of Avian Primordial Germ Cells: Implications for Germ Cell Culture and Transgenesis in Poultry 97
J.N. Petitte, S. D'Costa and L. Karagenç

9. Generation of Transgenic Poultry by Transfection of Primordial Germ Cells 117
E.A. Wong, A.L. Wentworth, B.C. Wentworth, J.A. Proudman and M.E. El Halawani

10. Expression of Insulin-like Growth Factor-I in Skeletal Muscle of Transgenic Swine 131
V.G. Pursel, R.J. Wall, A.D. Mitchell, T.H. Elsasser, M.B. Solomon, M.E. Coleman, F. DeMayo and R.J. Schwartz

11. Production and Analysis of Transgenic Pigs Containing a Metallothionein Porcine Growth Hormone Gene Construct 145
M.B. Nottle, H. Nagashima, P.J. Verma, Z.T. Du, C.G. Grupen, S.M. McIlfatrick, R.J. Ashman, M.P. Harding, C. Giannakis, P.L. Wigley, I.G. Lyons, D.T. Harrison, B.G. Luxford, R.G. Campbell, R.J. Crawford and A.J. Robins

12. The Utilization of Bacterial Genes to Modify Domestic Animal Biochemistry 157
K.A. Ward, Z. Leish, A.G. Brownlee, J. Bonsing, C.D. Nancarrow and B.W. Brown

13. Production of Transgenic Cattle Expressing a Recombinant Protein in Milk 177
W.H. Eyestone

14. Changing the Composition and Properties of Milk 193
J.D. Murray and E.A. Maga

15. Comparison of Traditional Breeding and Transgenesis in Farmed Fish with Implications for Growth Enhancement and Fitness 209
R.A. Dunham and R.H. Devlin

16. Direct and Correlated Responses to Short-term Selection of 8-week Body Weight in Lines of Transgenic (oMt1a-oGH) Mice 231
F. Siewerdt, E.J. Eisen ana J.D. Murray

17. Ethics, Animal Welfare and Transgenic Farm Animals 251
J.A. Mench

18. The Future of Transgenic Farm Animals 269
G.E. Seidel, Jr

Index **283**

List of Contributors

G.B. Anderson, Department of Animal Science, University of California, Davis, CA 95616, USA.

R.J. Ashman, BresaGen Ltd, PO Box 259, Rundle Mall, Adelaide, SA 5000, Australia.

C. Blackwell, Advanced Cell Technology Inc., Paige Laboratory, University of Massachusetts, Amherst, MA 01003, USA.

J. Bonsing, CSIRO Division of Animal Production, Clunies Ross Street, Prospect, NSW, Australia.

B.W. Brown, CSIRO Division of Animal Production, Clunies Ross Street, Prospect, NSW, Australia.

A.G. Brownlee, CSIRO Division of Animal Production, Clunies Ross Street, Prospect, NSW, Australia.

L. Caldovic, Department of Genetics and Cell Biology, University of Minnesota, 250 Biological Sciences Center, 1445 Gortner Avenue, St Paul, MN 55108, USA.

K.H.S. Campbell, PPL Therapeutics, Roslin, Midlothian EH25 9PP, UK.

R.G. Campbell, Bunge Meat Industries, PO Box 78, Corowa, NSW 2646, Australia.

J.B. Cibelli, Department of Veterinary and Animal Sciences and Advanced Cell Technology Inc., Paige Laboratory, University of Massachusetts, Amherst, MA 01003, USA.

M.E. Coleman, GeneMedicine Inc., The Woodlands, TX 77381, USA.

A. Colman, PPL Therapeutics, Roslin, Midlothian EH25 9PP, UK.

R.J. Crawford, BresaGen Ltd, PO Box 259, Rundle Mall, Adelaide, SA 5000, Australia.

S. D'Costa, Department of Poultry Science, North Carolina State University, Raleigh, NC 27695–7680, USA.

F. DeMayo, Baylor College of Medicine, Houston, TX 77030, USA.

R.H. Devlin, Fisheries and Oceans Canada, 4160 Marine Drive, West Vancouver, BC V7V 1N6, Canada.

E.S. Dickenson, Advanced Cell Technology Inc., Paige Laboratory, University of Massachusetts, Amherst, MA 01003, USA.

Z.T. Du, BresaGen Ltd, PO Box 259, Rundle Mall, Adelaide, SA 5000, Australia.

R.A. Dunham, Department of Fisheries and Allied Aquacultures, Alabama Agricultural Experiment Station, Auburn University, AL 36849, USA.

E.J. Eisen, Department of Animal Science, North Carolina State University, Raleigh, NC 27695, USA.

M.E. El Halawani, University of Minnesota, St Paul, MN 55108, USA.

T.H. Elsasser, US Department of Agriculture, ARS, Beltsville, MD 20705, USA.

W.H. Eyestone, PPL Therapeutics Inc., 1700 Kraft Drive, Blacksburg, VA 24060, USA.

C. Giannakis, BresaGen Ltd, PO Box 259, Rundle Mall, Adelaide, SA 5000, Australia.

P.G. Golueke, Advanced Cell Technology Inc., Paige Laboratory, University of Massachusetts, Amherst, MA 01003, USA.

C.G. Grupen, BresaGen Ltd, PO Box 259, Rundle Mall, Adelaide, SA 5000, Australia.

P.B. Hackett, Department of Genetics and Cell Biology, University of Minnesota, 250 Biological Sciences Center, 1445 Gortner Avenue, St Paul, MN 55108, USA.

M.P. Harding, BresaGen Ltd, PO Box 259, Rundle Mall, Adelaide, SA 5000, Australia.

D.T. Harrison, Bunge Meat Industries, PO Box 78, Corowa, NSW 2646, Australia.

Z. Ivics, Department of Genetics and Cell Biology, University of Minnesota, 250 Biological Sciences Center, 1445 Gortner Avenue, St Paul, MN 55108, USA.

Z. Izsvak, Department of Genetics and Cell Biology, University of Minnesota, 250 Biological Sciences Center, 1445 Gortner Avenue, St Paul, MN 55108, USA.

J. Jerry, Department of Veterinary and Animal Sciences, Paige Laboratory, University of Massachusetts, Amherst, MA 01003, USA.

J.J. Kane, Advanced Cell Technology Inc., Paige Laboratory, University of Massachusetts, Amherst, MA 01003, USA.

L. Karagenç, Department of Poultry Science, North Carolina State University, Raleigh, NC 27695–7680, USA.

A.J. Kind, PPL Therapeutics, Roslin, Midlothian EH25 9PP, UK.

Z. Leish, CSIRO Division of Animal Production, Clunies Ross Street, Prospect, NSW, Australia.

B.G. Luxford, Bunge Meat Industries, PO Box 78, Corowa, NSW 2646, Australia.

I.G. Lyons, BresaGen Ltd, PO Box 259, Rundle Mall, Adelaide, SA 5000, Australia.

E.A. Maga, Department of Animal Science, University of California, Davis, CA 95616–8421, USA.

S.M. McIlfatrick, BresaGen Ltd, PO Box 259, Rundle Mall, Adelaide, SA 5000, Australia.

J. McWhir, Roslin Institute, Roslin, Midlothian EH25 9PP, UK.

J.A. Mench, Department of Animal Science and Center for Animal Welfare, University of California, Davis, CA 95616, USA.

A.D. Mitchell, US Department of Agriculture, ARS, Beltsville, MD 20705, USA.

J.D. Murray, Department of Animal Science and Department of Population Health and Reproduction, School of Veterinary Medicine, University of California, Davis, CA 95616–8421, USA.

H. Nagashima, BresaGen Ltd, PO Box 259, Rundle Mall, Adelaide, SA 5000, Australia.

C.D. Nancarrow, CSIRO Division of Animal Production, Clunies Ross Street, Prospect, NSW, Australia.

M.B. Nottle, BresaGen Ltd, PO Box 259, Rundle Mall, Adelaide, SA 5000, Australia.

J.N. Petitte, Department of Poultry Science, North Carolina State University, Raleigh, NC 27695–7680, USA.

C.A. Pinkert, Department of Comparative Medicine and the UAB Transgenic Animal/ES Cell Resource, The University of Alabama at Birmingham, Birmingham, AL 35294–0019, USA.

F.A. Ponce de Leon, Department of Veterinary and Animal Sciences, Paige Laboratory, University of Massachusetts, Amherst, MA 01003, USA.

J.A. Proudman, US Department of Agriculture, ARS, Beltsville, MD 20705, USA.

V.G. Pursel, US Department of Agriculture, ARS, Beltsville, MD 20705, USA.

A.J. Robins, BresaGen Ltd, PO Box 259, Rundle Mall, Adelaide, SA 5000, Australia.

J.M. Robl, Department of Veterinary and Animal Sciences, Paige Laboratory, University of Massachusetts, Amherst, MA 01003, USA.

E. Schnieke, PPL Therapeutics, Roslin, Midlothian EH25 9PP, UK.

R.J. Schwartz, Baylor College of Medicine, Houston, TX 77030, USA.

G.E. Seidel Jr, Animal Reproduction and Biotechnology Laboratory, Colorado State University, Fort Collins, CO 80523, USA.

F. Siewerdt, Department of Animal Science, North Carolina State University, Raleigh, NC 27695, USA.

M.B. Solomon, US Department of Agriculture, ARS, Beltsville, MD 20705, USA.

E.J. Squires, Department of Animal and Poultry Sciences, University of Guelph, Guelph, Ontario, Canada, N1G 2W1.

S.L. Stice, Advanced Cell Technology Inc., Paige Laboratory, University of Massachusetts, Amherst, MA 01003, USA.

P.J. Verma, BresaGen Ltd, PO Box 259, Rundle Mall, Adelaide, SA 5000, Australia.

R.J. Wall, Gene Evaluation and Mapping Laboratory, Agricultural Research Service, US Department of Agriculture, Beltsville, MD 20705, USA.

K.A. Ward, CSIRO Division of Animal Production, Clunies Ross Street, Prospect, NSW, Australia.

K.D. Wells, Gene Evaluation and Mapping Laboratory, Agricultural Research Service, US Department of Agriculture, Beltsville, MD 20705, USA.

A.L. Wentworth, University of Wisconsin, Madison, WI 53706, USA.

B.C. Wentworth, University of Wisconsin, Madison, WI 53706, USA.

P.L. Wigley, BresaGen Ltd, PO Box 259, Rundle Mall, Adelaide, SA 5000, Australia.

I. Wilmut, Roslin Institute, Roslin, Midlothian EH25 9PP, UK.

E.A. Wong, Virginia Polytechnic Institute and State University, Blacksburg, VA 24061, USA.

Preface

This volume represents the majority of the papers on the application of transgenic animals for use in production agriculture that were presented at a meeting held during August 1997 at the Granlibakken Conference Center in Tahoe City, California. In many cases, the papers have been updated to reflect research published up to the time of writing (June 1998). The impetus for this meeting came from the realization that it had been several years since a meeting had been held that was both this focused, i.e. limited to transgenic applications for animal agriculture, yet inclusive of work on fish, birds and mammals. As many of the problems faced when doing research involved with the transgenic manipulations of vertebrates are universal, such as the attempts to isolate embryonic stem cells or constructing an efficient expression vector, much benefit was gained by the interactions of scientists working with different species who were able to attend this meeting. It is our hope that the sphere of individuals benefiting from this meeting will be greatly expanded by the publication of this volume.

Our colleague Dr Robert Wall (USDA-ARS Beltsville, Maryland) often states that the field of transgenic large animals is one of the few fields in modern science where there are more review papers than data papers. This volume may seem a bit uneven as well, as it is a mixture of review papers and primary data papers. However, in many cases the leading-edge research reported at this conference, such as nuclear transfer-based cloning (Chapters 5 and 6) or the work on the integration of a transgene into a selection experiment (Chapter 16), represents such recent advances that these papers are some of the first to be written in the area. Other papers, however, are more comprehensive reviews of the information pertaining either to specific technical areas, e.g. sperm-mediated gene transfer (Chapter 7), or the targeted application of transgenes to a specific species, such as the use of growth hormone constructs in fish (Chapter 15) or the attempts to transfer two genes in a biochemical pathway to alter the intermediary metabolism of mammals (Chapter 12). In each chapter, we hope to convey to the reader a better understanding of the possibilities and limitations of the current state of our art and the excitement of the participants at this meeting as we try to apply transgenic technology to help improve the animals used in agriculture.

Finally, as editors and organizers, we need to thank those people other than the speakers and authors who gave freely of their time to assist us in bringing together first the conference and now this volume. A brief list includes Robert Devlin, James Petitte, Carl Pinkert, Caird Rexroad Jr and George Seidel, who helped us to plan the scientific programme and to identify speakers. We also extend our thanks to a large number of anonymous referees who provided the peer review for each manuscript and in so doing provided valuable comments to both the authors and to us. The conference, and thus this volume, would not have been possible without financial support from the University of California, Davis, grants from the University of California Systemwide Biotechnology Research and Education Program and the USDA-CSREES, and contributions from the following corporate sponsors: Bayer Pharmaceuticals, Wyeth-Ayerst Laboratories, Perkin Elmer ABI, Biogenics, Metamorphix, PPL Therapeutics, Pharming B.V., Genzyme Transgenic Corporation and BioTime Corporation. This book would never have been finished without Randy Cook's unflagging assistance in helping to pull it all together. Finally, we would like to acknowledge Mr Tim Hardwick from CABI *Publishing*, both for his encouragement to undertake the preparation of this volume and his patience as time went by.

James D. Murray, Gary B. Anderson,
Martina M. McGloughlin and Anita M. Oberbauer
Davis, California
June 1998

Transgenic Farm Animals

1

Carl A. Pinkert[1] and James D. Murray[2]

[1] *Department of Comparative Medicine and the UAB Transgenic Animal/ES Cell Resource, The University of Alabama at Birmingham, Birmingham, Alabama, USA;*
[2] *Department of Animal Science and Department of Population Health and Reproduction, School of Veterinary Medicine, University of California, Davis, California, USA*

The last quarter-of-a-century has witnessed a rapid advance in the application of genetic engineering techniques to increasingly complex organisms, from bacteria and yeasts to mammalian species. In 1985, the first report on the production of genetically engineered farm animals described transgenic rabbits, sheep and pigs. Since that time, in addition to models for a number of mammalian species, transgenic fish and bird models have also been developed. Transgenic animals have provided us with a means of analysing developmental and regulatory mechanisms *in vivo*. Areas of research focusing on production characteristics have targeted growth and development, disease resistance, reproduction, lactational performance, feed efficiency, immune responsiveness and fibre production. Additionally, novel biomedical applications have forged ahead, using transgenic farm animals as research models and as bioreactors to produce biologically important proteins, tissues and organs for a host of specific applications. Today, transgenic animals embody one of the most potent and exciting research tools in the agricultural and biological sciences. These genetically engineered animals can be custom tailored to address specific scientific questions that were previously beyond our reach. Transgenic animal technology is of particular relevance in the rapid genetic modification of farm animal species, especially when one considers that selective breeding, which can be used to direct the modification of a specific phenotype, cannot be used to engineer a specific genetic trait in a directed fashion. As we enter the 21st century, novel methods to enhance the efficiency of transgenic animal production and to increase the utility of transgenic animal models in agriculture and society continue to evolve.

Introduction

The scientific breakthroughs that have enabled the current successes in the genetic engineering of animals occurred over the past century beginning

© CAB INTERNATIONAL 1999. *Transgenic Animals in Agriculture* (eds J.D. Murray, G.B. Anderson, A.M. Oberbauer and M.M. McGloughlin)

with the first attempts to culture and transfer embryos in the late 1800s (Table 1.1). While recent progress seems extremely rapid, it is still difficult to believe that, following the first published report of a microinjection method (Lin, 1966), 15 years passed before the first transgenic mice were created by Gordon *et al.* (1980). The first technological shift toward transgenic mouse production occurred in 1977, when Gurdon transferred mRNA and DNA into *Xenopus* embryos and observed that the transferred nucleic acids could function in an appropriate manner. Then, in 1980, Brinster and his colleagues reported on similar studies in the mouse. They demonstrated that an appropriate translational product was produced following transfer of a specific mRNA into mouse embryos. Sequentially, these studies laid the groundwork for the development of the first 'gain-of-function' transgenic mouse models.

From late 1980 through 1981, six research groups reported success in gene transfer and the development of transgenic mice. To describe animals carrying new genes (integrating foreign DNA into their genome), Gordon and Ruddle (1981) coined the term 'transgenic'. This definition has since been extended to include animals that result from the molecular manipulation of endogenous genomic DNA, including all techniques from DNA microinjection to embryonic stem (ES) cell transfer and 'knockout' mouse production.

Table 1.1. Transgenic animal milestones.

0000	genetic selection to improve animal productivity
1880	mammalian embryo cultivation attempted
1891	first successful embryo transfer
Early 1900s	*in vitro* embryo culture develops
1961	mouse embryo aggregation to produce chimeras
1966	zygote microinjection technology established
1973	foreign genes function after cell transfection
1974	development of teratocarcinoma cell transfer
1977	mRNA and DNA transferred into *Xenopus* eggs
1980	mRNA transferred into mammalian embryos
1980–1981	transgenic mice first documented
1981	transfer of ES cells derived from mouse embryos
1982	transgenic mice demonstrate an enhanced growth (GH) phenotype
1983	tissue-specific gene expression in transgenic mice
1985	transgenic domestic animals produced
1987	chimeric 'knock-out' mice described
1989	targeted DNA integration and germline chimeric mice
1993	germline chimeric mice produced using co-culture
1994	spermatogonia cell transplantation
1997	nuclear transfer using ES and adult cell nuclei in sheep
1998	nuclear transfer using ES cells to derive transgenic sheep
2000	????

Since the early 1980s, the production of transgenic mice by microinjection of DNA into the pronucleus of zygotes has been the most productive and widely used technique. Using transgenic technology in the mouse, such as antisense RNA encoding transgenes, it is now possible to add a new gene to the genome, increase the level of expression or change the tissue specificity of expression of a gene, and decrease the level of synthesis of a specific protein (see Sokol and Murray, 1996). Removal or alteration of an existing gene via homologous recombination required the use of ES cells and was limited to the mouse until the advent of nuclear transfer cloning procedures (Wilmut *et al.*, 1997; see also Chapter 5, this volume).

This review notwithstanding, there are now literally hundreds of excellent reviews that detail the production and utility of transgenic animals. (A number of reviews and texts are cited in the references in addition to a journal, *Transgenic Research*, which is dedicated to this field.) Yet, the most influential experimentation to impact on transgenic farm animal research was the work of Palmiter and Brinster in the early 1980s. Their studies related to growth, performance and the dramatic phenotype of mice transgenic for growth hormone (GH), influenced animal agriculture in dramatic fashion. In these pioneering studies 'Super Mice', which grew 100% larger than normal or littermate mice, were produced by redirecting GH production to the mouse's liver, using a liver-specific metallothionein promoter fused to a GH structural gene (e.g. Palmiter *et al.*, 1982).

During the past 15 years, transgenic technology has been extended to a variety of animal species beyond the mouse, including rats, rabbits, swine, ruminants (sheep, goats and cattle), poultry and fish (Table 1.2). With advances in the understanding of promoter-enhancer elements and transcription-regulatory proteins involved in the control of gene expression, the technology continues to evolve using different model systems (Box 1.1). In the systems explored to date, gene transfer technology is a proven asset in science as a means of dissecting gene regulation and expression *in vivo*.

Table 1.2. Genetically engineered vertebrate species.

Mammals	Birds	Fish
Mice	Chickens	Salmon
Rats	Japanese quail	Trout
Rabbits		Tilapia
Cattle		Carp
Pigs		Catfish
Sheep		Medaka
Goats		Zebrafish
		Loach
		Goldfish
		Pike

Box 1.1. Application and use of transgenic animal models.

Transgenic animals have provided models in agricultural, biomedical, biotechnological and veterinary disciplines in the study of gene expression and developmental biology, as well as for modelling:

- Increased efficiency of animal production.
- Genetic bases of animal and human diseases (leading to the design and testing of strategies for therapy).
- Gene therapy.
- Disease resistance in animals and in humans.
- Drug and product efficacy testing/screening.
- Novel or improved product development, 'molecular farming', ultimately targeting products or productivity of domestic animals. Models range from enhancing production traits of interest to 'foreign' protein production and human organ replacement (xenotransplantation).

As such, the primary questions that are addressed concern the roles of individual genes in development or in particular developmental pathways. With this caveat, considerations include the ramifications of gene activity, from intracellular to inter- and extracellular events within a given tissue or cell-type milieu.

Gene transfer has been used to produce both random and targeted insertion of discrete DNA fragments into the mouse genome. For targeted insertions, where the integration of foreign genes is based on a recombinational gene insertion with a specific homology to cellular sequences (termed homologous recombination), the efficiency at which DNA microinjection is effective is extremely low (Brinster *et al.*, 1989). In contrast, the use of ES cell transfer into mouse embryos has been quite effective in allowing an investigator to preselect a specific genetic modification, via homologous recombination, at a precise chromosomal position. This preselection has led to the production of mice: (i) incorporating a novel foreign gene in their genome, (ii) carrying a modified endogenous gene, or (iii) lacking a specific endogenous gene following gene deletion or 'knock-out' procedures (see Capecchi, 1989; Brinster, 1993).

Isolation and propagation strategies for ES cells in domestic species have proven elusive, with much of the effort now being directed towards the isolation of primordial germ (PG) cells. Techniques such as nuclear transfer might use donor nuclei from various sources (e.g. ES cell embryonic cell lines, PG cells or spermatogonia) to produce offspring. The utility of ES cells or related methodologies to provide efficient and targeted *in vivo* genetic manipulations offer the prospects of profoundly useful animal models for biomedical, biological and agricultural applications. The road to such success has been most challenging, but recent developments in this field are extremely encouraging.

Production of Transgenic Domestic Animals

The success of transgenic mouse experiments led a number of research groups to study the transfer of similar gene constructs into the germline of domestic animal species. With one exception, these efforts have been directed primarily toward either of two general goals: (i) improving the productivity traits of domestic food animal species, or (ii) developing transgenic lines for use as 'bioreactors'; i.e. as producers of recoverable quantities of medically or biologically important proteins. These studies revealed basic biological mechanisms as well as a need for precise regulation of gene expression. Since 1985, transgenic farm animals harbouring growth-related gene constructs have been created, although ideal growth phenotypes were not achieved because of an inability to coordinately regulate either gene expression or the ensuing cascade of endocrine events (see Pursel *et al.*, 1989; Pursel and Rexroad, 1993; Pinkert *et al.*, 1997).

Presently, DNA microinjection and now nuclear transfer (Schnieke *et al.*, 1998) are the only methods used to produce transgenic livestock successfully. Although involved and at times quite tedious, the steps in the development of transgenic models are relatively straightforward. For either DNA microinjection or nuclear transfer, once a specific fusion gene has been cloned and characterized, sufficient quantities are isolated, purified and tested in cell culture if possible. Once the appropriate gene construct has been identified, the fragment is linearized, purified and readied for preliminary mammalian gene transfer experiments. In contrast with nuclear transfer studies, DNA microinjection experiments are first performed in the mouse. While the transgenic mouse model will not always identify likely phenotypic expression patterns in domestic animals, we have not observed a single construct that would function in a pig when there was no evidence of transgene expression in mice. Therefore, preliminary experimentation in mice has been a crucial component of any gene transfer experiment in domestic animals.

With the exception of recently reported nuclear transfer experiments in sheep and cattle, there has been little change in the methods used to produce transgenic mammals, birds and fish over the last few years. For the sake of brevity, further discussion in this paper will be centred around the production of transgenic livestock in order to illustrate some points concerned with the production, utilization and limitations of transgenic animals in general. In practice, except for the nuclear transfer reports by Wilmut in sheep (Schnieke *et al.*, 1998) and Robl in cattle (Cibelli *et al.*, 1998), all other transgenic farm animals to date have been produced by pronuclear microinjection and in all cases the efficiency of producing transgenic animals is low (Table 1.3; also see Wall *et al.*, 1992). While nuclear transfer might be considered inefficient in its current form, we anticipate major strides in enhancing experimental protocols within the next few years, comparable perhaps with the early advances in DNA

Table 1.3. Efficiency of producing transgenic farm animals (percentage of transferred microinjected zygotes).

Species	Born	Transgenic
Pig	9.9	0.91
Sheep	10.6	0.88
Goat	14.3	0.99
Cattle*	16.2	0.79
Mice	15.0	≤3.5

* Based on transfer of morulae/blastocysts.
Modified from Pursel and Rexroad (1993).

microinjection technology. The added possibility of gene targeting through nuclear transplantation opens up a host of applications, particularly with regard to the use of transgenic animals to produce human pharmaceuticals (see Pinkert, 1997).

The current state of the art for the production of transgenic farm animals is still relatively unchanged from what it was 13 years ago; however, there are a host of procedures in development that may very well change 'state-of-the-art' technology very shortly. The only major technological advance since the initial production of transgenic farm animals has been the development of methods for the *in vitro* maturation of oocytes (IVM), *in vitro* fertilization (IVF) and subsequent culture of injected embryos prior to transfer to recipient females at some point up to, and including, the early blastocyst stage (Gordon and Lu, 1990). IVM and IVF have made the production of transgenic cattle economically feasible, even though the overall efficiency is low. Considerable effort has been expended towards establishing ES cells for cattle, sheep, chickens and pigs, but to date without success. While the techniques currently used to produce transgenic animals are inefficient, a variety of species can be, and are, routinely genetically engineered. This suggests that 'new' types of transgenic farm animals will continue to be produced for some time.

The major limiting factor in the production of transgenic mammals is the rate at which the microinjected DNA is integrated into the recipient genome (Wall *et al.*, 1992). However, to date, there has been virtually no research done to ascertain the mechanism(s) responsible for integration. Once the mechanism of integration is known, it may be possible to develop techniques to enhance the rate of transgene incorporation and thus gain significant efficiencies in the overall rate at which transgenic mammals can be produced.

Using DNA microinjection, the types of genes and regulatory sequences introduced into livestock species become important considerations. Pursel and Rexroad (1993) provided a comprehensive list of gene constructs used in the production of transgenic cattle, goats, pigs and sheep that has not changed significantly over the last 4 years. Table 1.4 summarizes their data

Table 1.4. Number of genes transferred into livestock.

Species	Growth factors	Milk genes	Total
Pig	15	2	23
Sheep	5	2	11
Goat	—	2	2
Cattle	4	1	6

with respect to the total number of genes transferred into each species and the two principal functional types of coding sequences. As can be seen, the types of transgenes used fall into two main types: those encoding growth factors and those encoding proteins for expression in the mammary gland.

The work with growth factors was carried out in an attempt to alter the efficiency of meat production and alter the partitioning of nutrient resources towards increased lean production; i.e. these projects were intended to alter animals for use in production agriculture. To date, these attempts have failed to result in the production of genetically superior livestock (sheep and pigs) due to a variety of undesirable side effects in these animals, although in general the transgenic animals have been more feed efficient and leaner (Pursel *et al.*, 1989; Nancarrow *et al.*, 1991). In addition to the work with livestock transgenic for growth factor, considerable effort has been directed towards increasing the efficiency of wool growth in Australian sheep by insertion of the two bacterial or yeast genes required for sheep to synthesize *de novo* the sulphur amino acid cysteine (see Rogers, 1990; Ward and Nancarrow, 1991; Chapter 12, this volume).

Work on the directed expression of new proteins with pharmaceutical value to the mammary gland of cattle, goats, pigs and sheep has been more successful. A number of pharmaceutically important proteins have been expressed in the mammary gland, with human α_1-antitrypsin being expressed in sheep milk (Archibald *et al.*, 1990; for review see Maga and Murray, 1995; Pinkert, 1997) at levels high enough for consideration for commercial extraction. While pharmaceutical-producing farm animals will continue to be developed, they will not have a direct effect on agriculture and, as there is high value in the protein being produced, it would not even be necessary for these animals to ever enter the human food chain. Thus, the value of this work to agriculture is in the knowledge gained concerning the control of mammary gland gene expression and the potential development of new techniques to increase the efficiency of producing transgenic farm animals.

Yet, the major scientific limitations to the wide-scale application of transgenic technology to improve farm animals basically have not changed since 1986 (Ward *et al.*, 1986). Those limitations include:

1. Lack of knowledge concerning the genetic basis of factors limiting production traits.

2. Identification of tissue- and developmentally specific regulatory sequences for use in developing gene constructs, expression vectors and in gene targeting.
3. Establishment of novel methods to increase the efficiency of transgenic animal production.

The production of transgenic farm animals is not undertaken lightly due to the high costs associated with obtaining and maintaining these animals. Thus it is prudent to confirm transgene expression in mice before it becomes cost effective to initiate DNA microinjection experiments in other species. In mouse experiments, less than 2 months is required from the time the purified construct is ready for microinjection until the weaning of founder pups. In contrast, for pig experiments, 1 month to a year is required for a sufficient number of DNA injections and recipient transfers to ensure the likelihood of success. Experimental efficiencies coupled with a long generational interval (i.e. 114 day gestation period, 21–28 day lactation and onset of puberty at 6–9 months of age) reflect the efforts necessary to identify and characterize transgenic pigs and illustrate the extended time-lines associated with the production of any transgenic livestock model. In addition, the time-frame from birth of a founder transgenic animal to the establishment of lines can be 1–2 years for pigs, sheep and goats to 4–5 years for cattle (while also dependent on the sex of founders). Hence, there is an obvious advantage to characterizing transgenic mouse models to expedite what will ultimately be a lengthy undertaking.

More recently, protocols were developed to permit removal of individual blastomeres from microinjected pre-implantation embryos maintained in culture prior to transfer to recipient females followed by PCR analysis of DNA purified from individual blastomeres to identify those embryos that bear the transgene of interest. Use of such methods has the potential to greatly increase the efficiency associated with production of transgenic farm animals and to thereby significantly reduce the associated costs. However, to date the potential increase in efficiency due to the identification of embryos carrying the transgene prior to embryo transfer is offset by a loss of viability of the biopsied embryo and the occurrence of false negatives and false positives in the PCR analysis (e.g. Behboodi *et al.*, 1993; Horvat *et al.*, 1993).

Strain and Species Considerations

Transgenic techniques have been developed for a variety of vertebrate species in addition to the mouse (Table 1.2). However, the most informative system is encountered in the production of transgenic mice, simply because so much work has been done with this species. In mice, differences in reproductive productivity, behaviour, related husbandry requirements and responses to various experimental procedures that affect overall production efficiency are well documented. Additionally, strain differences may have significant influences on modifying gene expression; e.g. gene expression and tumour formation in lines of transgenic mice harbouring human oncogenes (or with tumour suppressor genes 'knocked out') vary when these mice are backcrossed to different inbred or outbred strains (Harris *et al.*, 1988; Chisari *et al.*, 1989; Cho *et al.*, 1989; Donehower *et al.*, 1995).

DNA microinjection protocols developed in mice have been modified to accommodate production of other transgenic species. Differences between these species and mice in the embryo quality and physical response to microinjection, requirements for embryo culture, quantity of embryos needed for embryo transfer and pregnancy maintenance, as well as differences in general husbandry practices, are well documented.

To this point, we have not mentioned the production methods used to produce transgenic poultry and fish. In both instances, genetic selection is an exceedingly slow process. Since DNA microinjection into pronuclei of embryonic cells in poultry is not feasible, transfection methodologies using replication-competent and replication-compromised retroviruses has taken centre-stage (Shuman, 1991; Perry and Sang, 1993; Cioffi *et al.*, 1994). As described, methods have included transfection of genes into cells of embryonic blastoderm; insertion of genes using replication-competent retroviruses; the use of replication-defective retroviruses; and sperm-mediated gene transfer. While the latter method has come under critical dispute, the other methods have led to the development of experimental models.

In contrast with poultry studies, work with fish has moved ahead with far greater speed. The principal area of research has focused on growth performance, and initial transgenic GH fish models have demonstrated accelerated and beneficial phenotypes (Fletcher and Davies, 1991; Houdebine and Chourrout, 1991; Cioffi *et al.*, 1994). DNA microinjection methods have propelled the many studies reported and have been most effective due to the relative ease of working with fish embryos. Ideally, efforts at developing 'mass transfer' techniques (e.g. electroporation, sperm binding and lipofection-mediated transfer) would aid in commercializing transgenic fish for the aquaculture industry.

Stem Cells and Alternative Methods for Gene Transfer

The development of ES cell technologies emanated from efforts of the early cell biologists. Teratocarcinoma cell transfer and cell aggregation work in the 1970s evolved from the earlier characterization and studies of teratocarcinoma cells (Pierce, 1975; see also Brinster, 1993; Pinkert, 1997). This led to work with the '129' mouse strain and pluripotential teratocarcinoma cells, and then ultimately to the basis for work with embryonic carcinoma and stem cells in 1981 (Evans and Kaufman, 1981; Martin, 1981). By 1985, purified mouse ES cells were characterized, and by 1987 homologous recombination, gene targeting and the production of chimeric 'knock-out' mice ushered in a new era of 'loss-of-function' mutants to accompany existing techniques (Thomas and Capecchi, 1987; see also Capecchi, 1989; Brinster, 1993). Then, in a relatively brief period, the ability to target DNA integration (as opposed to *random integration* of microinjected genes) and to produce germline-competent chimeric mice was demonstrated. Within a few more years experimental efficiency was enhanced by the development of co-culture techniques, where blastocyst injection was not the only route for ES cell transfer. With co-culture, host embryos could be cultured on a lawn of ES cells, with the ES cells preferentially being incorporated into the embryo proper. Yet, in all of these cases, techniques continuously improved in incremental steps. Thus, the recent successful 'cloning' of a sheep (Wilmut *et al.*, 1997) has captured the imagination of researchers around the world. This technological breakthrough should play a significant role in the development of new procedures for genetic engineering in a number of mammalian species. It should be noted that nuclear cloning, with nuclei obtained from either mammalian stem cells or differentiated 'adult' cells, is an especially important development in 'non-mouse' species. This is because, until the report by Schnieke *et al.* (1998), germline-competent transgenics had only been produced in mammalian species, other than mice, using DNA microinjection.

In contrast with progress in embryo manipulation, a completely different tack was taken with the advent of sperm-related transfer procedures. In 1989, sperm-mediated gene transfer was reported but hotly disputed when many laboratories around the world were unable to duplicate the procedures. Yet, by 1994, the sperm-mediated story generated interest that resulted in the development of spermatogonial cell transplantation procedures as a potentially feasible alternative for gene transfer experimentation (Brinster and Avarbock, 1994; Brinster and Zimmerman, 1994). With embryo- and sperm-related procedures leading the way, as we move into the 21st century, many of our existing procedures will continue to evolve and become more practicable (Box 1.2). However, whole-animal and somatic cell techniques (including liposome-mediated gene transfer, particle bombardment and jet injection), coupled with novel vectors and vector design, will continue in their evolution and in enhancing our gene-transfer capabilities.

Gene Transfer and Gene Regulation

The various strategies for producing genetically engineered animals extend from the mechanistic (e.g. DNA microinjection, ES cell- or retroviral-mediated transfer) to the requisite gene cloning and modelling techniques. However, our understanding of promoter-enhancer sequences and external

Box 1.2. Gene transfer methodologies.

Mouse modelling techniques have evolved from procedures for non-specific (whole genome) transfer, as in aggregation and teratocarcinoma studies, to the transfer of discrete genes and the modification of endogenous genes.

- Blastomere/embryo aggregation.
- Teratocarcinoma cell transfer.
- Retroviral infection.
- Microinjection.
- Electrofusion.
- Nuclear transplantation.
- Embryonic stem (ES) cell transfer.
- Spermatozoa- and spermatogonial cell-mediated transfer.
- Particle bombardment and jet injection.

transcription-regulatory proteins involved in the control of gene expression continues to advance using different model systems. In the systems explored to date, gene transfer technology is a proven asset in science as a means of dissecting gene regulation and expression *in vivo*. However, the primary type of question that is addressed in these systems still concerns the particular role of a single gene in development or in a given developmental pathway.

The three major factors that influence gene expression in all animals, but are particularly relevant to transgenic animals, include: *cis*-acting elements, *trans*-acting factors and the specific gene location (insertion site) within the genome. *Cis*-acting elements determine the state of chromosomal accessibility and consequently the tissue distribution and developmental timing of gene expression. *Cis*-acting elements act in proximity to a given gene and include both promoters and enhancers. Promoters are location-dependent regions of DNA involved in the binding of RNA polymerase to initiate gene transcription. Enhancers are location-independent sequences (they function in either orientation, upstream or downstream of a promoter), and increase the utilization of promoters. In contrast, *trans*-acting factors interact with genes in open domains and stimulate transcription. Normally, gene function is influenced by both *cis*-acting elements and *trans*-acting factors. For transferred genes, the *cis*- and *trans*-activators work in conjunction with the gene integration/insertion event. The chromosomal environment is a major factor that influences gene expression (i.e. the insertion site may alter expression of an endogenous gene), as seen when a gene fails to function (express) in one or more lines of transgenic animals, while it is active in other lines. Using genes that code for reporter proteins (e.g. GH or *lacZ* constructs), analysis of transgenic animals has revealed the importance of these three factors in determining developmental timing, efficiency and tissue distribution of gene expression. Additionally, transgenic animals have proven quite useful in unravelling *in vivo* artefacts of other non-transgenic model systems and techniques. Interestingly, regulation of specific genes in one species does not always correspond to species-specific homologues or the regulation seen in other species.

Traits Affecting Domestic Animal Productivity

Interest in modifying traits that determine the productivity of domestic animals was greatly stimulated by early experiments in which body size and growth rates were dramatically affected in transgenic mice expressing GH transgenes driven by a metallothionein (MT) enhancer/promoter (Palmiter *et al.*, 1982). From that starting point, similar attempts followed in swine and sheep studies to enhance growth by introduction of various GH gene constructs under control of a number of different regulatory promoters (see Pursel and Rexroad, 1993; Table 1.4). Use of these constructs was intended to allow for tight regulation of individual transgene expression by dietary supplementation. However, although resulting phenotypes included altered fat composition, feed efficiency and rate of gain, and lean:fat body composition, they were accompanied by undesirable side-effects, e.g. joint pathology, skeletal abnormalities, increased metabolic rate, gastric ulcers and infertility (Pursel *et al.*, 1989; Nancarrow *et al.*, 1991). Such problems were attributed to chronic overexpression or aberrant expression of the growth-related transgenes and could be mimicked, in several cases, in normal animals by long-term treatment with elevated doses of GH.

Subsequent efforts to genetically alter growth rates and patterns have included production of transgenic swine and cattle expressing a foreign *c-ski* oncogene, which targets skeletal muscle, and studies of growth in lines of mice and sheep that separately express transgenes encoding growth hormone-releasing factor (GRF) or insulin-like growth factor I (IGF-I). Cumulatively, it has become apparent from these studies that greater knowledge of the biology of muscle growth and development will be required in order to genetically engineer lines of domestic animals with these desired characteristics. However, recent work on IGF-I and GH transgenic pigs reported in this volume (Chapters 10 and 11) indicate that progress is being made.

Other productivity traits that are major targets for genetic engineering include altering the properties or proportions of caseins, lactose or butterfat in milk of transgenic cattle and goats, more efficient wool production, and enhanced resistance to viral and bacterial diseases (including development of 'constitutive immunity' or germline transmission of specific, rearranged antibody genes).

Domestic Animals as Bioreactors

The second general area of interest has been the development of lines of transgenic domestic animals for use as bioreactors. One of the main targets of these so-called 'gene farming' efforts has involved attempts to direct expression of transgenes encoding biologically active human proteins. In such a strategy, the goal is to recover large quantities of functional proteins that have therapeutic value, from serum or from the milk of lactating females. To date, expression of foreign genes encoding α_1-antitrypsin, tissue plasminogen activator, clotting factor IX and protein C were successfully targeted to the mammary glands of goats, sheep, cattle and/or swine (Table 1.5).

Similarly, lines of transgenic swine and mice have been created that produce human haemoglobin or specific circulating immunoglobulins. The ultimate goal of these efforts is to harvest proteins from the serum of transgenic animals for use as important constituents of blood transfusion substitutes, or for use in diagnostic testing.

Commercialization

While transgenic animal technology continues to open new and unexplored agricultural frontiers, it also raises questions concerning regulatory and commercialization issues, as demonstrated by molecular farming efforts. A number of major regulatory and public perception hurdles exists that may affect the time to commercialization of transgenic animals. These include perceptions of genetic engineering motives, ethical considerations including animal welfare issues, and product uniformity and economic production (scale-up) issues. A further issue is the potential environmental impact

Table 1.5. Molecular farming projects: a host of commercial projects are underway using transgenic farm animals as bioreactors to produce important biomedical products (from Pinkert, 1997).

Product	Use	Commercializing firm(s)
α_1-antitrypsin	Hereditary emphysema/cystic fibrosis	PPL
α-glucosidase	Glycogen storage disease	Pharming
Antibodies	Anti-cancer	CellGenesys, Genzyme, Ligand
Antithrombin III	Emboli/thromboses	Genzyme
Collagen	Rheumatoid arthritis	Pharming
CFTR	Ion transport/cystic fibrosis	Genzyme
Factor IX	Blood coagulation/haemophilia	Genzyme, PPL
Fibrinogen	Tissue sealant development	ARC, PPL
Haemoglobin	Blood substitute development	Baxter
Lactoferrin	Infant formula additive	Pharming
Protein C	Blood coagulation	ARC, PPL
Serum albumin	Blood pressure, trauma/burn treatment	Pharming
tPA	Dissolve fibrin clots/heart attacks	Genzyme
Tissues/organs	Engineered for xenotransplantation	Alexion, Baxter, CTI, Novartis

following the 'release' of transgenic animals, particularly fish. These societal issues will exist and will continue to influence the development of value-added animal products through transgenesis until transgenic products and foodstuffs have been proven safe for human use and are accepted by a wide cross-section of society.

Conclusions

The use of transgenic animal models for the study of gene regulation and expression has become commonplace in the biological sciences. However, contrary to the early prospects related to commercial exploitation in agriculture, there are numerous societal challenges regarding potential risks that still lie ahead. The risks at hand can be defined not only by scientific evidence but also in relation to public concern (whether perceived or real). Therefore, the central questions will revolve around the proper safeguards to employ and the development of a coherent and unified regulation of the technology. Will new animal reservoirs of fatal human diseases be created? Will more virulent pathogens be artificially created? What is the environmental impact of the 'release' of genetically engineered animals? But perhaps most importantly, we have to ask the question 'do the advantages of a bioengineered product outweigh potential consequences of its use?'

In spite of the inherent limitations in existing methodologies, transgenic livestock will continue to hold great promise for the agricultural industry. The rate of progress to date has, to a certain degree, been limited by the resources available to the scientific community. The cost of producing transgenic farm animals is high and, thus, it is of no surprise that most efforts are carried out in laboratories receiving large amounts of direct government funding. This is, at least in part, changing as venture capital and industry money is now been put into the development of transgenic livestock to produce pharmaceuticals in transgenic animal bioreactors.

Our role as scientists, consumers and regulators is, in part, to decide at what levels or stages and to what degree the development of agriculturally important transgenic animals must be monitored and regulated to ensure consumer safety and animal well-being, and address societal concerns. A further corollary to this responsibility is to ensure that the consuming public understands the processes to the extent that they can accept government approval of such animals in the food chain.

References

Archibald, A.L., McClenaghan, M., Hornsey, V., Simons, J.P. and Clark, A.J. (1990) High-level expression of biologically active human α1-antitrypsin in the milk of transgenic mice. *Proceedings of the National Academy of Sciences USA* 87, 5178–5182.

Behboodi, E., Anderson, G.B., Horvat, S., Medrano, J.F., Murray, J.D. and Rowe, J.D. (1993) Survival and development of bovine embryos microinjected with DNA and detection of the exogenous DNA at the blastocyst stage. *Journal of Dairy Science* 76, 3392–3399.

Brinster, R.L. (1993) Stem cells and transgenic mice in the study of development. *International Journal of Developmental Biology* 37, 89–99.

Brinster, R.L. and Avarbock, M.R. (1994) Germline transmission of donor haplotype following spermatogonial transplantation. *Proceedings of the National Academy of Sciences USA* 91, 11303–11307.

Brinster, R.L. and Zimmerman, J.W. (1994) Spermatogenesis following male germ-cell transplantation. *Proceedings of the National Academy of Sciences USA* 91, 11298–11302.

Brinster, R.L., Chen, H.Y., Trumbauer, M.E. and Avarbock, M.R. (1980) Translation of globin messenger RNA by the mouse ovum. *Nature* 283, 499–501.

Brinster, R.L., Sandgren, E.P., Behringer, R.R. and Palmiter, R.D. (1989) No simple solution for making transgenic mice. *Cell* 59, 239–241.

Cappechi, M.R. (1989) Altering the genome by homologous recombination. *Science* 244, 1288–1292.

Chisari, F.V., Klopchin, K., Moriyama, T., Pasquinelli, C., Dunsford, J.A., Sell, S., Pinkert, C.A., Brinster, R.L. and Palmiter, R.D. (1989) Molecular pathogenesis of hepatocellular carcinoma in hepatitis B virus transgenic mice. *Cell* 59, 1145–1156.

Cho, H.J., Seiberg, M., Georgoff, I., Teresky, A.K., Marks, J.R. and Levine, A.J. (1989) Impact of the genetic background of transgenic mice upon the formation and timing of choroid plexus papillomas. *Journal of Neuroscience Research* 24, 115–122.

Cibelli, J.B., Stice, S.L., Golueke, P.J., Kane, J.J., Jerry, J., Blackwell, C., Ponce de León, F.A. and Robl, J.M. (1998) Cloned transgenic calves produced from nonquiescent fetal fibroblasts. *Science* 280, 1256.

Cioffi, L.C., Kopchick, J.J. and Chen, H.Y. (1994) Production of transgenic poultry and fish. In: Pinkert, C.A. (ed.) *Transgenic Animal Technology: a Laboratory Handbook*. Academic Press, San Diego, pp. 279–313.

Donehower, L.A., Harvey, M., Vogel, H., McArthur, M.J., Montgomery, C.A., Jr, Park, S.H., Thompson, T., Ford, R.J. and Bradley, A. (1995) Effects of genetic background on tumorigenesis in p53-deficient mice. *Molecular Carcinogenesis* 14, 16–22.

Evans, M.J. and Kaufman, M.H. (1981) Establishment in culture of pluripotential cells from mouse embryos. *Nature* 292, 154–156.

Fletcher, G.L. and Davies, P.L. (1991) Transgenic fish for aquaculture. *Genetic Engineering* 13, 331–370.

Furth, P.A., Shamay, A. and Hennighausen, L. (1995) Gene transfer into mammalian cells by jet injection. *Hybridoma* 14, 149–152.

Gordon, I. and Lu, K.H. (1990) Production of embryos *in vitro* and its impact on livestock production. *Theriogenology* 33, 77–87.

Gordon, J.W. and Ruddle, F.H. (1981) Integration and stable germline transmission of genes injected into mouse pronuclei. *Science* 214, 1244–1246.

Gordon, J.W., Scangos, G.A., Plotkin, D.J., Barbosa, J.A. and Ruddle, F.H. (1980) Genetic transformation of mouse embryos by microinjection of purified DNA. *Proceedings of the National Academy of Sciences USA* 77, 7380–7384.

Gurdon, J.B. (1977) Nuclear transplantation and gene injection in amphibia. *Brookhaven Symposia in Biology* 29, 106–115.

Hammer, R.E., Pursel, V.G., Rexroad, C.E., Jr, Wall, R.J., Bolt, D.J., Ebert, K.M., Palmiter, R.D. and Brinster, R.L. (1985) Production of transgenic rabbits, sheep and pigs by microinjection. *Nature* 315, 680–683.

Harris, A.W., Pinkert, C.A., Crawford, M., Langdon, W.Y., Brinster, R.L. and Adams, J.M. (1988) The Eμ-*myc* transgenic mouse: a model for high-incidence spontaneous lymphoma and leukemia of early B cells. *Journal of Experimental Medicine* 167, 353–371.

Hogan, B., Beddington, R., Costantini, F. and Lacy, E. (1994) *Manipulating the Mouse Embryo: a Laboratory Manual.* Cold Spring Harbor Press, Cold Spring Harbor, New York.

Horvat, S., Medrano, J.F., Behboodi, E., Anderson, G.B. and Murray, J.D. (1993) Sexing and detection of gene construct in microinjected bovine blastocysts using the polymerase chain reaction. *Transgenic Research* 2, 134–140.

Houdebine, L.M. and Chourrout, D. (1991) Transgenesis in fish. *Experientia* 47, 891–897.

Irwin, M.H., Moffatt, R.J. and Pinkert, C.A. (1996) Identification of transgenic mice by PCR analysis of saliva. *Nature Biotechnology* 14, 1146–1149.

Larsson, S., Hotchkiss, G., Andang, M., Nyholm, T., Inzunza, J., Jansson, I. and Ahrlund-Richter, L. (1994) Reduced β2-microglobulin mRNA levels in transgenic mice expressing a designed hammerhead ribozyme. *Nucleic Acid Research* 22, 2242–2248.

Lin, T.P. (1966) Microinjection of mouse eggs. *Science* 151, 333–337.

Maga, E.A. and Murray, J.D. (1995) Mammary gland expression of transgenes and the potential for altering the properties of milk. *Bio/Technology* 13, 1452–1457.

Martin, G.R. (1981) Isolation of a pluripotent cell line from early mouse embryos cultured in medium conditioned by teratocarcinoma stem cells. *Proceedings of the National Academy of Sciences USA* 78, 7634–7638.

Monastersky, G.M. and Robl, J.M. (1995) *Strategies in Transgenic Animal Science.* American Society for Microbiology Press, Washington, DC.

Nagy, A., Rossant, J., Nagy, R., Abramow-Newerly, W. and Roder, J.C. (1993) Derivation of completely cell culture-derived mice from early-passage embryonic stem cells. *Proceedings of the National Academy of Sciences USA* 90, 8424–8428.

Nancarrow, C.D., Marshall, J.T.A., Clarkson, J.L., Murray, J.D., Millard, R.M., Shanahan, C.M., Wynn, P.C. and Ward, K.A. (1991) Expression and physiology of performance regulating genes in transgenic sheep. *Journal of Reproduction and Fertility* 43 (Suppl.), 277–91.

Palmiter, R.D., Brinster, R.L., Hammer, R.E., Trumbauer, M.E., Rosenfeld, M.G., Birnberg, N.C. and Evans, R.M. (1982) Dramatic growth of mice that develop from eggs microinjected with metallothionein-growth hormone fusion genes. *Nature* 300, 611–615.

Palmiter, R.D., Norstedt, G., Gelinas, R.E., Hammer, R.E. and Brinster, R.L. (1983) Metallothionein–human GH fusion genes stimulate growth of mice. *Science* 222, 809–814.

Perry, M.M. and Sang, H.M. (1993) Transgenesis in chickens. *Transgenic Research* 2, 125–133.

Pierce, G.B. (1975) Teratocarcinoma and perspectives. In: Sherman, M.I. and Solter, D. (eds) *Teratomas and Differentiation.* Academic Press, New York, pp. 3–12.

Pinkert, C.A. (1994a) *Transgenic Animal Technology: a Laboratory Handbook.* Academic Press, San Diego.

Pinkert, C.A. (1994b) Transgenic pig models for xenotransplantation. *Xeno* 2, 10–15.

Pinkert, C.A. (1997) The history and theory of transgenic animals. *Laboratory Animal* 26, 29–34.

Pinkert, C.A. and Stice, S.L. (1995) Embryonic stem cell strategies: beyond the mouse model. In: Monastersky, G.M. and Robl, J.M. (eds) *Strategies in Transgenic Animal Science.* American Society for Microbiology Press, Washington, DC, pp. 73–85.

Pinkert, C.A., Irwin, M.H. and Moffatt, R.J. (1997) Transgenic animal modeling. In: Meyers, R.A. (ed.) *Encyclopedia of Molecular Biology and Molecular Medicine* Vol. 6. VCH Publishers, New York, pp. 63–74.

Pursel, V.G. and Rexroad, C.E., Jr (1993) Status of research with transgenic farm animals. *Journal of Animal Science* 71 (Suppl. 3), 10–19.

Pursel, V.G., Pinkert, C.A., Miller, K.F., Bolt, D.J., Campbell, R.G., Palmiter, R.D., Brinster, R.L. and Hammer, R.E. (1989) Genetic engineering of livestock. *Science* 244, 1281–1288.

Rogers, G.E. (1990) Improvement of wool production through genetic engineering. *Trends in Biotechnology* 8, 6–11.

Schnieke, A.E., Kind, A.J., Ritchie, W.A., Mycock, K., Scott, A.R., Ritchie, M., Wilmut, I., Colman, A. and Campbell, K.H.S. (1998) Human factor IX transgenic sheep produced by transfer of nuclei from transfected fetal fibroblasts. *Science* 278, 2130–2133.

Shuman, R.M. (1991) Production of transgenic birds. *Experientia* 47, 897–905.

Sokol, D.L. and Murray, J.D. (1996) Antisense and ribozyme constructs in transgenic animals. *Transgenic Research* 5, 363–371.

Thomas, K.R. and Capecchi, M.R. (1987) Site-directed mutagenesis by gene-targeting in mouse embryo-derived stem cells. *Cell* 51, 503–512.

Van Brunt, J. (1988) Molecular farming: transgenic animals as bioreactors. *Bio/Technology* 6, 1149–1154.

Wagner, E.F., Stewart, T.A. and Mintz, B. (1981a) The human β-globin gene and a functional viral thymidine kinase gene in developing mice. *Proceedings of the National Academy of Sciences USA* 78, 5016–5020.

Wagner, T.E., Hoppe, P.C., Jollick, J.D., Scholl, D.R., Hodinka, R.L. and Gault, J.B. (1981b) Microinjection of a rabbit β-globin gene into zygotes and its subsequent expression in adult mice and their offspring. *Proceedings of the National Academy of Sciences USA* 78, 6376–6380.

Wall, R.J., Hawk, H.W. and Nel, N. (1992) Making transgenic livestock: Genetic engineering on a large scale. *Journal of Cellular Biochemistry* 49, 113–120.

Ward, K.A. and Nancarrow, C.D. (1991) The genetic engineering of production traits in domestic animals. *Experientia* 47, 913–922.

Ward, K.A., Franklin, I.R., Murray, J.D., Nancarrow, C.D., Raphael, K.A., Rigby, N.W., Byrne, C.R., Wilson, B.W. and Hunt, C.L. (1986) The direct transfer of DNA by embryo microinjection. In: *Proceedings of the 3rd World Congress on Genetics Applied to Livestock Breeding, Lincoln,* Vol. 12, pp. 6–21. Lincoln, Nebraska.

Wilmut, I., Schnieke, A.E., McWhir, J., Kind, A.J. and Campbell, K.H. (1997) Viable offspring derived from fetal and adult mammalian cells. *Nature* 385, 810–813.

Wood, S.A., Pascoe, W.S., Schmidt, C., Kemler, R., Evans, M.J. and Allen, N.D. (1993) Simple and efficient production of embryonic stem cell-embryo chimeras by coculture. *Proceedings of the National Academy of Sciences USA* 90, 4582–4585.

Wright, G., Carver, A., Cottom, D., Reeves, D., Scott, A., Simons, P., Wilmut, I., Garner, I. and Colman, A. (1991) High level expression of active human α-1-antitrypsin in the milk of transgenic sheep. *Bio/Technology* 9, 830–834.

2

Development of Genetic Tools for Transgenic Animals

P.B. Hackett, Z. Izsvak, Z. Ivics and L. Caldovic

Department of Genetics and Cell Biology, University of Minnesota, St Paul, Minnesota, USA

There is a chronic need to develop transgenic fish for aquaculture and genetically engineered farm animals for various agricultural and medical purposes. To meet this need we have developed new lines of vectors based on transposable elements and border elements for genetically engineering animals in an efficient, cost-effective manner. The principles of two newly developed tools for fish are described. First, a transposable element system, based on the *Tc1/mariner* family of transposons that is active in animals from fish to mammals, is described. Second, the effectiveness of border elements, derived from insects and birds, to insulate transgenes from position effects is presented. Fish are used as a model system and as an example of the type of needs that can be met because these species are especially convenient and inexpensive for use in the development of laboratory procedures.

Introduction

The aquatic resources of the world are being exhausted by over harvesting of finfish and other aquatic organisms. As the world's population grows, its fisheries are being depleted at increasing rates. The USA suffers from a staggering international trade imbalance in fisheries products, about US$3 billion per year, the third largest contributor to its annual imbalance of payments. The US consumption of about 20 kg/person requires the harvest of a total of about 6 Mt per year (Parfit, 1995). To meet this demand, the US fishing fleet is harvesting more fish and depleting the fish stocks required to maintain fish populations, which has led in the last 4 years to a decline in wild fisheries off the coasts of the USA. As a result, in future years we will not be able to produce sufficient quantities of fish for our national needs. The situation will get worse if we continue with a 'fishing as usual' policy.

© CAB INTERNATIONAL 1999. *Transgenic Animals in Agriculture* (eds J.D. Murray, G.B. Anderson, A.M. Oberbauer and M.M. McGloughlin)

Placing long-term moratoria on fishing in some regions is an economically and politically difficult choice, and the strategy will not result in increased harvests in the long run. Another way to increase yields of fish is through aquaculture of genetically superior stocks of fish that can be farmed at a faster rate for lower cost.

There are two methods for achieving improved stocks of fish and other commercially important animals. The first is classical breeding, which has worked well for land animals, but which takes decades to bring about major changes. For example, by controlled breeding, growth rates in coho salmon (*Oncorhynchus kisutch*) increased by 60% over four generations (Herschberger *et al.*, 1990) and body weights of two strains of channel catfish (*Ictalurus punctatus*) were increased by 21–29% over three generations (Dunham *et al.*, 1994). The second method is genetic engineering, a selective process by which genes whose behaviours we think we understand are introduced into the chromosomes of animals or plants to give these organisms a new trait or characteristic, such as improved growth or greater resistance to disease. The results of genetic engineering have exceeded those of breeding in some cases. In a single generation, increases in body weight of 58% were obtained in common carp (*Cyprinus carpio*) with extra rainbow trout growth hormone I genes (Chen *et al.*, 1993), with increases of more than 1000% in salmon with extra salmon growth hormone genes, and less in trout (Agellon *et al.*, 1988; Du *et al.*, 1992; Devlin *et al.*, 1994).

The advantage of genetic engineering in fish is that an organism can be altered directly in a very short period of time if the appropriate gene has been identified (Fletcher and Davis, 1991; Hackett, 1993; Iyengar *et al.*, 1996). The disadvantage of genetic engineering in fish is that few of the many genes that are involved in growth and development have been identified and the interactions of their protein products is poorly understood. At present, we cannot produce certain changes that might be possible through breeding programmes; however, as we identify genes in fish genomes, our ability to introduce new characteristics into fish will improve and allow us to avoid the long process of breeding. Well-developed procedures that work well for genetic analysis in other organisms are lacking in fish and other economically important animals. These include methods for insertional mutagenesis (gene tagging) and efficient procedures for producing transgenic animals. Transgenic DNA is not efficiently incorporated into chromosomes. Only about one in a million of the foreign DNA molecules integrates into the cellular genome, generally several cleavage cycles into development. Consequently, most transgenic fish are mosaic (Hackett, 1993). As a result, fish raised from embryos into which transgenic DNA has been delivered must be cultured until gametes can be assayed for the presence of integrated foreign DNA. The screening is extremely tedious, time-consuming and therefore relatively expensive. Many transgenic fish fail to express the transgene due to position effects. Consequently, transgenic fish are not routinely made as frequently as they are needed. A simple, reliable procedure that directs early integration of exogenous DNA into the chromosomes of animals at the one-cell stage is needed. Moreover, we need a method for getting the genes to express reliably and continuously once they are in animal chromosomes.

Here, we present the results of work in our laboratory over the past 5 years to improve transgenesis in animals. The experimental animal we use is the fish because: (i) they are easy to raise; (ii) embryos can be obtained daily in large numbers; (iii) embryos develop outside the mother, which greatly reduces the expense and effort of raising transgenic individuals; (iv) development is rapid, about 2 days to hatching in zebrafish; (v) development takes place in an optically clear chorion, allowing visual inspection of development as it occurs following experimental genetic manipulation of the embryos; and (vi) the results appear to be applicable to all other vertebrates which we have tested. The initial drive for transgenic fish came from attempts to enhance production of fish in Minnesota. Our goal was to use genetic engineering to improve the characteristics of several species of fish so as to increase their commercial value. To do this, several genetic tools were developed which work as well in land vertebrates as in fish.

Results

Early results of genetic engineering in fish

The first goal of the Minnesota Transgenic Fish Group (MTFG, formed by Drs Kevin Guise, Anne Kapuscinski, Anthony Faras and Perry Hackett) was growth enhancement of commercially valuable fish, with the understanding that if we succeeded in this area, we could continue improvement of fish species for Minnesota aquaculture. More than 60,000 embryos of walleye, northern pike and rainbow trout were microinjected with DNA constructs that contained a growth hormone gene and the required genetic switches to ensure expression of the transgene. From many thousands of embryos that were microinjected, only about 1000 fish survived to adulthood. The low survival rate was due to many factors including constant stress of moving from one fish facility to another in various regions of the state, the lack of knowledge of indoor rearing of wild game fish, the natural cannibalism practiced by the fish at early stages, and possible lowered fitness of growth-enhanced, transgenic fish.

Although the number of surviving founder fish was low, many were transgenic. Only one or two actively expressing fish are required to serve as broodstock for future generations. Initially we used a construct composed of the Rous sarcoma virus long terminal repeat sequence (which harbours enhancers and a semi-constitutive promoter) juxtaposed with the bovine growth hormone gene (RSV/bGH) for test purposes in the northern pike. In later studies we switched to an 'all-fish' construct composed of the carp β-actin enhancers plus promoter driving a chinook salmon growth hormone gene (β-act/csGH) (Liu *et al.*, 1990) for northern pike, walleye, rainbow trout and Atlantic salmon (Gross *et al.*, 1992). About 10,000 northern pike embryos were injected with either of the two transgenic constructs. Of several thousand embryos injected with the RSV/bGH construct 1218 were examined by radioimmunoassay and 36 (3%) had elevated levels of bGH in their blood. Of the several thousand embryos that were injected with β-act/csGH, 1398 were screened to yield 88 (6%) with elevated csGH in the blood (MTFG, unpublished observations). This was in the range we expected.

The initial results were encouraging. There was an almost 40% increase in size of the fish, similar to what has been seen with other experiments of this type. However, these results did not take into consideration the sex of the fish, females are larger than males, or problems that could occur due to fish crowding. At later times, when several of the confounding parameters were considered, we found that only microinjected males showed a low but consistent increase (about 25%) in size over a 16 month time interval (Gross *et al.*, 1992). Upon further consideration of our results compared with those of others who obtained larger fish (Agellon *et al.*, 1988; Zhang *et al.*, 1990; Du *et al.*, 1992; Chen *et al.*, 1993; Devlin *et al.*, 1994; Dunham *et al.*, 1994), we realized that we had not removed the prokaryotic vector sequences from our transgenic DNA. This was a serious omission because prokaryotes do not have a prejudice against CpG dinucleotide base pairs in their DNAs as do animal cells, where these base pairs are sites for DNA methylation (an activity associated with inhibition of gene expression). Thus, it is very possible that although the transgenes were present in a mosaic fashion in our fish, they were not active.

Our studies and those of others indicated that mosaicism was common, i.e. that the transgenic DNA integrated into the fish chromosomes after the initial cleavages, yielding fish that had the transgene in some tissues but not others. This determination came from Southern blotting of various tissues including muscle, fin, blood, kidney, spleen, heart and brain (Hallerman *et al.*, 1990). Thus, by using PCR techniques on small samples of fin from the experimental fish, we could determine whether or not the transgenic construct was present in the fin but not necessarily in the most important tissue, the gonads. Presence in the gonads is important for passage of the trait to offspring. To determine further our abilities to get the transgenic DNA into fish chromosomes, we sacrificed a sample of the fish and carried out Southern blotting analysis of tissues. Thirty per cent of the fish had the transgene in one or more of the tissues analysed, but only about 12% had the gene in fin tissue. Thus, a positive signal in the fin samples represented only about 40% of the transgenic fish.

Why did only 3–6% of the fish show elevated levels of transgenic GH when 30% of the fish had the transgene in one or more tissues? The answer is not known. Our best speculation, which is fairly well founded on other systems, is that expression of the transgenes is dependent not only on the accompanying genetic control elements, but also on regulatory sequences in the fish chromatin near the site of integration of the construct. There are about 2,000,000,000 potential sites of entry of the transgenic material, and the regulatory units around these sites play a role in transgene expression. Likewise, the problem of mosaicism is widespread in every laboratory that is attempting to make transgenic fish (Hackett, 1993). Screening transgenic fish would be much easier, and the results of the procedure far more predictable, if mosaicism were reduced. Accordingly, we initiated studies to improve the rate of early integration of transgenes into fish and to reduce the effects of neighbouring chromatin sequences on expression of integrated transgenes.

Efficacy of co-delivery of transgenes plus integrase to produce transgenic animals

Our first goal was to enhance integration by using a recombinase protein, an enzyme that is associated with restructuring DNA sequences. There are many types of recombinase proteins found in nature. We elected to examine recombinase proteins that were known to mediate integration of DNA sequences into chromosomal DNA. For this, specific recombinase-binding DNA sequences had to be added to both sides of the transgenic DNA construct (Fig. 2.1). However, some recombinases require host-specific cofactors for efficient activity; the key was to find a recombinase that either had no such requirement or required a factor(s) that was ubiquitous in a wide range of animal cells. We first used murine retroviral integrase synthesized from a baculovirus expression vector and from extracts of retrovirus-packaging cells (Ivics *et al.*, 1993). We achieved a 10- to 40-fold enhancement of transgene expression, although we had hoped to get a much higher level of activity. There may have been several factors contributing to the low activity: (i) the integrase preparations made in the baculovirus-infected cells may have had low specific activities as a result of denaturation or poor solubility; (ii) the chromatin may have needed to be 'activated' for integration (a site in the chromosome may have needed to be 'opened' or cleaved), possibly by an endonuclease that produces ends complementary to those of the transgene; (iii) integrase may have to be packaged together with target DNA (as happens with reverse transcriptase and viral RNA during virus assembly in retrovirus-infected cells) to form an 'integration kit'; moreover, (iv) host factors probably are necessary for efficient function of integrase, and therefore mixing purified integrase with substrate DNA may not have mimicked the natural retroviral process faithfully. Nevertheless, the most important conclusion of this early work was that recombinases could be employed to enhance integration of transgenic DNA. To avoid the problem of host factors and find an alternative to using

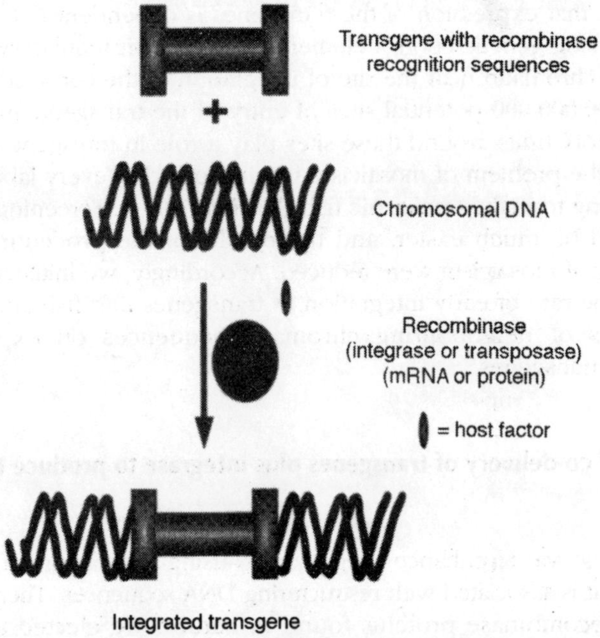

Fig. 2.1. Recombinase-directed integration of transgenes into chromatin. The schematic illustrates the principle of using either a transposase or integrase protein (large oval), with or without the assistance of cofactors (small, narrow oval) to mediate the integration of transgenic DNA (horizontal bar) into chromatin (double helix). The transgene is shown with recombinase recognition sequences (vertical rectangles flanking the transgene).

oncoviruses (or their products), we investigated repetitive elements in fish genomes (Izsvák *et al.*, 1995; Izsvák *et al.*, 1997). This was done in order to find other types of sequences that might harbour recombinases to catalyse efficient integration of DNA sequences into animal chromosomes. We especially scrutinized Tc1-type transposable elements, hereafter referred to as TcEs.

Adaptation of transposable elements as genetic tools in animal genetic engineering

The wide distribution of TcEs suggested that these elements require few, if any, species-specific host factors. In contrast, other mobile DNAs such as *P* elements, which appear to have requirements for specific host cofactors, do not function in fish (our unpublished observations and Gibbs *et al.*, 1994). Transposase-deficient TcEs carrying marker genes can be mobilized by transposase provided in *trans*. These features suggested that TcEs would be suitable for many genetic applications. However, because species-specific constraints of TcE transposition never had been evaluated rigorously, it was important to transfer elements from a species that was relatively close in evolution to the species in which they were to be applied. Indeed, we were unsuccessful in mobilizing the *Tc1* element from *Caenorhabditis elegans*, in fish.

Fish transposable elements belonging to the Tc1 family were discovered 3 years ago (Radice *et al.*, 1994). Since then, we have identified TcEs in a dozen species of fish, indicating that TcEs are prevalent components of many fish genomes (Izsvák *et al.*, 1995). Due to a variety of mutations in their transposase genes (Radice *et al.*, 1994; Izsvák *et al.*, 1995; Ivics *et al.*, 1996), none of the identified fish TcEs encoded an active transposase. We found that the majority of fish TcEs can be classified into two major subfamilies, zebrafish- and salmonid-type elements. Although the two subfamilies of fish TcEs appear to have a common ancestor, they are characteristically different in their encoded transposases and their flanking sequences (Ivics *et al.*, 1996). These findings suggested that a heterologous, *salmonid* transposon could be revived for use in developmental genetic studies in fish and other animals.

Structural and functional features of TcEs

TcEs contain a single gene encoding a transposase flanked by inverted repeats (IRs). Transposons spread when the transposase is expressed and their flanking IR sequences are exposed. The transposase catalyses the excision of the transposon from its original location and promotes its reintegration elsewhere in the genome (a 'cut-and-paste' mechanism). In leaving a site, the transposon leaves behind a gap in chromosomal DNA, which is often repaired by a mechanism that can regenerate a portion (a 'footprint') of the transposon at its original site of insertion. The prototype *Tc1* transposon has short, 54-bp IRs flanking its transposase gene. In contrast, most of the fish TcEs have long, 210–250 bp, IRs at their termini and directly repeated (DR) DNA sequence motifs at the ends of each IR. In this respect, fish TcEs are similar to other transposable elements from flies. We proposed that these IR/DR elements form a group of TcEs on the basis of the organization of their IRs, and that they transpose by a similar mechanism (Izsvák *et al.*, 1995; Ivics *et al.*, 1996). The direct repeats in the IR/DR flanks are the cores of the binding sites for transposase. These observations suggested that the number of putative transposase binding sites in the IR/DR-transposons is twice that of most known TcEs; therefore, these may be sites for regulation of mobility. An important observation is that *Minos*, a TcE from the fly *Drosophila hydei*, is active in *Ceratitis capita*, a non-drosophilid species of fly (Loukeris *et al.*, 1995). This suggested to us that, unlike *P*-elements, TcEs of the IR/DR class could be active beyond the species in which they were found. Multiple sequence alignment of TcE transposases allowed us to identify highly conserved amino acid domains of functional importance (Fig. 2.2) (Ivics *et al.*, 1996). Phylogenetic sequence comparisons show that the C-terminal halves of TcE transposase proteins which make up the catalytic centre and contain the DDE box, are highly conserved and shared by other recombinase proteins, suggesting similar mechanisms for DNA cleavage and joining. A nuclear localization signal (NLS), and casein kinase II (CK-II) phosphorylation sites which mediated nuclear translocation were identified. Tc1 transposase has a bipartite DNA-binding domain overlapping the NLS motif at a cluster of basic amino acids. The N-terminal region (marked as a string of leucines) of the fish transposases has a striking sequence similarity to the bipartite *paired* domain. Our identification of functional domains in TcE transposases was important because it allowed us to improve the efficiency of the system.

An active TcE vector system for vertebrate animals

The two components of any transposon system are an active transposase and the DNA sequences that are mobilized. We searched for a transposase-producing IR/DR element in a number of fish species; however, *all* the elements that we examined appeared to be inactive due to deletions and other mutations. Therefore, we used the accumulated sequence data to reconstruct a salmonid transposase gene from sequence alignments of TcEs found in 11 fish species. Since parsimony analysis could not resolve the phylogenetic relationships among salmonid-type TcEs (Ivics *et al.*, 1996), we engineered a consensus transposon with an intact transposase gene from salmonid elements which we call *Sleeping Beauty* (SB). A series of ten

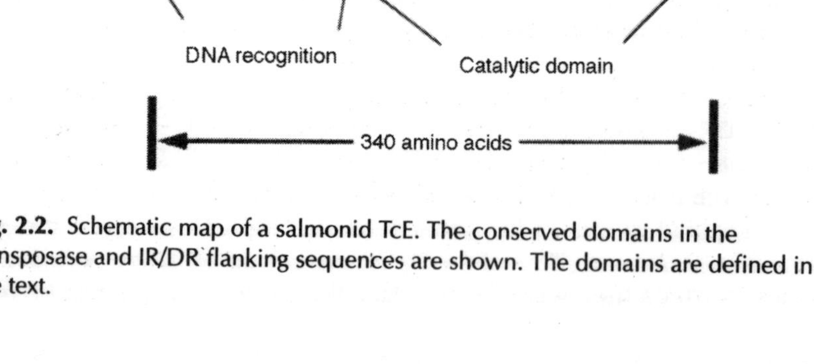

Fig. 2.2. Schematic map of a salmonid TcE. The conserved domains in the transposase and IR/DR flanking sequences are shown. The domains are defined in the text.

constructs (Fig. 2.3) were made by a PCR-mutagenesis strategy to step-by-step produce a synthetic gene encoding a putative salmonid transposase protein of 340 amino acids that is nearly identical to the consensus, and that possesses all the most conserved domains (Ivics *et al.*, 1997). By this method, selected nucleotides were changed in codons to restore the amino acids that were in the putatively active transposase gene many millions of years ago. We did not blindly use a 'majority-rule' consensus sequence; e.g. at some loci it appeared that C→T mutations had been fixed where deamination of $5^{m}C$ residues had occurred (which leads to C being converted to T which in turn can lead to the 'repair' of the mismatched G residue to an A). We could test for various expected activities of the resurrected transposase, and thus the accuracy of our engineering, by examining several

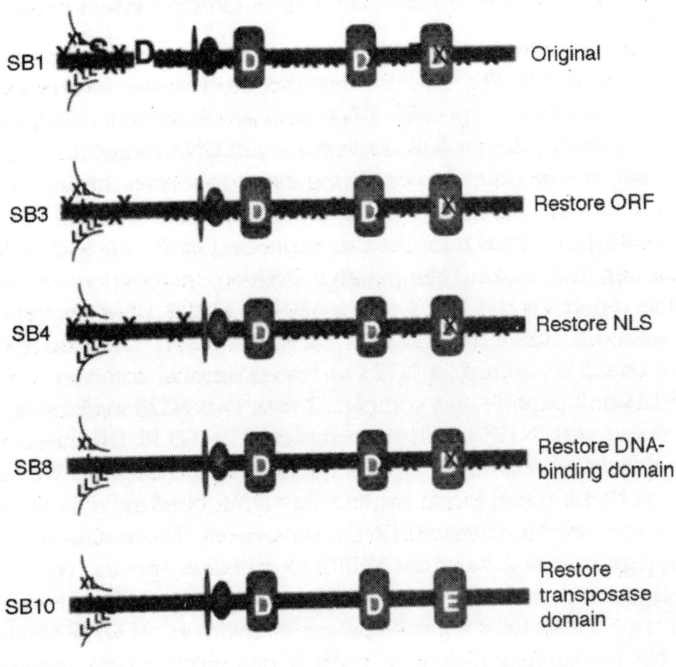

Fig. 2.3. Molecular reconstruction of the SB transposase gene. Several types of site-specific changes were made: black-D (a major deletion was filled in) and S (translational termination codons) and F (frameshift mutations) were replaced. Residues marked by Xs were changed to the consensus. In the right margin, the net results, based on various functional tests, are indicated. SB1 is the initial salmonid TcE transposase gene with the conserved domains indicated. The first two stages resulted in a complete open reading frame for the transposase (SB3). Systematic replacement of specific amino acids restored the bipartite nuclear localization signal (SB4), the DNA-binding domain specific for salmonid but not zebrafish TcEs (SB8) and catalytic domain (SB10). The methods used for the site-specific mutagenesis are described in detail in Ivics *et al.* (1997).

specific functions which are found in active transposase proteins (see right margin of Fig. 2.3).

We have four sources of synthetic SB transposase protein:

1. Extracts of tissue-cultured zebrafish or carp EPC cells transfected with a eukaryotic expression plasmid carrying the SB transposase gene driven by the human cytomegalovirus (CMV) promoter.
2. Purified SB transposase protein, or portions thereof, obtained from extracts of *Escherichia coli* transformed with the SB transposase gene in a pET (Novagen) bacterial expression vector; this vector directs the addition of a histidine-tag for purification and an epitope-tag for detection of the recombinant protein.
3. Extracts of 5-h old zebrafish embryos microinjected at the one-cell stage with SB transposase mRNA synthesized *in vitro*.
4. SB transposase expressed in bacculovirus-infected insect cells.

We examined some of the biochemical activities of the SB protein. One such test is shown in Fig. 2.4, the mobility-shift assay. In this assay, if a protein can bind to a specific DNA sequence, which is crucial for a transposase protein, then it will convert a small DNA molecule into a larger complex that will migrate slower during electrophoresis through a gel. We tested the DNA-binding activity of the amino-terminal fragment (123 amino acids, called N123) of SB transposase, expressed in *E. coli* and isolated via a histidine tag, that contains the putative DNA-recognition motif (Ivics *et al.*, 1997). The target for the N123 was a 320-bp IR/DR DNA sequence from either a salmonid TcE or the zebrafish Tdr1 transposon. The salmonid IR/DR sequence could be shifted by N123 to two positions, complex 1 with one bound N123 polypeptide and complex 2 with two N123 molecules. The SB IR complexed with N123 could be completed by SB IR/DR fragments, but not with closely related IR/DR DNA sequences from zebrafish. We conclude that our synthetic transposase protein has DNA-binding activity, and this binding is specific for salmonid IR/DR sequences. These data suggest that the SB transposase will have the ability to mobilize specific vectors based on our SB transposon without disturbing endogenous TcEs of the host.

Two crucial tests for a transposase are its ability to cut DNA precisely out of one DNA molecule and then to insert it into another DNA sequence. We tested excision and integration activities of our SB transposase by an inter-plasmid marker-transfer assay. The indicator (donor) plasmids for monitoring transposon excision and/or integration had two features: (i) a marker gene that, when recovered in *E. coli* or in fish cells, can be screened by virtue of either the loss or the gain of a function, and (ii) transposase-recognition sequences in the IRs flanking the marker gene. Care was taken during these constructions to keep the total size of the marked transposons around 1.6 kb, the natural size of TcEs found in teleost genomes. Using the assay system, we examined the integration activity of SB transposase. Table 2.1 shows that the number of recombinant plasmids (doubly resistant to the antibiotics

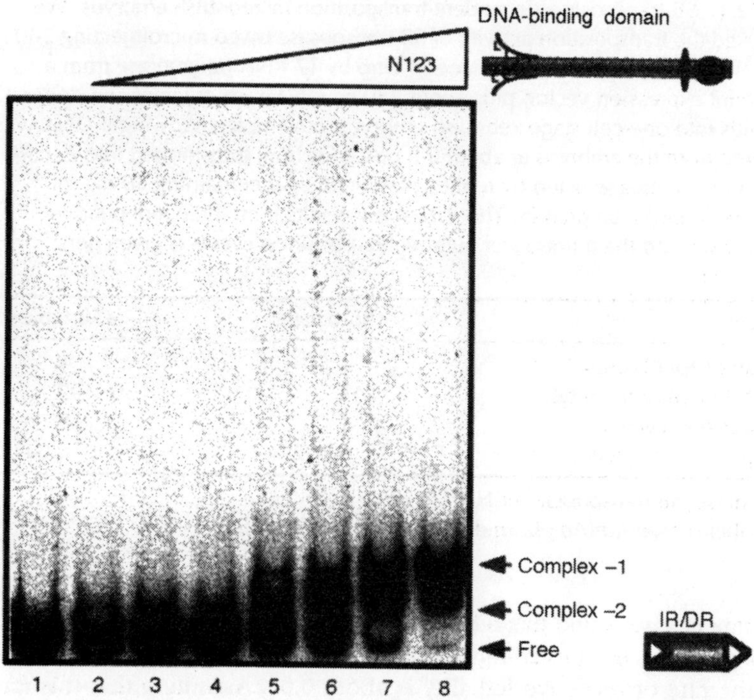

Fig. 2.4. DNA-binding activities of SB transposase. Mobility-shift analysis of the ability of the amino-terminal third (first 123 amino acids; N123) of SB transposase to bind to the inverted repeats of fish transposons. The N123 peptide was purified from *E. coli* in which it was cloned and expressed. ^{32}P-labelled IR sequences were combined with the N123 peptide in varying amounts plus unlabelled poly(dI:dC) non-specific competitor DNA. Lane 1: probe only; lanes 2–8: 100,000-, 50,000-, 20,000-, 10,000-, 5000-, 2500- and 1000-fold dilutions of the N123 preparation. The binding of two N123 peptides to each IR/DR DNA segment did not display cooperative binding kinetics.

ampicillin and kanamycin) was about 20-fold higher than control transformations lacking SB transposase mRNA. The physical translocation of the *Ap*-marked TcE was confirmed by Southern hybridization; we did not find any evidence that double antibiotic resistance was due to co-transformation by two original plasmids. Based on the estimated average number of plasmids injected per embryo, the activity shown in Table 2.1 corresponds to an average of 20–50 insertions per zebrafish embryo. Approximately 5×10^5 3-kb plasmids were injected per embryo. This corresponds to about 1.5×10^9 bp of total injected DNA, nearly the same as the zebrafish genome (about 1.6×10^9 bp). Assuming that the plasmid DNA formed nucleosomes, this assay is an excellent test of the system *in vivo*. In the course of these

Table 2.1. SB transposase-dependent transposition in zebrafish embryos. We evaluated the transposition activity of SB transposase by co-microinjecting 200 ng ml^{-1} of SB transposase mRNA, made *in vitro* by T7 RNA polymerase from a Bluescript expression vector, plus about 250 ng ml^{-1} each of target and donor plasmids into one-cell stage zebrafish embryos. Low molecular weight DNA was prepared from the embryos at about 5 h post-injection, transformed into *E. coli* cells, and colonies selected by replica plating on agar containing 50 mg ml^{-1} kanamycin and/or ampicillin. The data show results from three independent experiments and the numbers of putative transposition events is given per experiment.

Reagent	− SB txpase	+SB txpase[a]
Indicator (*Ap*) plasmid	+	+
Target (*Km*) plasmid (total)	43,858	46,492
Transposition events[b]	1	19
Transposition frequency	0.002%	0.041%

[a] SB txpase, SB transposase mRNA.
[b] Number of true *Km/Ap* plasmids found in 5 h zebrafish embryos.

experiments, we found that a high percentage of the embryos did not survive beyond 4 days. Insertional mutagenesis studies in the mouse have suggested that the rate of recessive lethality is about 0.05. Assuming that this rate is applicable to zebrafish, the approximate level of mortality suggests that with the microinjection conditions we used there were, coincidentally, about 20 insertions per genome, which is in the range of that suggested by the inter-plasmid assays.

Using an assay to monitor transfer of a gene from a plasmid to a chromosome (rather than an inter-plasmid assay as with the zebrafish embryos) in transformed human cells and cultured fish EPC cells, we found a comparable 12-fold stimulation of integration of a transposon-like substrate DNA by SB transposase (Table 2.2). HeLa cells were transfected with SB transposase and an SV40–neomycin–phosphotransferase-II expression vector flanked by IR/DR repeats. The transfected cells were selected with G418 and stained with methylene blue to indicate survivors. The staining patterns clearly demonstrated that the integration of an SV40-*neo* construct flanked by IR/DR repeats into HeLa cells was significantly increased when an active transposase construct was coinjected when compared with control constructs with an inverted SB coding region or a defective transposase gene lacking the DDE-catalytic domain but containing the DNA-binding and NLS functions. These data suggested that the SB system is active in vertebrates other than fish and that the activity of the SB transposase was due to its catalytic ability rather than merely binding to the vector DNA and facilitating its movement into the nucleus. Thus, the data in Tables 2.1 and 2.2 demonstrate that the SB transposon system is about as active for inserting DNA into chromatin in humans as in fish.

Table 2.2. SB transposase-dependent transposition in vertebrate chromosomes. Two plasmids were constructed, one encoding the SB transposase behind a cytomegalovirus promoter and the other with a transposon carrying an SV40-*neo* construct flanked by the SB IR sequences. The plasmid with the *neo* transposon was delivered to cultured cells by lipofection; the results show the relative levels of transformation when the plasmid with the SB transposase gene was co-transfected or not.

Cell line	– SB txpase	+ SB txpase	Increase
Carp EPC	362	868	2.4X
Mouse LMTK$^-$	66	308	4.7X
Human HeLa	34	417	12.3X

SB txpase, plasmid encoding SB transposase.

Construction of expression vectors for use with the transposon system in fish

Parallel to our development of transposition vectors, we initiated development of expression vectors that can be used either to overexpress, ectopically express or block expression of genes. Optimally, expression vectors will have a predictable level of gene expression and not be regulated by nearby *cis*-acting chromosomal sequences (position effects). We have found that border elements from flies and chickens function in zebrafish and confer not only position-independent expression, but also maintain expression levels through at least the F3 generation (Caldovic and Hackett, 1995; Caldovic and Hackett, 1998). In many transgenic experiments, it is desirable to have expression of the transgene that is not subject to position–effect variegation (i.e. the expression is not modulated by chromatin sequences flanking the integration site). We tested chick lysozyme *A* elements and *Drosophila* heat shock *scs/scs′* elements in zebrafish and found that they confer position-independent expression of a CAT transgene in every case where we found transmission of the genes (Caldovic and Hackett, 1995; Caldovic and Hackett, 1998) (Table 2.3). The data in Table 2.4 show that overall, our efficiency of producing transgenic fish is low, about 2.4%. Rates of integration do not appear to be influenced by the presence of border elements.

Table 2.3 shows that whereas the FV3CAT expression vector has non-uniform expression without border elements, with either of the two border elements it confers position-independent expression that is maintained through three generations of fish. Moreover, within a factor of two in zebrafish, the expression per gene per cell at 1, 2 and 5 days of development is constant for transgenes flanked with either *scs/scs′* or *A* elements. These data suggest that the border elements can confer reliable levels of expression of a transgene regardless of cell type or stage of development.

Table 2.3. Frequency of transgenic fish in F1, F2, and F3 generations.

Transgenic line	Expression pattern in F1	Total F1 fish	F1 fish	F2 fish	F3 fish
FV3CAT[a]	Non-uniform	522	13 (2.5%)	None found	Not done
scs(FV3CAT)scs[b]	Uniform	1278	46 (3.6%)	44/96 (46%)	Not done
A(FV3CAT)A[c]	Uniform	190	16 (8.4%)	166/317 (52%)	138/191 (72%)[d]

[a] Four lines of FV3CAT without border elements are totalled.
[b] Seven lines of FV3CAT flanked by *scs/scs'* border elements are totalled.
[c] Three lines of FV3CAT flanked by *A* elements are totalled.
[d] These F3 generation were interbred, leading to the higher percentage transmission of transgenes.

Table 2.4. Frequency of transgenic fish in F0 generation. One- or two-cell zebrafish embryos were injected with approximately 10^6–10^7 copies of the constructs shown in the table. Transgenic fish were identified by taking a small sample of fin tissue (fin clip) and screening for the presence of the CAT transgene by PCR. Fish that gave positive results were further analysed by Southern blotting to ensure that integration of the transgenic DNA had occurred (Caldovic and Hackett, 1998).

Construct	Total of F0 fish	Transgenic fish
FV3CAT	180	4 (2.2%)
scs(FV3CAT)scs	204	8 (4%)
A(FV3CAT)A	286	4 (1.4%)

Discussion

The results reported here demonstrate that it is now possible to move exogenous DNA more efficiently into the chromosomes of fish and other vertebrates using the SB transposon system. The mobilization and integration of exogenous, transgenic DNA depends only on the presence of transposase and inverted repeats of the correct sequence. Thus, we expect that the SB transposon will be useful in mammals, as well as in fish and birds, for delivering transgenic constructs into the chromosomes of animals (and perhaps plants) for biotechnological exploitation. Moreover, we have found that border elements flanking transgenes can be used to stabilize the expression of the exogenous genes such that their expression is consistent from line to line and is maintained during vertical transmission through the germline.

The results will have applications to many areas of aquacultural biotechnology. Development of transposable elements for vectors in animals will permit the following: (i) efficient insertion of genetic material into animal chromosomes; (ii) identification, isolation and characterization of genes involved with growth and development; (iii) identification, isolation and characterization of transcriptional regulatory sequences controlling growth and development; (iv) use of marker constructs for quantitative trait loci (QTL) analysis; (v) identification of genetic loci of economically important traits, besides those for growth and development, i.e. disease resistance; (vi) site-specific gene inactivation for obtaining stocks with specific inactivated genes. This could be useful for producing sterile transgenic fish so that broodstock with inactivated genes could be mated to produce sterile offspring for either biological containment or for maximizing growth rates in aquacultured fish. We have participated in the construction of a genetic vaccine for fish that protects them against a killer rhabdovirus, infectious haematopoietic necrosis virus (Anderson *et al.*, 1996). Use of our vectors in this area should be especially important to worldwide aquaculture.

Acknowledgements

We thank J. Essner, S. Fahrenkrug, C. Kaufman, D. Mohn and M. Simmons for discussions and reading the manuscript. This work was supported by grants from USDA (92–37205–7842), NIH (RO1-RR06625) and SeaGrant (USDOC/NA46RG O101–04).

References

Agellon, L.B., Emery, C.J., Jones, J.M., Davies, S.L., Dingle, A.D. and Chen, T.T. (1988) Promotion of rapid growth of rainbow trout (*Salmo gairdneri*) by a recombinant fish growth hormone. *Canadian Journal of Fisheries and Aquatic Science* 45, 146–151.

Anderson, E.D., Mourich, D.V., Fahrenkrug, S.C., LaPatra, S., Shepard, J. and Leong, J.-A.C. (1996) Genetic immunization of rainbow trout (*Oncorhynchus mykiss*) against infectious hematopoietic necrosis virus. *Molecular Marine Biology and Biotechnology* 5, 105–113.

Caldovic, L. and Hackett, P.B. (1995) Development of position-independent expression vectors and their transfer into transgenic fish. *Molecular Marine Biology and Biotechnology* 4, 51–61.

Caldovic, L. and Hackett, P.B. (1998) Position-independent expression and germline transmission of transgenic DNA in zebrafish. *Transgenic Research* (in press).

Chen, T.T., Kight, K., Lin, C.M., Powers, D.A., Gayat, M.H., Chatakondi, N., Ramboux, A.C., Duncan, P.L. and Dunham, R.A. (1993) Expression and inheritance of RSVLTR-rtGH1 complementary DNA in the trangenic common carp, *Cyprinus carpio. Molecular Marine Biology and Biotechnology* 2, 88–95.

Devlin, R.H., Yesaki, T.Y., Biagi, C., Donaldson, E.M., Swanson, P. and Chan, W.K. (1994) Extraordinary salmon growth. *Nature* 371, 209–210.

Du, S.J., Gong, Z., Fletcher, G.L., Shears, M.A., King, M.J., Idler, D.R. and Hew, C.L. (1992) Growth enhancement in transgenic Atlantic salmon by the use of an 'all fish' chimeric growth hormone gene construct. *Bio/Technology* 10, 176–181.

Dunham, R.A., Smitherman, R.O. and Rezk, M. (1994) Response and correlated responses to three generations of selection for increased body weight in channel catfish (*Ictalurus punctatus*). In: *World Aquaculture '94, Book of Abstracts*, p. 188.

Fletcher, G.L. and Davies, P.L. (1991) Transgenic fish for aquaculture. *Genetic Engineering* 13, 331–370.

Gibbs, P.D.L., Gray, A. and Thorgaard, G. (1994) Inheritance of *P* element and reporter gene sequences in zebrafish. *Molecular Marine Biology and Biotechnology* 3, 317–326.

Gross, M.L., Kapuscinski, A.R., Schneider, J.F., Liu, Z., Moav, N., Moav, B., Myster, S., Hew, C., Guise, K.S., Hackett P.B. and Faras, A.J. (1992) Growth evaluation of northern pike (*Esox lucius*) injected with growth hormone genes. *Aquaculture* 103, 253–273.

Hackett, P.B. (1993) The molecular biology of transgenic fish. In: Hochachka, P.W. and Mommsen, T.P. (eds) *Biochemistry and Molecular Biology of Fishes*, Vol. 2. Elsevier, Amsterdam, pp. 207–240.

Hallerman, E.M., Schneider, J.F., Gross, M., Liu, Z., Yoon, S.J., He, L., Hackett, P.B., Faras, A.J., Kapuscinski, A.R. and Guise, K.S. (1990) *Animal Biotechnology* 1, 79–93.

Herschberger, W.K., Meyers, J.M., Iwamoto, R.N., McAuley, W.C. and Saxon, A.M. (1990) Genetic changes in the growth of coho salmon (*Oncorhynchus kisutch*) in marine net-pens produced by ten years of selection. *Aquaculture* 85, 187–198.

Ivics, Z., Izsvák, Z. and Hackett, P.B. (1993) Enhanced incorporation of transgenic DNA into zebrafish chromosomes by a retroviral integration protein. *Molecular Marine Biology and Biotechnology* 2, 162–173.

Ivics, Z., Izsvák, Z., Minter, A. and Hackett, P.B. (1996) Identification of functional domains and evolution of Tc1 family of transposable elements. *Proceedings of the National Academy of Sciences USA* 93, 5008–5013.

Ivics, Z., Izsvák, Z. and Hackett, P.B. (1997) Molecular reconstruction of *Sleeping Beauty*, a *Tc1*-like transposon from fish, and its transposition in human cells. *Cell* 91, 501–510.

Iyengar, A., Müller, F. and Maclean, N. (1996) Expression and regulation of transgenes in fish. *Transgenic Research* 5, 147–165.

Izsvák, Z., Ivics Z. and Hackett, P.B. (1995) Characterization of a Tc1-like transposable element in zebrafish (*Danio rerio*). *Molecular and General Genetics* 247, 312–322.

Izsvák, Z., Ivics, Z., Garcia-Estefania, D., Fahrenkrug, S.C. and Hackett, P.B. (1996) DANA elements: a family of composite, short interspersed, DNA elements associated with mutational activities in zebrafish (*Danio rerio*). *Proceedings of the National Academy of Sciences USA* 93, 1044–1048.

Izsvák, Z., Ivics, Z. and Hackett, P.B. (1997) Repetitive elements and their genetic applications in zebrafish (review). *Biochemistry and Cell Biology* 75, 507–523.

Liu, Z., Moav, B., Faras, A.J., Guise, K.S., Kapuscinski, A.R. and Hackett, P.B. (1990) Development of expression vectors for gene transfer into fish. *BioTechnology* 8, 1268–1272.

Loukeris, T.G., Livadaras, I., Arca, B., Zabalou, S. and Savakis, C. (1995) Gene transfer into the medfly, *Ceratitis captiata*, with a *Drosophila hydei* transposable element. *Science* 270, 2002–2005.

Parfit, M. (1995) Diminishing returns: exploiting the ocean's bounty. *National Geographic* 185 (5), 2–37.

Radice, A.D., Bugaj, B., Fitch, D. and Emmons, S.W. (1994) Widespread occurence of the Tc1 transposon family: Tc1-like transposons from teleost fish. *Molecular and General Genetics* 244, 606–612.

Zhang, P., Hayat, M., Joyce, C., Gonzalez-Villasenor, L.I., Lin, C.M., Dunham, R.A., Chen, T.T. and Powers, D.A. (1990) Gene transfer, expression and inheritance of pRSV-rainbow trout-GH cDNA in the common carp, *Cyprinus carpio* (Linnaeus). *Molecular Reproduction and Development* 25, 3–13.

3

One Gene is Not Enough: Transgene Detection, Expression and Control

K.D. Wells and R.J. Wall

Gene Evaluation and Mapping Laboratory, USDA-ARS, Beltsville, Maryland, USA

Introduction

It is not surprising that recently reported successes with sheep and cattle nuclear transfer have rekindled interest in examining new methodologies for producing transgenic animals. The proponents of nuclear transfer and other new approaches in part justify their efforts by pointing to the inefficiency of producing transgenic livestock. The techniques for producing transgenic livestock have remained basically unchanged since the first transgenic farm animals were reported in 1985 (Hammer *et al.*, 1985). A review of the transgenic livestock literature suggests that the inefficiency of producing transgenic animals, by pronuclear microinjection, can be attributed to poor embryo survival, poor transgene integration rates and unpredictable transgene behaviour (Wall, 1996). At a minimum, performing nuclear transfer with genetically engineered nuclei will eliminate the inefficiency related to poor transgene integration rates. The impact of nuclear transfer on embryo/fetal survival and on the ability to select integrations with predictable transgene expression is still unknown. Therefore, until the new approaches are validated and are shown to increase the overall efficiency, we believe it is prudent to review some of the proposed strategies for improving the efficiency of current technology. Furthermore, no matter which way genes are introduced, their behaviour will be dictated by still ill-defined transgene design criteria. Though we are unable to offer a resolution to the vagaries of transgene design, it is our hope that some of the molecular approaches discussed will stimulate readers to re-evaluate their transgene design guidelines.

The title of this chapter was chosen in recognition of that fact that from the very beginning of transgenic animal technology, transgenes (fusion

© CAB INTERNATIONAL 1999. *Transgenic Animals in Agriculture* (eds J.D. Murray, G.B. Anderson, A.M. Oberbauer and M.M. McGloughlin)

genes) were generated from sequences borrowed from more than one gene. Also, many of the more promising techniques proposed to improve efficiency rely on using two genes: a selectable reporter gene and a gene of interest. Furthermore, the strategies for more precisely controlling transgene behaviour, discussed in this review, utilize several pieces of DNA from a variety of sources.

Identification of Transgenic Embryos as a Means of Improving Efficiency

No validated tool has been developed to precisely measure the frequency at which microinjected transgenes integrate into mammalian genomes. In the absence of a direct measure, we have been forced to estimate that parameter based on the proportion of transgenic offspring born. Published estimates of the percentage of transgenics born range from about 4 to 30%. The true integration frequency may be higher if one assumes that a greater proportion of transgenic embryos or fetuses is lost during development than their non-transgenic counterparts (Canseco *et al.*, 1994). Even though these estimates suggest that transgene integration is not a particularly rare event, maintaining an adequate recipient population (especially for monotocous species) is a costly proposition. In an attempt to reduce recipient costs a significant effort has been placed on identification of transgenic embryos before they are transferred to recipients. Primarily, two tactics have been employed, one based on direct detection of the transgene and the other based on detection of transgene expression. Unfortunately, there are rational theoretical grounds to question the wisdom of most approaches tested to date. A third approach, evaluating the methylation status of transgenes has received less attention.

Direct transgene detection

Direct transgene detection methods are, by necessity, PCR-based (there is simply not enough DNA in an embryo, or part thereof, to employ other techniques). The most serious drawback of PCR in this context is its inability to distinguish between transgenes that have integrated into the genome and copies of the transgenes that have not. The lack of distinction is problematic because it appears that unintegrated DNA can persist in the developing embryo for a significant time (Burdon and Wall, 1992; Cousens *et al.*, 1994). Modified PCR strategies that are dependent on ligation are capable of distinguishing between integrated and unintegrated transgenes (Ninomiya *et al.*, 1990; MacGregor and Overbeek, 1991; Jones and Winistorfer, 1992). Unfortunately, such techniques cannot be applied to this problem because the amount of DNA available in a few cells is insufficient.

However, it has been shown that following microinjection, transgenes frequently segregate during the early stages of development, resulting in a mosaic distribution. The mosaic nature of microinjected embryos has been demonstrated by indirectly monitoring transgene expression (Hattman *et al.*, 1978) and more directly by detection of transgenes in individual blastomeres (Burdon and Wall, 1992; Cousens *et al.*, 1994). Due to this mosaicism, detection of a transgene in two or three cells harvested from a developing eight- to 16-cell embryo could easily generate a misleading result.

It would appear that the inadequacy of PCR for this application, coupled with the mosaic distribution of transgenes in the developing embryo, has thus far thwarted attempts to select transgenic embryos based on direct detection of transgenes. None of the published accounts have succeeded in increasing the proportion of transgenic offspring or have not conducted the necessary experiments to fully validate this approach (Ninomiya *et al.*, 1989; Horvat *et al.*, 1993; Bowen *et al.*, 1994; Sparks *et al.*, 1994; Page *et al.*, 1995; Hyttinen *et al.*, 1996; Seo *et al.*, 1997).

Detection of transgene expression

Techniques based on monitoring transgene expression in preimplantation embryos are also hindered by lingering unintegrated transgenes which are likely expressed as well as, if not more strongly than, integrated copies. The phenomenon of diminished expression levels of integrated transgenes has clearly been demonstrated in tissue culture where levels of expression of stably transfected (integrated genes) cells are usually much lower than those of transient (unintegrated genes) transfected cells. Therefore, an expression-based selection system requires the potentially interfering background expression of unintegrated DNA to be eliminated or rendered non-detectable at the time of screening. If a regulatory element could be found that 'turns on' a reporter gene after unintegrated DNA is degraded, or if a transgene could be designed that only expressed after integration, this selection approach might work. However, even if unintegrated DNA has been degraded or rendered non-functional, background expression might still be a problem unless the mRNA and protein generated by the unintegrated transgenes have relatively short half-lives.

A number of different strategies to monitor transgene expression in preimplantation embryos have been proposed. They all depend on two genes, a reporter (selectable gene) and the gene of interest. One could devise non-destructive methods for detecting expression in the whole embryo or analyse expression in blastomere biopsies. Wisely, the expression-monitoring experiments that have been published thus far are based on evaluating the whole embryo. Each of the three approaches that have been tested relies on monitoring the expression of a reporter gene, but methods of detection differ.

In one case, expression monitoring was based on detecting a secreted gene product, luciferase (Thompson *et al.*, 1995). Surprisingly, the authors claimed that the expression of their reporter construct increased upon integration. This promising approach has not yet been used to demonstrate that selection based on high luciferase expression will yield a higher proportion of transgenic animals.

A selection system based on resistance to the toxic effects of the neomycin analogue G418, commonly used in tissue culture transfection studies, has been evaluated in mouse (Tada *et al.*, 1995) and bovine (Bondioli and Wall, 1996) embryos. G418 blocks translation, but embryos are protected if they express the neomycin resistance gene (*neo*). In the mouse study, the authors reported a 48% survival rate to the blastocyst stage after microinjection with the *neo* gene and 2 days of culture in G418 at a concentration which otherwise would kill 100% of non-injected embryos. However, these authors did not present evidence to demonstrate that *neo* selection increased the proportion of transgenic mice born. In the bovine embryo study the *neo* gene was fused to the *LacZ* gene in a bicistronic construct. Since bicistronic constructs generate a single mRNA, it can be reasonably assumed that β-galactosidase-positive blastomeres (those expressing the *LacZ* gene) also were producing the neomycin resistance protein. Staining for LacZ after 8 days of culture to the expanded blastocyst stage revealed that 15% of microinjected, but unselected, embryos were LacZ-positive (contained 80% or more β-galactosidase-positive blastomeres). Of the microinjected embryos treated with G418, 60% were LacZ-positive. It remains to be seen if the enrichment of expression-positive embryos translates into more transgenic calves. However, preliminary results using the same construct in mouse embryos suggest that G418 compromises development beyond the blastocyst stage (K.R. Bondioli, K.D. Wells and R.J. Wall, unpublished data).

Possibly the most obvious and straightforward approach for identifying transgenic embryos based on expression of a transgene, was the utilization of green fluorescent protein (GFP) as the reporter (Takada *et al.*, 1997). In that very encouraging study an elongation factor promoter ligated to GFP was injected into mouse embryos and green fluorescence was monitored at the morula/blastocyst stage. Eight out of 12 fetuses or pups resulting from green fluorescing embryos were transgenic as assessed by Southern blot analysis. The authors also reported that GFP could be detected in bovine embryos injected with the same construct. This is the first report that we are aware of that clearly demonstrates an increase in the proportion of transgenic animals based on a selection system. This approach warrants further investigation.

Even if GFP selection or another method can be found to enhance the proportion of transgenic animals born, the use of a selectable reporter gene could introduce some problems. Expression of the reporter could compromise embryo viability, thus reducing pregnancy rates. It is also possible that expression of the reporter transgene might interfere with the function of the co-selected transgene of interest. There is precedence for one transgene influencing expression of another following co-injection. John Clark and colleagues have shown that constructs that are expressed poorly can be rescued by co-injecting them with the ovine β-lactoglobulin gene (Clark *et al.*, 1992). However, they have also demonstrated that co-injected gene constructs can silence or diminish expression of otherwise highly expressed sequences (Clark *et al.*, 1997). The implication of those observations is that it might be prudent to isolate the reporter construct from the co-selected transgene by using DNA sequences such as matrix attachment regions (MAR) or scaffold attachment regions (SAR) which will be discussed later.

A selection concept that has yet to be tested in transgenic animals is based on designing a transgene that can only be expressed after integration. It may be possible to take advantage of the unique characteristics of peptide nucleic acids (PNA) to achieve this goal (Nielsen *et al.*, 1991; Almarsson *et al.*, 1993; Betts *et al.*, 1995; Norton *et al.*, 1995; Corey, 1997). The PNA molecule has a peptide backbone with purines and pyrimidines as the functional side groups (Nielsen and Haaima, 1997). The peptide backbone maintains the same distance between bases as nucleic acids but is uncharged. In the absence of the charged phosphodiester backbone, PNA annealed to DNA through normal base pairing is much more stable than other nucleic acids hybridized in the same fashion (Egholm *et al.*, 1993). The PNA/DNA interaction is sufficiently strong to prevent polymerases from displacing the PNA. The annealed PNA can therefore block progression of transcription. In theory, this ability to block expression could be used in microinjected embryos to silence transient expression of the transgene. Upon replication (assumed for this argument to be integration-dependent), the PNA/DNA hybrid is removed by the editing machinery of the cell. Once this block to expression is removed, an appropriately designed transgene can be detected by its expression.

Detection of methylation status

One approach to selection that does not neatly fit in either the direct detection or the expression monitoring paradigms is based on changes in the methylation state of specific bases in transgenes. Methylation at the N^6 of the adenine within the nucleotide sequence GATC is a characteristic of *dam⁺ Escherichia coli*. Mammalian cells do not normally methylate this adenine and consequently the absence of methylation following introduction of DNA of bacterial origin (transgenes isolated from plasmids) into mammalian cells is taken to indicate that the DNA has been replicated within the host cell (Hattman *et al.*, 1978; Peden *et al.*, 1980). Since replication would be expected to occur following integration in dividing cells, the absence of methylation would be indicative of transgene integration. The recognition site of the restriction enzyme *Dpn*I includes an adenine, but *Dpn*I only cuts if the adenine within the recognition site is methylated. Attempts to take advantage of the changing state of methylation pattern to monitor integration of transgenes in embryos have failed (Burdon and Wall, 1992; Cousens *et al.*, 1994). The failure might result from some specific inhibitory substance in embryos that blocks the action of *Dpn*I, modification of the methylation state of the *Dpn*I site independent of replication or possibly because transgenes are replicated before they are integrated (not a commonly accepted hypothesis).

Transgene Behaviour

Several strategies have been employed to regulate transgene expression patterns. Initial attempts utilized promoter elements from cloned mammalian genes to control tissue specificity and temporal expression. This strategy has been invaluable for targeting expression to particular tissues. However, it has proven to be of little use for temporal control due to limited availability of inducible promoters with low basal activity and the common occurrence of ectopic expression. To overcome this difficulty, strategies have been developed that utilize ligand-dependent sequence-specific DNA binding proteins to initiate or perturb transcription. The 'switches' thus far designed require a minimum of two genes. One gene encodes the 'switch protein' and the second gene responds to the switch protein (Fig. 3.1). Generally the switches fall into three categories. Those that inhibit expression, those that induce expression and those that utilize a 'double switch.'

Inhibition of basal expression

To inhibit expression of a transgene that would otherwise be expressed at some basal level, a strategy has been devised which is based on conditional binding of a switch protein to a specific DNA sequence within the transgene. The binding of the switch protein interferes with initiation or elongation of transcription. For example, transgenes can be designed that allow conditional binding of the *E. coli lac* repressor. The repressor binding sequences are placed in untranslated regions of the transgene and interfere with the production of functional transgene mRNA (Fig. 3.2a; Biard *et al.*, 1992). Similarly, the coding region of a silencing domain from a transcription factor such as KRAB (Margolin *et al.*, 1994) can be fused to the coding region of the DNA binding domain of *E. coli tet* repressor to produce a synthetic transcription factor which silences expression when bound to the transgene promoter (Fig. 3.2b; Deuschle *et al.*, 1995). In either of these cases, to provide complete repression, expression of the switch protein must be in every cell

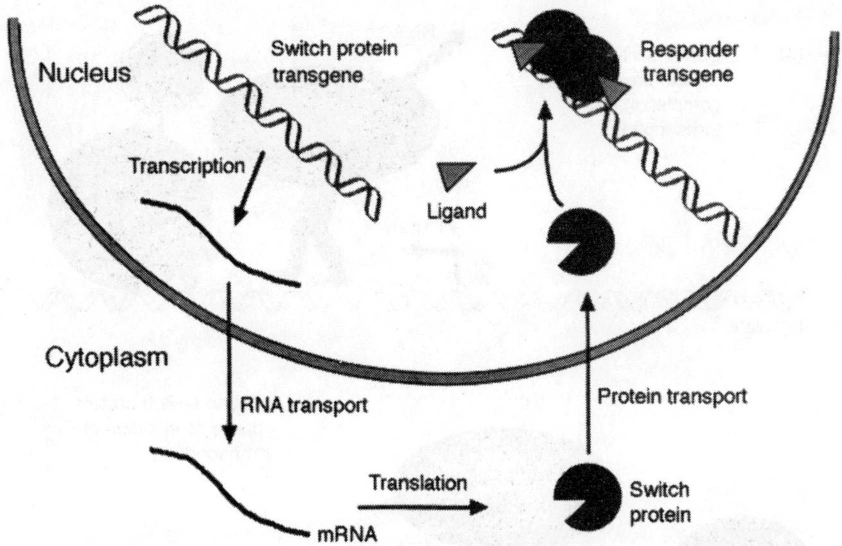

Fig. 3.1. A typical two-gene transgene switch. The transgene encoding the switch protein is controlled by a tissue-specific promoter sequence as necessary. Upon transcription of the switch protein gene and subsequent translation of its mRNA, the switch protein enters the nucleus. In this example, a ligand must interact with the switch protein for the protein switch to bind to its target sequence on the responder transgene. Supplying and withdrawing ligand affords temporal control. Switch systems can be designed that are ligand independent. They can be controlled by developmentally regulated promoters.

that would otherwise express the transgene. Expression would be induced by administration of a ligand (i.e. galactose or tetracycline analogues) which prevents sequence-specific binding of the switch protein and therefore allows the transgene to be expressed. One could predict that these strategies would have limited success in animals due to different ectopic expression patterns between the two transgenes. Switch protein would be required to express not only in the tissue targeted for transgene control but also in every cell that may otherwise express the gene of interest ectopically.

This type of switch may be useful in any situation that requires external control of transgene expression. For example, if one wanted to express a circulating hormone, such as somatotropin, and a dependable promoter with no ectopic expression was available, then that promoter could be used to drive expression of both the switch protein and somatotropin. The promoter driving expression of somatotropin would also include a binding site for the switch protein. When the switch protein was bound to the somatotropin transgene promoter, then only endogenous somatotropin would be produced. In breeding animals or very young animals this state would be desirable. When the switch protein was not bound to the somatotropin transgene promoter, the

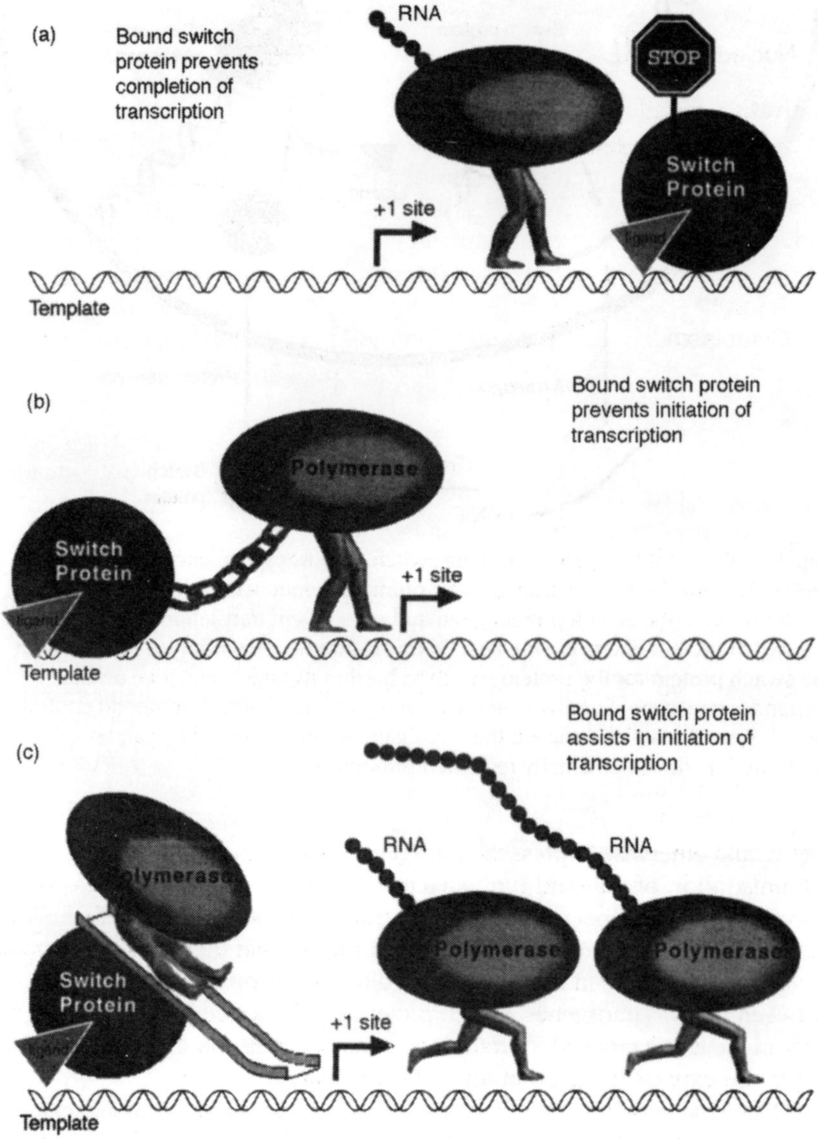

Fig. 3.2. Three ligand-dependent ways to control gene expression. (a) Ligand-dependent DNA binding proteins can perturb progression of RNA polymerase to prevent functional mRNA from being produced. In the presence of bound 'switch protein', the transgene is down-regulated. In the absence of bound switch protein, the transgene would be regulated by the transgene promoter. (b) Ligand-dependent DNA binding proteins can be fused to repressor domains to prevent initiation of transcription. The repressor domain interacts with a series of co-repressors to induce

peptide hormone would be made and secreted into the circulation. This state would be desirable during the growth phase of meat production animals.

Activation of induced expression

A second strategy attempts to mimic mammalian transcription factors to induce expression from an otherwise silent promoter. Two genes and an inducing molecule are required. One gene encodes the switch protein and the other responds when it binds the switch protein. The switch protein in these cases bind unique DNA sequences located upstream of a minimal promoter (Fig. 3.2c). The minimal promoter is typically composed of little more than a TATA box and the switch protein binding sequence. Minimal promoters can support initiation of transcription but require the switch protein to be bound to promote elevated expression. These switch proteins are again ligand-dependent conditional binding proteins but now include transactivation domains. In the first scenario the intention was to turn off a gene, whereas this strategy is designed to turn genes on. For example, a mutant *E. coli tet* repressor that binds a specific DNA sequence only in the presence of a tetracycline analogue, doxycycline, can be fused to the transactivation domain of herpes VP16 to make a protein referred to as rtTA (Gossen *et al.*, 1995). In the absence of doxycycline, the switch protein, rtTA, does not specifically bind to the responder transgene minimal promoter and the transgene is silent. Upon administration of doxycycline, the switch protein binds to the DNA included in the minimal promoter of the responder transgene and transgene expression is induced.

Similarly, transcription factors from other eukaryotes such as the *Drosophila melanogaster* ecdysone receptor can be used as the switch protein. The ecdysone receptor normally functions as a heterodimer with the

---

presumptive conformational changes represented here as a chain. In the presence of bound switch protein and endogenous co-repressor protein, the transgene is down-regulated. In the absence of the bound switch protein, the transgene is regulated by the transgene promoter. (c) Ligand-dependent DNA binding proteins can be fused to transactivation domains to enhance initiation of transcription. The transactivation domain interacts with a series of co-activators to stimulate assembly and initiation shown here as a slide. In transgenes with no other enhancers, the minimal promoter is insufficient to drive expression. In the presence of bound switch protein, the transgene is induced to express from a minimal promoter.

gene product ultraspiracle (Yao *et al.*, 1992). Fortunately, the mammalian homologue, retenoic acid receptor X (RXR), can serve in the place of ultraspiracle to heterodimerize with the ecdysone receptor to produce a functional transcription factor (Thomas *et al.*, 1993). Thus, the ecdysone receptor can function as a switch protein in cells that express RXR (Christopherson *et al.*, 1992). A mutant progesterone receptor has also been used as a switch protein (Wang *et al.*, 1997). This mutant does not induce expression in the presence or absence of endogenous steroids but is active upon administration of very low levels of the progesterone antagonist RU486. To prevent induction of endogenous progesterone responsive genes by the switch protein, the native DNA binding domain was replaced with the yeast Gal4 DNA binding domain (Wang *et al.*, 1994). Similar modifications have also been made to the ecdysone receptor (No *et al.*, 1996).

This type of switch may be useful in same scenarios as the first strategy. However, when low background is a major concern this switch is less dependent on a predictable promoter for the switch protein. Instead, this type of switch is most dependent on the construction of a responder transgene that has very low background expression. The level of induction is dependent on both the promoter of the switch protein and the inducibilty of an otherwise silent responder gene.

Cross-breed for expression

The third category of switches does not perturb or mimic mammalian transcription. Instead, this strategy finds an alternative method for producing an active gene. It still requires a switch protein gene in addition to the transgene of interest. One example would be to use a phage RNA polymerase to drive expression of a transgene (Lieber *et al.*, 1989). In this strategy, the gene of interest need not contain any mammalian promoter element thus ensuring little or no basal expression. In the presence of the phage RNA polymerase the gene is active. In the absence of the phage RNA polymerase the gene is inactive. The tissue specificity is provided by the promoter used to drive the phage RNA polymerase expression. The timing of the expression is also provided by the promoter used to drive polymerase expression. Therefore, a second switching mechanism is required to control the polymerase (cross-breeding serves as the second 'switch' in this example). The use of a double switch has been demonstrated to provide tighter control (Aubrecht *et al.*, 1996) but this binary strategy is primarily applicable to situations where transgene expression needs to be regulated in a particular generation and requires the production of two types of transgenic animals. For example, one might want to express a transgene that could prevent sexual maturation or gonadal development as an alternative to castration. If an animal carried both the phage RNA polymerase gene and the responder gene, those individuals could not reproduce. However, if breeding stock were developed so that one line carried the responsive gene of interest and another line harboured the switch protein gene, then only the offspring generated by crossing those lines would have this phenotype. Similar arguments could be made for transgenic fish farms to prevent a phenotype from being transferred to wild populations even if the transgenes themselves are transferred to wild populations. Like the phage RNA polymerase, recombinases which remove inhibitory sequence or invert a transgene coding sequence, can be used to control gene expression. Some success using both the yeast FLP recombinase and the bacterial Cre recombinase has been demonstrated in tissue culture and transgenic animals (O'Gorman *et al.*, 1991; Barlow *et al.*, 1997).

The strategy utilizing synthetic transcription factors as the switch protein will continue to be improved. Already the tetracycline repressor switch has incorporated DNA binding mutants such that the wild-type version can be induced by ligand withdrawal and the mutant version by ligand administration. In many livestock scenarios breeding strategies required to use the binary approach are impractical. However, for livestock breeding companies that want to control proprietary genetics, a binary system may provide a means of protecting their unique genetic stock. For other situations where efficient transmission of transgenes is important, the genes of the binary system may be combined into a single gene construct incorporating both coding regions in the same construct (Schultze *et al.*, 1996). However, these systems are greatly affected by integration position. Integration sites that provide adequate transgene induction tend to have greater basal expression. If the effect of integration site can be removed or reduced, the strategy may become very useful.

Integration Site Effects

In most studies about half of the transgenic founder animals do not express their transgene. The efficiency of producing transgenic animals could be doubled if a way could be found to correct that problem. Most transgenes are composite sequences originating from a variety of genes, species or even kingdoms. The regulatory elements are often less than fully characterized and the structural gene can contain sequences that interfere with the intended regulation. In addition, the sequences surrounding the integration may be as varied as the genome itself. The surround sequence may therefore also interfere with the intended regulation of the transgene. Several strategies have been used to overcome some of these problems.

Insulators

The location in which a transgene integrates can contain enhancers, silencers or other elements that may interact with the transgene promoter. Transgene expression therefore can be greatly affected by site of integration. In

D. melanogaster, sequences have been identified on the basis of their ability to functionally separate enhancers from their corresponding promoters (Cal and Levine, 1995; Gdula and Corces, 1997). These regions therefore insulate the sequence on one side of the insulating element from effectors located on the other side of the element. One example of a *D. melanogaster* insulator is located on a P-element, *Gypsy* (Roseman *et al.*, 1995). The *Gypsy* insulator has been demonstrated to reduce the effect of integration site on transgenes. A vertebrate sequence has been shown to be functionally similar to *Gypsy* by insulating transgenes from position effects in transgenic flies (Chung *et al.*, 1993). This vertebrate sequence was characterized by its location as the boundary of the DNase hypersensitive region of the expressing locus of chicken lysozyme gene (Stief *et al.*, 1989). This boundary element was further demonstrated to have sequences which bind to the nuclear matrix (MAR) and are thought to define the border of the lysozyme locus. In transgenic mice, this sequence was also shown to insulate heterologous transgenes from position effects, to allow more appropriate developmental timing of expression, and to reduce ectopic expression (McKnight *et al.*, 1992, 1996; Greenberg *et al.*, 1994). Sequences from other genes have been described as locus control regions (LCRs) (Stamatoyannopoulos *et al.*, 1997), as chromosomal SARs (Allen *et al.*, 1996) and MARs. At some level these DNA fragments have related functions since they all can decrease position effects. However, these sequences are not identical nor are the effects of these regions universal among different transgenes (Barash *et al.*, 1996a).

MARs, as mentioned above, have sequences that can bind the nuclear matrix and/or chromosomal scaffold. As such, these sequences are thought to anchor chromatin loops to the central organizing matrix of each chromosome. This function alone cannot explain the insulating ability of MARs. This function however is associated with maintenance of an open chromatin region once a gene is transactivated (Ciejek *et al.*, 1983; Tewari *et al.*, 1996). Regions within some MARs may assist in transactivator access to promoters. The chicken lysozyme MAR does contain intrinsically bent DNA – a structure associated with transactivator accessibility (Phi-Van and Stratling, 1996).

An insulating function such as that found in the globin LCR or the *Gypsy* insulator may not be associated with matrix binding yet can interact with the basal transcription machinery to prevent interactions with distal enhancers. Such a function has been implicated in the endogenous regulation of globin genes to prevent fetal globin enhancers from activating adult globin promoters (Boulikas, 1994). Given the functional similarity between *Gypsy* and the chicken lysozyme MAR, sequences within or near some MARs may be binding sites for insulator proteins such as those described for *Gypsy*, suppressor of *Hairy-wing* (Smith and Corces, 1992).

The mechanism of protection from integration position effects is unknown. In many cases MAR-like sequences may provide some protection from position effects. Providing locus boundaries, scaffold attachment, topoisomerase substrates or direct interaction between enhancers and the basal transcription machinery may be involved in the action of MARs (Boulikas, 1995). The use of MARs in specific scenarios has proven to be less than predictable (Barash *et al.*, 1996a). However, as another tool in the arsenal of DNA elements available for transgene design, MARs will prove to be very useful in many cases.

Rescue

In the above discussions, the availability of tissue-specific promoters has been assumed to be present. However, very few loci have been sufficiently described as to assure that a particular promoter region contains sufficient elements to provide the predicted expression pattern. As the globin locus has illustrated, regulatory elements can be located at distances impractical for inclusion in most plasmids (Slightom *et al.*, 1997). Regulatory elements have been described upstream and downstream of transcriptional start sites, within coding regions and within introns. When spatial relationships between regulatory elements are important it may not be possible to create a transgene which maintains those relationships. As an alternative, non-manipulated configurations have been coinjected with transgene constructs to increase the fidelity of regulation (Barash *et al.*, 1996b). This strategy has been called 'rescue' and has, in fact, rescued correct expression patterns from a transgene which otherwise does not express as predicted. Again, the mechanism and universal application is unknown. However, in the absence of full understanding of the elements required for correct expression, this strategy allows a spatially correct pattern of regulatory elements to be included within a transgene locus.

Native regulation

As an alternative to transgene rescue, inclusion of transgenes with native context offers the potential to generate spatially correct transgenes with all or most regulatory elements for the endogenous gene. One such strategy is the insertion, by homologous recombination, of the coding region of interest within the native gene whose expression pattern meets all requirements. This strategy is similar to the proven technologies of enhancer trapping and promoter trapping but relies on precise placement of the transcoding region within an endogenous locus. If the disturbed allele is required for some other function, this strategy may create unanticipated problems. These unknowns may be circumvented by including an internal ribosome entry site for the transcoding region or the endogenous coding region. One still risks perturbing expression at the targeted allele by changing spatial relationships or by inserting within important sequences. Likewise the introduction of regulatory elements within the transcoding region or its introns may also create an unpredicted expression pattern. However unproven, this strategy remains potentially useful.

When the transgene needs to retain its endogenous expression pattern, very large fragments of DNA such as those found in yeast or bacterial artificial chromosomes (YACs or BACs, respectively) may prove useful (Peterson *et al.*, 1996). By using such large regions of the genome, inclusion of the entire locus is more likely. However, this strategy will generally prevent known sequence from being integrated unless the entire artificial chromosome has been sequenced. The lack of complete sequence information would be of serious concern for commercial projects in the US because the Federal Drug Agency currently requires such information.

Codon bias

As suggested above, transgene coding regions may include regulatory elements that generate unpredicted expression patterns when combined with well described promoters. The unexpected pattern of expression may be no detectable expression at all. Regulatory elements have been described within both coding regions and introns. However, since prokaryotes have had no genetic selection against sequence which have functions in eukaryotes, it is inevitable that silencers, enhancers, splice sites and degradation signals will be present in some genes of bacterial origin. Such sequences have been described (Artelt *et al.*, 1991; Kushner *et al.*, 1994). In addition, the base composition of DNA between divergent species is sometimes very different. Those differences are reflected in the codons that are favoured by particular species. Some tRNAs are relatively rare in species that have a codon bias and the codon for those tRNAs are also rare. The extreme sequence differences that cause codon bias between species may cause inefficiencies in translation, but those differences may also represent sequences that have been selected against in the target organism. For example, attempts to express an insect resistance gene (*Bacillus thuringiensis*, cryIA) in transgenic plants failed until the coding region was chemically synthesized to generate a coding region that statistically represented the DNA sequence most likely to encode the desired amino acid sequence (Perlak *et al.*, 1991). By altering the codon bias, three poly(A) cleavage signals were removed. The resulting coding region, now codon biased for the host plant, expressed at very high levels. This strategy remains to be proven in transgenic vertebrates but codon bias has been shown to be important in many other systems (Perlak *et al.*, 1993; Alexandrova *et al.*, 1995).

Conclusions

The search for alternative approaches for producing transgenic animals is well justified in light of the low efficiency of the current process which results in high-cost scientific projects that are often beyond the capacity of granting agencies to fund. Recent developments in nuclear transfer promise to eliminate at least one of the low efficiency steps, namely poor integration rates. The heightened interest in nuclear transfer may also result in identifying new embryo culture systems that address another of the low efficiency steps, that of poor embryo/fetal survival. Nevertheless, it may not be time to abandon microinjection as a method for producing transgenic animals, especially in light of the encouraging results with at least one scheme for selecting transgenic embryos. If GFP expression can be used as effectively for selection of transgenic livestock embryos as it was in mice, producing 50–60% transgenic founders would make current technology more appealing.

Furthermore, new approaches to gene transfer are going to have little immediate impact on correcting transgene function. Even if genes can be readily introduced by homologous recombination, the transgenic mouse literature suggests that homologous recombination will not always be an appropriate strategy and will not be a panacea for achieving precise control of gene behaviour. Use of isolating elements and transgene switches hold the best promise for controlling the action of transgenes. Transgene switches have already been used to control formation and regression of carcinomas (Furth *et al.*, 1994), control gene expression in embryonal carcinoma cells (Miller and Rizzino, 1995) and heart muscle (Yu *et al.*, 1996) and in plants (Weinmann *et al.*, 1994). It is probably time for these switches to be tested in transgenic livestock.

References

Alexandrova, R., Eweida, M., Georges, F., Dragulev, B., Abouhaidar, M.G. and Ivanov, I. (1995) Domains in human interferon α-1 gene containing tandems of arginine codons AGG play the role of translational initiators in *Escherichia coli*. *International Journal of Biochemistry and Cell Biology* 27, 469–473.

Allen, G.C., Hall, G.J., Michalowski, S., Newman, W., Spiker, S., Weissinger, A.K. and Thompson, W.F. (1996) High-level transgene expression in plant cells: effects of a strong scaffold attachment region from tobacco. *Plant Cell* 8, 899–913.

Almarsson, O., Bruice, T.C., Kerr, J. and Zuckermann, R.N. (1993) Molecular mechanics calculations of the structures of polyamide nucleic acid DNA duplexes and triple helical hybrids. *Proceedings of the National Academy of Sciences USA* 90, 7518–7522.

Artelt, P., Grannemann, R., Stocking, C., Friel, J., Bartsch, J. and Hauser, H. (1991) The prokaryotic neomycin-resistance-encoding gene acts as a transcriptional silencer in eukaryotic cells. *Gene* 99, 249–254.

Aubrecht, J., Manivasakam, P. and Schiestl, R.H. (1996) Controlled gene expression in mammalian cells via a regulatory cascade involving the tetracycline transactivator and lac repressor. *Gene* 172, 227–231.

Barash, I., Ilan, N., Kari, R., Hurwitz, D.R. and Shani, M. (1996a) Co-integration of β-lactoglobulin/human serum albumin hybrid genes with the entire β-lactoglobulin gene or the matrix attachment region element: repression of human serum albumin and β-lactoglobulin expression in the mammary gland and dual regulation of the transgenes. *Molecular Reproduction and Development* 45, 421–430.

Barash, I., Nathan, M., Kari, R., Ilan, N., Shani, M. and Hurwitz, D.R. (1996b) Elements within the β-lactoglobulin gene inhibit expression of human serum albumin cDNA and minigenes in transfected cells but rescue their expression in the mammary gland of transgenic mice. *Nucleic Acids Research* 24, 602–610.

Barlow, C., Schroeder, M., Lekstrom-Himes, J., Kylefjord, H., Deng, C.X., Wynshaw-Boris, A., Spiegelman, B.M. and Xanthopoulos, K.G. (1997) Targeted expression of *Cre* recombinase to adipose tissue of transgenic mice directs adipose-specific excision of LoxP-flanked gene segments. *Nucleic Acids Research* 25, 2543–2545.

Betts, L., Josey, J.A., Veal, J.M. and Jordan, S.R. (1995) A nucleic acid triple helix formed by a peptide nucleic acid–DNA complex. *Science* 270, 1838–1841.

Biard, D.S., James, M.R., Cordier, A. and Sarasin, A. (1992) Regulation of the *Escherichia coli lac* operon expressed in human cells. *Biochimica Biophysica Acta* 1130, 68–74.

Bondioli, K.R. and Wall, R.J. (1996) Positive selection of transgenic bovine embryos in culture. *Theriogenology* 45, 345.

Boulikas, T. (1994) Transcription factor binding sites in the matrix attachment region (MAR) of the chicken alpha-globulin gene. *Journal of Cellular Biochemistry* 55, 513–529.

Boulikas, T. (1995) Chromatin domains and prediction of MAR sequences. *International Review of Cytology* 162A, 279–288.

Bowen, R.A., Reed, M.L., Schnieke, A., Seidel, G.E., Jr, Stacey, A., Thomas, W.K. and Kajikawa, O. (1994) Transgenic cattle resulting from biopsied embryos: expression of *c-ski* in a transgenic calf. *Biology of Reproduction* 50, 664–668.

Burdon, T.G. and Wall, R.J. (1992) Fate of microinjected genes in preimplantation mouse embryos. *Molecular Reproduction and Development* 33, 436–442.

Cal, H. and Levine, M. (1995) Modulation of enhancer–promoter interactions by insulators in the *Drosophila* embryo. *Nature* 376, 533–536.

Canseco, R.S., Sparks, A.E., Page, R.L., Russell, C.G., Johnson, J.L., Velander, W.H., Pearson, R.E., Drohan, W.N. and Gwazdauskas, F.C. (1994) Gene transfer efficiency during gestation and the influence of co-transfer of non-manipulated embryos on production of transgenic mice. *Transgenic Research* 3, 20–25.

Christopherson, K.S., Mark, M.R., Bajaj, V. and Godowski, P.J. (1992) Ecdysteroid-dependent regulation of genes in mammalian cells by a *Drosophila* ecdysone receptor and chimeric transactivators. *Proceedings of the National Academy of Sciences USA* 89, 6314–6318.

Chung, J.H., Whiteley, M. and Felsenfeld, G. (1993) A 5′ element of the chicken *beta-globin* domain serves as an insulator in human erythroid cells and protects against position effect in *Drosophila*. *Cell* 74, 505–514.

Ciejek, E.M., Tsai, M.J. and O'Malley, B.W. (1983) Actively transcribed genes are associated with the nuclear matrix. *Nature* 306, 607–609.

Clark, A.J., Cowper, A., Wallace, G., Wright, G. and Simons, J.P. (1992) Rescuing transgene expression by co-injection. *Bio/Technology* 10, 1450–1454.

Clark, A.J., Harold, G. and Yull, F.E. (1997) Mammalian cDNA and prokaryotic reporter sequences silence adjacent transgenes in transgenic mice. *Nucleic Acids Research* 25, 1009–1014.

Corey, D.R. (1997) Peptide nucleic acids: expanding the scope of nucleic acid recognition. *Trends in Biotechnology* 15, 224–229.

Cousens, C., Carver, A.S., Wilmut, I., Colman, A., Garner, I. and O'Neill, G.T. (1994) Use of PCR-based methods for selection of integrated transgenes in preimplantation embryos. *Molecular Reproduction and Development* 39, 384–391.

Deuschle, U., Meyer, W.K. and Thiesen, H.J. (1995) Tetracycline-reversible silencing of eukaryotic promoters. *Molecular and Cellular Biology* 15, 1907–1914.

Egholm, M., Buchardt, O., Christensen, L., Behrens, C., Freier, S.M., Driver, D.A., Berg, R.H., Kim, S.K., Norden, B. and Nielsen, P.E. (1993) PNA hybridizes to complementary oligonucleotides obeying the Watson–Crick hydrogen-bonding rules. *Nature* 365, 566–568.

Furth, P.A., St Onge, L., Boger, H., Gruss, P., Gossen, M., Kistner, A., Bujard, H. and Hennighausen, L. (1994) Temporal control of gene expression in transgenic mice by a tetracycline-responsive promoter. *Proceedings of the National Academy of Sciences USA* 91, 9302–9306.

Gdula, D.A. and Corces, V.G. (1997) Characterization of functional domains of the su(Hw) protein that mediate the silencing effect of mod(mdg4) mutations. *Genetics* 145, 153–161.

Gossen, M., Freundlieb, S., Bender, G., Muller, G., Hillen, W. and Bujard, H. (1995) Transcriptional activation by tetracyclines in mammalian cells. *Science* 268, 1766–1769.

Greenberg, N.M., DeMayo, F.J., Sheppard, P.C., Barrios, R., Lebovitz, R., Finegold, M., Angelopoulou, R., Dodd, J.G., Duckworth, M.L., Rosen, J.M. and Matusik, R.J. (1994) The rat *probasin* gene promoter directs hormonally and developmentally regulated expression of a heterologous gene specifically to the prostate in transgenic mice. *Molecular Endocrinology* 8, 230–239.

Hammer, R.E., Pursel, V.G., Rexroad, C.E. Jr, Wall, R.J., Bolt, D.J., Ebert, K.M., Palmiter, R.D. and Brinster, R.L. (1985) Production of transgenic rabbits, sheep and pigs by microinjection. *Nature* 315, 680–683.

Hattman, S., Brooks, J.E. and Masurekar, M. (1978) Sequence specificity of the P1 modification methylase (*M.Eco P1*) and the DNA methylase (*M.Eco dam*) controlled by *Escherichia coli* dam gene. *Journal of Molecular Biology* 126, 367–369.

Horvat, S., Medrano, J.F., Behboodi, E., Anderson, G.B. and Murray, J.D. (1993) Sexing and detection of gene construct in microinjected bovine blastocysts using the polymerase chain reaction. *Transgenic Research* 2, 134–140.

Hyttinen, J.M., Peura, T., Tolvanen, M., Aalto, J. and Jänne, J. (1996) Detection of microinjected genes in bovine preimplantation embryos with combined DNA digestion and polymerase chain reaction. *Molecular Reproduction and Development* 43, 150–157.

Jones, D.H. and Winistorfer, S.C. (1992) Sequence specific generation of a DNA panhandle permits PCR amplification of unknown flanking DNA. *Nucleic Acids Research* 3, 595–600.

Kushner, P.J., Baxter, J.D., Duncan, K.G., Lopez, G.N., Schaufele, F., Uht, R.M., Webb, P. and West, B.L. (1994) Eukaryotic regulatory elements lurking in plasmid DNA: the activator protein-1 site in pUC. *Molecular Endocrinology* 8, 405–407.

Lieber, A., Kiessling, U. and Strauss, M. (1989) High level gene expression in mammalian cells by a nuclear T7-phase RNA polymerase. *Nucleic Acids Research* 17, 8485–8493.

MacGregor, G.R. and Overbeek, P.A. (1991) Use of a simplified single-site PCR to facilitate cloning of genomic DNA sequences flanking a transgene integration site. *PCR Methods and Applications* 2, 129–135.

Margolin, J.F., Friedman, J.R., Meyer, W.K., Vissing, H., Thiesen, H.J. and Rauscher, F.J. (1994) Kruppel-associated boxes are potent transcriptional repression. *Proceedings of the National Academy of Sciences USA* 91, 4509–4513.

McKnight, R.A., Shamay, A., Sankaran, L., Wall, R.J. and Hennighausen, L. (1992) Matrix-attachment regions can impart position-independent regulation of a tissue-specific gene in transgenic mice. *Proceedings of the National Academy of Sciences USA* 89, 6943–6947.

McKnight, R.A., Spencer, M., Wall, R.J. and Hennighausen, L. (1996) Severe position effects imposed on a 1 kb mouse whey acidic protein gene promoter are overcome by heterologous matrix attachment regions. *Molecular Reproduction and Development* 44, 179–184.

Miller, K. and Rizzino, A. (1995) The function of inducible promoter systems in F9 embryonal carcinoma cells. *Experimental Cell Research* 218, 144–150.

Nielsen, P.E. and Haaima, G. (1997) Peptide nucleic acid (PNA). A DNA mimic with a pseudopeptide backbone. *Chemical Society Reviews* 26, 73–78.

Nielsen, P.E., Egholm, M., Berg, R.H. and Buchardt, O. (1991) Sequence-selective recognition of DNA by strand displacement with a thymine-substituted polyamide. *Science* 254, 1497–1500.

Ninomiya, T., Hoshi, M., Mizuno, A., Nagao, M. and Yuki, A. (1989) Selection of mouse preimplantation embryos carrying exogenous DNA by polymerase chain reaction. *Molecular Reproduction and Development* 1, 242–248.

Ninomiya, T., Iwabuchi, T., Soga, Y. and Yuki, A. (1990) Direct sequencing of flanking regions of a transgene amplified by inverted PCR. *Agricultural and Biological Chemistry* 7, 1869–1872.

No, D., Yao, T.P. and Evans, R.M. (1996) Ecdysone-inducible gene expression in mammalian cells and transgenic mice. *Proceedings of the National Academy of Sciences USA* 93, 3346–3351.

Norton, J.C., Waggenspack, J.H., Varnum, E. and Corey, D.R. (1995) Targeting peptide nucleic acid-protein conjugates to structural features within duplex DNA. *Bioorganic and Medicinal Chemistry* 3, 437–445.

O'Gorman, S., Fox, D.T. and Wahl, G.M. (1991) Recombinase-mediated gene activation and site-specific integration in mammalian cells. *Science* 251, 1351–1355.

Page, R.L., Canseco, R.S., Russell, C.G., Johnson, W.H., Velander, W.H. and Gwazdauskas, F.C. (1995) Transgene detection during early murine embryonic development after pronuclear microinjection. *Transgenic Research* 4, 12–17.

Peden, K.W.C., Pipas, J.M., Pearson-White, S. and Nathan, D. (1980) Isolation of mutants of an animal virus in bacteria. *Science* 209, 1392–1396.

Perlak, F.J., Fuchs, R.L., Dean, D.A., McPherson, S.L. and Fischhoff, D.A. (1991) Modification of the coding sequence enhances plant expression of insect control protein genes. *Proceedings of the National Academy of Sciences USA* 88, 3324–3328.

Perlak, F.J., Stone, T.B., Muskopf, Y.M., Petersen, L.J., Parker, G.B., McPherson, S.A., Wyman, J., Love, S., Reed, G. and Biever, D. (1993) Genetically improved potatoes: protection from damage by Colorado potato beetles. *Plant Molecular Biology* 22, 313–321.

Peterson, K.R., Clegg, C.H., Navas, P.A., Norton, E.J., Kimbrough, T.G. and Stamatoyannopoulos, G. (1996) Effect of deletion of 5′HS3 or 5′HS2 of the human *beta-globin* locus control region on the developmental regulation of *globin* gene expression in *beta-globin* locus yeast artificial chromosome transgenic mice. *Proceedings of the National Academy of Sciences USA* 93, 6605–6609.

Phi-Van, L. and Stratling, W.H. (1996) Dissection of the ability of the chicken lysozyme gene 5′ matrix attachment region to stimulate transgene expression and to dampen position effects. *Biochemistry* 35, 10735–10742.

Roseman, R.R., Johnson, E.A., Rodesch, C.K., Bjerke, M., Nagoshi, R.N. and Geyer, P.K. (1995) A P element containing suppressor of hairy-wing binding regions has novel properties for mutagenesis in *Drosophila melanogaster. Genetics* 141, 1061–1074.

Schultze, N., Burki, Y., Lang, Y., Certa, U. and Bluethmann, H. (1996) Efficient control of gene expression by single step integration of the tetracycline system in transgenic mice. *Nature Biotechnology* 14, 499–503.

Seo, B.B., Kim, C.H., Tojo, H., Tanaka, S., Yamanouchi, K., Takahashi, M., Sawasaki, T. and Tachi, C. (1997) Efficient selection of preimplantation transgenic embryos by an improved procedure using *Dpn I-Bal 31* digestion and the polymerase chain reaction. *Reproduction, Fertility and Development* 9, 263–269.

Slightom, J.L., Bock, J.H., Tagle, D.A., Gumucio, D.L., Goodman, M., Stojanovic, N., Jackson, J., Miller, W. and Hardison, R. (1997) The complete sequences of the galago and rabbit *beta-globin* locus control regions: extended sequence and functional conservation outside the cores of DNase hypersensitive sites. *Genomics* 39, 90–94.

Smith, P.A. and Corces, V.G. (1992) The suppressor of *Hairy-wing* binding region is required for *Gypsy* mutagenesis. *Molecular and General Genetics* 233, 65–70.

Sparks, A.E.T., Canseco, R.S., Russell, C.G., Johnson, J.L., Moll, H.D., Velander, W.H. and Gwazdauskas, F.C. (1994) Effects of time of deoxyribonucleic acid microinjection on gene detection and *in vitro* development of bovine embryos. *Journal of Dairy Science* 77, 718–724.

Stamatoyannopoulos, J.A., Clegg, C.H. and Li, Q. (1997) Sheltering of *gamma-globin* expression from position effects requires both an upstream locus control region and a regulatory element 3′ to the A *gamma-globin* gene. *Molecular and Cellular Biology* 17, 240–247.

Stief, A., Winter, D.M., Stratling, W.H. and Sippel, A.E. (1989) A nuclear DNA attachment element mediates elevated and position-independent gene activity. *Nature* 341, 343–345.

Tada, N., Sato, M., Hayashi, K., Kasai, K. and Ogawa, S. (1995) *In vitro* selection of transgenic mouse embryos in the presence of G-418. *Transgenics* 1, 535–540.

Takada, T., Iida, K., Awaji, T., Itoh, K., Takahashi, R., Shibui, A., Yoshida, K., Sugano, S. and Tsujimoto, G. (1997) Selective production of transgenic mice using green fluorescent protein as a maker. *Nature Biotechnology* 15, 458–461.

Tewari, R., Gillemans, N., Harper, A., Wijgerde, M., Zafarana, G., Drabek, D., Grosveld, F. and Philipsen, S. (1996) The human beta-globin locus control region confers an early embryonic erythroid-specific expression pattern to a basic promoter driving the bacterial *lacZ* gene. *Development* 122, 3991–3999.

Thomas, H.E., Stunnenberg, H.G. and Stewart, A.F. (1993) Heterodimerization of the *Drosophila* ecdysone receptor with retinoid X receptor and ultraspiracle. *Nature* 362, 471–475.

Thompson, E.M., Adenot, P., Tsuji, F.I. and Renard, J.P. (1995) Real time imaging of transcriptional activity in live mouse preimplantation embryos using a secreted luciferase. *Proceedings of the National Academy of Sciences USA* 92, 1317–1321.

Wall, R.J. (1996) Transgenic livestock: progress and prospects for the future. *Theriogenology* 45, 57–68.

Wang, Y., O'Malley, B.W.J , Tsai, S.Y. and O'Malley, B.W. (1994) A regulatory system for use in gene transfer. *Proceedings of the National Academy of Sciences USA* 91, 8180–8184.

Wang, Y., DeMayo, F.J., Tsai, S.Y. and O'Malley, B.W. (1997) Ligand-inducible and liver-specific target gene expression in transgenic mice. *Nature Biotechnology* 15, 239–243.

Weinmann, P., Gossen, M., Hillen, W., Bujard, H. and Gatz, C. (1994) A chimeric transactivator allows tetracycline-responsive gene expression in whole plants. *Plant Journal* 5, 559–569.

Yao, T.P., Segraves, W.A., Oro, A.E., McKeown, M. and Evans, R.M. (1992) *Drosophila* ultraspiracle modulates ecdysone receptor function. *Cell* 71, 63–72.

Yu, Z., Redfern, C.S. and Fishman, G.I. (1996) Conditional transgene expression in the heart. *Circulation Research* 79, 691–697.

4

Embryonic Stem Cells in Agricultural Species

G.B. Anderson

Department of Animal Science, University of California, Davis, California, USA

Introduction

Embryonic stem (ES) cells are undifferentiated, pluripotent cells derived in culture from early embryos. When used to describe ES cells, 'undifferentiated' means that their fate has not been restricted to a particular cell lineage. 'Pluripotent' means that the cells have the capacity to develop into many cell types; in fact, when used to refer to ES cells, pluripotent has come to mean the capacity to contribute to both somatic- and germ-cell lineages. By definition, an ES cell usually is expected to have the capacity to produce both gametes and all somatic-cell lineages.

Embryonal Carcinoma Cells

Prior to when ES cells were first isolated and cultured from preimplantation embryos, developmental biologists studied stem cells with ES-like properties found in spontaneously occurring tumours. Certain strains of laboratory mice were known to develop gonadal tumours called teratocarcinomas, which contain numerous differentiated cell types in addition to undifferentiated embryonal carcinoma (EC) cells. EC cells were known to proliferate and to maintain their malignancy in culture. If EC cells were injected under the skin of a histocompatible host, a tumour developed from the injected cells. If EC cells were injected into the blastocoel of a blastocyst, and the blastocyst was transferred to the reproductive tract of a recipient female mouse, the injected cells could combine with cells of the host embryo to differentiate and develop into tissues and organs of a normal mouse pup. These pups, chimeras because their cells had two distinct embryonic origins, had no higher incidence of tumour formation than other mice. Gametes produced by some of these chimeras were derived not only from the host embryo but also from the EC cells.

Discovery and Use of Embryonic Stem Cells

The similarities between EC cells and early embryonic cells did not go unnoticed, and researchers searched for methods to produce EC-like cells directly from embryos. In 1981, two laboratories independently accomplished this feat (Evans and Kaufman, 1981; Martin, 1981). The embryo-derived cell lines isolated in culture were given the name embryonic stem cells or ES cells to distinguish them from EC cells. ES cells were shown to survive indefinitely in culture, to survive freezing and thawing, and to be capable of tumorigenesis when injected under the skin of a histocompatible host. These cells also retained the capacity to develop into normal tissues, including gametes, after blastocyst injection. In fact, ES cells were shown to have a higher capacity than EC cells to contribute to both somatic and germ cell lineages (Bradley *et al.*, 1984).

ES cells are typically isolated in culture from blastocyst-stage embryos. More specifically, they develop from the inner cell mass (ICM) of the blastocyst, at least some cells of which are thought to be undifferentiated. The other cells of the blastocyst, the trophectoderm, appear to have undergone differentiation and are incapable of producing ES cells. ES cells have become an important experimental model for research involving embryogenesis, cell-lineage studies and the study of stage- and cell-specific gene expression. A great deal of interest in ES cells, and why they are a subject for presentation at this conference, has developed from their capacity to integrate foreign DNA. Unlike the random integration into the genome that occurs with microinjection of DNA into embryos, ES cells can be used to target a specific site in the genome for genetic engineering. Low-frequency events such as homologous recombination, leading to gene replacement and gene knockout, can be exploited in ES cells because of their capacity to proliferate into large numbers in culture. Even if only a few cells of the millions growing in a culture plate undergo the appropriate genetic change, various enrichment techniques allow the desired cells to survive and grow in culture, while cells without the appropriate genetic change have a reduced ability to survive. The surviving colonies are then screened for a homologous recombination event.

Despite the potential to genetically engineer ES cells in ways not possible directly with embryos, the genetic modification can be introduced into an animal only if the engineered ES cells retain their capacity for normal differentiation and development into somatic and germ cells. Two methods are possible to produce a living animal from ES cells. The most common method is blastocyst injection whereby ES cells are injected into the blastocoel of a blastocyst. The injected cells can incorporate into the ICM of the host blastocyst and contribute to development of a chimera. If the chimeric animal produces gametes from the ES-cell line, a genetic change introduced into the ES cells can become established in a line of animals. A technique that has been developed more recently, and that has certain advantages over blastocyst injection, is nuclear transfer. Here the nucleus of a donor ES cell is introduced into an enucleated oocyte to produce a diploid embryo whose nuclear DNA is derived solely from the ES cell. An advantage of nuclear transfer over blastocyst injection for producing an animal from an ES-cell line is that nuclear transfer does not involve the uncertainty of germline transmission of the ES-cell genotype. With blastocyst injection, the chimeric animal might produce gametes only from the host embryo, or from the ES cells only at low frequency. A nuclear-transfer offspring will have the ES-cell genome in all cells of its body, ensuring that all gametes contain the ES-cell genotype. Despite rapid advances in nuclear-transfer technology, term development of nuclear-transfer embryos using ES cells as nuclear donors has not yet been reported.

Embryonic Stem Cell Technology and Species Specificity

To date, research with ES cells has been conducted almost exclusively with the laboratory mouse. The explanation for this limited application is quite simple: isolation of ES cells has not been accomplished unequivocally in other species, including in domestic livestock species. This fact could go unnoticed if one were to rely on the titles of many research articles published over the past decade in which the subject of the paper is proclaimed to be ES cells of one non-murine species or another. In fact, few of these papers provide results that adequately document that the cells under study meet the criteria to be called ES cells. Species for which the isolation of putative ES cells has been reported include the mouse (Evans and Kaufman, 1981; Martin, 1981), sheep (Handyside *et al.*, 1987), hamster (Doetschman *et al.*, 1988), pig (Piedrahita *et al.*, 1990a,b), cattle (Evans *et al.*, 1990), mink (Sukoyan *et al.*, 1992), rabbit (Giles *et al.*, 1993; Graves *et al.*, 1993), rat (Iannaccone *et al.*, 1994), monkey (Thompson *et al.*, 1995) and goat (Meinecke-Tillmann and Meinecke, 1996). This chronological listing of reports for the various species does not in all cases include papers that might be considered to contain the most convincing data, only papers that were among the first to be published for that species. If only species are listed for which chimeric offspring have been obtained after injection of putative ES cells into blastocysts, the list is considerably shorter. These species include the mouse (Evans and Kaufman, 1981; Martin, 1981), pig (Wheeler, 1994) and rabbit (Giles *et al.*, 1993; Schoonjans *et al.*, 1996). This already short list is unusual in the fact that it recently became shorter instead of longer. In 1994 the isolation of rat ES cells that produced chimeras after blastocyst injection was reported (Iannaccone *et al.*, 1994). Recently, the paper was retracted due to concern that the rat 'ES cells' had become contaminated with mouse ES cells, and the chimeric animals were really rat–mouse interspecies chimeras (Iannaccone *et al.*, 1997). The list of species in which ES-cell isolation has been documented by chimeras with demonstrated germline chimerism (i.e. transmission of the ES-cell genome in the gametes) after blastocyst injection is even shorter; only for the laboratory mouse has this criterion been met.

What accounts for the discrepancy between the lists of species for which ES-cell isolation has been reported versus the single species for which ES-cell isolation has been fully documented with germline chimerism? The discrepancy results from the different criteria researchers have used to describe their 'ES cells' from various species. One criterion used to identify cultured embryonic cells as ES cells is a determination of their having morphology and *in vitro* developmental characteristics similar to those of murine ES cells. In other words, if the cells were cultured from an early embryo, and they looked and behaved in culture like murine ES cells, they sometimes have been described as being ES cells. Some investigators, especially among the earliest publications on the subject, used only these morphological criteria to classify their cells as ES cells. A more stringent evaluation of a cell line would include not only morphological features but developmental criteria as well. Mouse ES cells are known to undergo differentiation *in vitro*, which can either occur spontaneously or be induced. Many investigators now include *in vitro* differentiation in their descriptions of how their cell lines resemble what we know to be ES cells. The most stringent criterion for identifying ES cells is their capacity to differentiate *in vivo*. As described in the previous paragraph, for only a few species have ES-like cells been shown to be capable of *in vivo* differentiation resulting in a chimera after blastocyst injection; only in the mouse has chimerism from ES cells been shown to extend to the germline.

Conditions for Isolation of Embryonic Stem Cells in Culture

Several conditions must be met in order for ES cells to be isolated *in vitro*:

1. Undifferentiated, pluripotent cells must be present in the embryo at the time of culture. This requirement may seem obvious, but for most species the precise stage of development appropriate for isolation of ES cells has not been determined.
2. The pluripotent cells must be deprived of differentiation signals in culture.
3. The cells must be stimulated, or at least be allowed, to proliferate.

We conducted experiments to determine whether, as in the mouse, transplantable undifferentiated cells could be isolated from porcine blastocysts. We isolated porcine ICM by immunosurgery and injected them into host blastocysts for embryo transfer and development to term. Chimeric pigs were born having characteristics similar to murine chimeras (Anderson *et al.*, 1994). Some of these chimeric pigs were shown to be germline chimeras, a requirement for ES cell technology to be useful for genetic modification. Our results confirmed that day-8 blastocysts were a source of undifferentiated embryonic cells; stimulation of proliferation and inhibition of differentiation of these cells in culture were the remaining challenges.

Culture conditions for early experiments aimed at defining conditions under which ICM-derived cells from livestock species will survive *in vitro* were based on culture systems previously defined for isolation of mouse ES cells, namely Dulbecco's modified Eagle's medium supplemented with fetal calf serum plus other additives and cultured over a monolayer of murine STO cells. Cell lines could be established from ICM culture, and these cell lines frequently maintained a morphology similar to mouse ES cells. Some cell lines were demonstrated to be pluripotent by virtue of their capacity to differentiate *in vitro* into various cell types (Piedrahita *et al.*, 1990a), but attempts to demonstrate *in vivo* differentiation were unsuccessful. These early results, though encouraging, fell short of actually documenting the isolation of ES cells. Experiments followed in which various homologous and heterologous feeder layers, or other slight modifications to the culture system, were tested (Piedrahita *et al.*, 1990b). Generally, few of these modifications to the culture system yielded results beyond those possible with culture conditions widely used for isolation of murine ES cells. Even today, in times of scarce resources, these early experiments frequently are being repeated and are yielding similar results as new laboratories become involved with livestock ES-cell research.

Some levels of success have been achieved with *in vitro* maintenance of pluripotent cells from agricultural species. Sims and First (1993) cultured bovine embryonic cells in low-density suspension culture and produced four nuclear-transfer calves after using these cells as nuclear donors. A substantially larger number of nuclear-transfer pregnancies was established, but most failed prior to term. Stice *et al.* (1996) used culture conditions more typical of those for murine ES cells to isolate ES-like cells from both *in vitro*- and *in vivo*-derived bovine embryos. In a large experiment involving more than 3000 nuclear-transfer embryos, between 12 and 40%, depending on the cell line, of the nuclear-transfer embryos developed to the blastocyst stage in culture. Up to 30% of embryo transfer recipients for these blastocysts were diagnosed pregnant at approximately 30 days of gestation, but only 20 days later at approximately 50 days gestation the percentage of pregnant recipients was only 0–15% for the various cell lines. None of the nuclear-transfer pregnancies developed to term. The authors speculated that failure of placental development was responsible for loss of pregnancy. A possible explanation for abnormal placental development in nuclear-transfer embryos from ES cells is the stage of imprinting in the ICM and hence in ES cells, but this theory has not been proven.

Absent from this discussion of the more successful attempts at ES-cell isolation in agricultural species are the nuclear-transfer results of Campbell *et al.* (1996) and Wells *et al.* (1997) describing the birth of lambs after use of embryo-derived cell lines as nuclear donor cells. These exciting results will not be discussed here, because they are included in Chapter 5 by Dr Ian Wilmut, and also because the cell lines used by these investigators likely were not ES cells. In the first paper in which cultured embryonic cell lines were used to produce viable nuclear-transfer embryos, the authors stated:

> The (cell) line was established from early passage colonies with a morphology like that of ES cells. By the second and third passages, the cells had a more epithelial, flattened morphology ... At passage 6, unlike murine ES cells, they expressed cytokeratin and nuclear lamin A/C, which are markers associated with differentiation (Campbell *et al.*, 1996).

Wells and colleagues (1997) also described their cell lines as having undergone some degree of differentiation in culture prior to nuclear transfer. Although some readers might view these distinctions as trivial given that, like ES cells, these embryo-derived cells (and differentiated somatic cells as well) could be useful for introducing genetic modifications into agricultural species, I believe that we should hesitate to refer to differentiated embryo-derived cells lines as ES cells. The biology of murine ES cells has a rich history, and researchers newly working with embryo-derived cell lines from livestock species should refrain from redefining terms based on how the cells might be used and for now should rely on definitions originally based on morphology, expression of ES-specific markers and developmental capacity.

Culture of Primordial Germ Cells and Isolation of Embryonic Germ Cells

Using information gained from research with murine embryos and ES cells, some laboratories (including ours) have worked for nearly a decade to establish conditions for isolation of ES cells in livestock species. Progress continues to be made but, while awaiting the significant breakthroughs that will facilitate isolation of ES cells in these species, alternative approaches must be considered. In 1992, two groups independently reported that pluripotent stem cells could be isolated from cultured primordial germ cells (PGC). PGC are embryonic cells that are the precursors to gametes in the adult (i.e. those cells all of whose surviving descendants form gametes). They are thought to be derived from extra-embryonic layers and to migrate into the embryo to the genital ridge. Matsui *et al.* (1992) and Resnick *et al.*

(1992) observed that when murine PGC are cultured in medium containing a cocktail of growth factors including stem cell factor, leukaemia inhibitory factor and basic fibroblast growth factor, the cells will survive and proliferate beyond the point at which they normally undergo mitotic (male germ cells) or meiotic (female germ cell) arrest in the gonad. With continued culture, the PGC lost their migratory capacity, attached to a monolayer of feeder cells and formed colonies resembling murine ES cells. The cells, designated embryonic germ (EG) cells to distinguish them from ES cells of ICM origin, were shown to share many characteristics with ES cells, including morphology, cellular markers and the capacity to differentiate *in vitro*. When injected into blastocysts, like ES cells, EG cells were shown to be capable of differentiating into normal cells of a chimeric mouse, including germ cells (Stewart *et al*., 1994). Available results from research with ES cells far exceed those with EG cells, but EG cells appear to be potentially useful for manipulations otherwise currently limited to ES cells. Debate continues over whether or not EG and ES cells, despite their different origins, are identical cell types.

In an effort to isolate porcine EG cells, we were able to collect, on average, approximately 15,000 PGC from each day-25 porcine embryo (Shim and Anderson, 1998). The cells resembled murine PGC in morphology and stained positive for alkaline phosphatase activity, a marker for murine PGC as well as for undifferentiated murine ES, EC and EG cells. When porcine PGC were cultured over a STO feeder monolayer in medium with or without growth factor supplementation, porcine PGC survived and proliferated without addition of growth factors required by murine PGC (Shim and Anderson, 1998). With prolonged culture, these cells attached to the monolayer of STO feeder cells and formed colonies similar in appearance to porcine ES-like colonies (Piedrahita *et al*., 1990a,b). Upon injection into host blastocysts, the PGC-derived cells were shown to have the capacity to differentiate into normal tissues of a chimeric piglet (Shim *et al*., 1997). Analysis of microsatellite DNA revealed that the EG cells had contributed to most somatic tissues tested. Germline transmission of porcine EG cells has yet to be demonstrated. Our results demonstrate that, as in the mouse, porcine PGC can be cultured to produce pluripotent stem cells. We are currently evaluating these cells for their ability to integrate foreign DNA as a vehicle for introducing changes into the porcine genome. One explanation for success in isolation of porcine EG cells is the substantially larger number of cells available to initiate culture from the genital ridge of day-25 porcine embryos compared with the ICM of day-8 blastocysts.

Another success at PGC culture in an agricultural species has been published in the popular press. A private company has been reported as having isolated and cultured PGC from day-30 bovine embryos. The cultured PGC ultimately were used as nuclear donors for nuclear-transfer experiments. Like most nuclear-transfer results from having used cultured bovine ICM and ES cells as nuclear donors (Sims and First, 1993; Stice *et al*.,

1996), pregnancies from nuclear transfer using EG cells were lost prior to term. When cells from nuclear-transfer embryos were retransplanted to produce second-series nuclear-transfer embryos, resulting pregnancies did survive to term. If these procedures prove to be reproducible, cultured PGC and EG cells could be useful for introducing genetic changes into the bovine genome. A description of these results is not yet available in the scientific literature, but information can be obtained over the Internet at www.absglobal.com.

Summary and Conclusions

ES cells have been demonstrated to be a powerful vehicle for targeting a specific site in the genome. To date, the technology has been developed and used almost exclusively in the laboratory mouse. Numerous embryo-derived cell lines have been established and described for agricultural species, but as yet none has yielded germline chimerism; however, some of these cell lines have yielded somatic chimerism. As an alternative to ES-cell culture, PGC can be cultured to produce EG cells with developmental capabilities similar to ES cells. The significant advances recently made in nuclear transfer using differentiated embryonic cells and somatic cells as nuclear donors have not been widely tested using ES and EG cells, and past experiments with disappointing results (e.g. low embryo and fetal survival) probably should be repeated. To end on an encouraging tone – despite lingering unanswered questions, steady progress continues toward isolation of the elusive undifferentiated stem cells in agricultural species.

References

Anderson, G.B., Choi, S.J. and BonDurant, R.H. (1994) Survival of porcine inner cell masses in culture and after injection into blastocysts. *Theriogenology* 42, 204–212.

Bradley, A., Evans, M., Kaufman, M.H. and Robertson, E. (1984) Formation of germline chimeras from embryo-derived teratocarcinoma cell lines. *Nature* 309, 255–256.

Campbell, K.H.S., McWhir, J., Ritchie, W.A. and Wilmut, I. (1996) Sheep cloned by nuclear transfer from a cultured cell line. *Nature* 380, 64–66.

Doetschman, T., Williams, P. and Maeda, N. (1988) Establishment of hamster blastocyst-derived embryonic stem (ES) cells. *Developmental Biology* 127, 224–227.

Evans, M.J. and Kaufman, M.H. (1981) Establishment in culture of pluripotential cells from mouse embryos. *Nature* 292, 154–156.

Evans, M.J., Notarianni, E., Laurie, S. and Moor, R.M. (1990) Derivation and preliminary characterization of pluripotent cell lines from porcine and bovine embryos. *Theriogenology* 33, 125–128.

Giles, J.R., Yang, X., Mark, X. and Foote, R.H. (1993) Pluripotency of cultured rabbit inner cell mass cells detected by isozyme analysis and eye pigmentation of fetuses following injection into blastocysts or morulae. *Molecular Reproduction and Development* 36, 130–138.

Graves, K.H. and Moreadith, R.W. (1993) Derivation and characterization of putative pluripotential embryonic stem cells from preimplantation rabbit embryos. *Molecular Reproduction and Development* 36, 424–433.

Handyside, A.H., Hooper, M.L., Kaufman, M.H. and Wilmut, I. (1987) Towards the isolation of embryonal stem cell lines from the sheep. *Roux's Archives* 196, 185–197.

Iannaccone, P.M., Taborn, G.U., Garton, R.L., Caplice, M.D. and Brenin, D. (1994) Pluripotent embryonic stem cells from the rat are capable of producing chimeras. *Developmental Biology* 163, 288–292.

Iannaccone, P.M., Taborn, G.U., Garton, R.L., Caplice, M.D. and Brenin, D. (1997) Pluripotent embryonic stem cells from the rat are capable of producing chimeras. *Developmental Biology* 185, 124–125.

Martin, G.R. (1981) Isolation of pluripotent cell lines from early mouse embryos cultured in medium conditioned with teratocarcinoma stem cells. *Proceedings of the National Academy of Sciences USA* 72, 1441–1445.

Matsui, Y., Zsebo, K. and Hogan, B.L. (1992) Derivation of pluripotential embryonic stem cells from murine primordial germ cells in culture. *Cell* 70, 841–847.

Meinecke-Tillmann, S. and Meinecke, B. (1996) Isolation of ES-like cell lines from ovine and caprine pre-implantation embryos. *Journal of Animal Breeding and Genetics* 113, 413–426.

Piedrahita, J., Anderson, G.B. and BonDurant, R.H. (1990a) Influence of feeder layer type on the efficiency of isolation of porcine embryo-derived cell lines. *Theriogenology* 34, 865–877.

Piedrahita, J., Anderson, G.B. and BonDurant, R.H. (1990b) On the isolation of embryonic stem cells: comparative behaviour of murine, porcine and ovine embryo. *Theriogenology* 34, 879–901.

Resnick, J.L., Bixter, L.S., Cheng, L. and Donovan, P.J. (1992) Long-term proliferation of mouse primordial germ cells in culture. *Nature* 359, 550–551.

Schoojans, L., Albright, G.M., Li, J.L., Collen, D. and Moreadith, R.W. (1996) Pluripotential rabbit embryonic stem (ES) cells are capable of forming overt coat color chimeras following injections into blastocysts. *Molecular Reproduction and Development* 45, 439–443.

Shim, H. and Anderson, G.B. (1998) *In vitro* survival and proliferation of porcine primordial germ cells. *Theriogenology* 49, 521–528.

Shim, H., Gutierrez-Adan, A., Chen, L.-R., BonDurant, R.H., Behboodi, E. and Anderson, G.B. (1997) Isolation of pluripotent stem cells from cultured porcine primordial germ cells. *Biology of Reproduction* 57, 1089–1095.

Sims, M.M. and First, N.L. (1993) Production of calves by transfer of nuclei from cultured inner cell mass cells. *Proceedings of the National Academy of Sciences USA* 90, 6143–6147.

Stewart, C.L., Gadi, I. and Bhatt, H. (1994) Stem cells from primordial germ cells can reenter the germline. *Developmental Biology* 161, 626–628.

Stice, S.L., Strelchenko, N.S., Keefer, C.L. and Matthews, L. (1996) Pluripotent bovine embryonic cell lines direct embryonic development following nuclear transfer. *Biology of Reproduction* 54, 100–110.

Sukoyan, M.A., Golubitsa, A.N., Zhelezova, A.I., Shilov, A.G., Vatolin, S.Y., Maximovsky, L.P., Andreeva, L.E., McWhir, J., Pack, S.D., Bayborodin, S.I., Kerkis, A.Y., Kizilova, H.I. and Serov, O.L. (1992) Isolation and cultivation of blastocyst-derived stem cell lines from American mink (*Mustela vison*). *Molecular Reproduction and Development* 33, 418–431.

Thompson, J.A., Kalishman, J., Golos, T.G., Durning, M., Harris, C.P. and Hearn, J.P. (1996) Pluripotent cell line derived from common marmoset (*Callithrix jacchus*) blastocysts. *Biology of Reproduction* 55, 254–259.

Wells, D.N., Misica, P.M., Day, A.M. and Tervit, H.R. (1997) Production of cloned lambs from an established embryonic cell line: a comparison between *in vivo*- and *in vitro*-matured cytoplasts. *Biology of Reproduction* 57, 385–393.

Wheeler, M.B. (1994) Development and validation of swine embryonic stem cells: a review. *Reproduction, Fertility and Development* 6, 1–6.

5

Nuclear Transfer in the Production of Transgenic Farm Animals

I. Wilmut[1], E. Schnieke[2], J. McWhir[1], A.J. Kind[2], A. Colman[2] and K.H.S. Campbell[3]

[1]*Roslin Institute, Roslin, UK;* [2]*PPL Therapeutics, Roslin, UK;* [3]*present address: PPL Therapeutics, Roslin, UK*

Introduction

The aim with any procedure for genetic modification is to be able to introduce the change which is desired, precisely and with confidence that there will be no other unintentional change. This has, so far, not been achieved in livestock, but recent results suggest that nuclear transfer from donor cells in which genetic change has been made by site-specific recombination will become possible and meet these aims. This chapter will discuss the limitations of the present procedures for genetic change, describe nuclear transfer and the factors which influence development of embryos produced by nuclear transfer, outline the approaches to gene transfer and modification before nuclear transfer, and consider some applications of genetic modification in livestock.

Methods for genetic modification of mammals

Direct injection

At present, the only method for the production of transgenic farm animals is to inject several hundred copies of the gene into a nucleus of an early embryo. While this approach has been useful, and has provided some commercial opportunities (e.g. for the production of therapeutic proteins in the milk of livestock), it has several serious limitations which are well known (Pursel *et al.*, 1989; Wilmut and Clark, 1990). It is believed that injection of the fluid causes breaks in the chromosomes, so stimulating repair mechanisms which inadvertently include some of the injected DNA into the chromosome. This sequence of events would certainly account for the facts that most of the injected embryos fail to develop to term and that there is a low frequency of transgenesis. The site of integration is apparently random and it is believed that the influence of the neighbouring sequences accounts for the great variability in the level of expression of the same transgene in different lines. In addition, it has been estimated that direct injection causes mutation to an endogenous gene in between 5% and 10% of mouse lines, and there is no reason to expect the frequency to be different in other species. The final, and most profound limitation of this approach is that it can only be used to add a gene and cannot be used to modify an endogenous gene.

Embryonic stem cells

The limitations of direct injection can be contrasted with those observed with the use of embryonic stem (ES) cells in mice (see Hooper, 1992). In the mouse, but so far only in the mouse, culture methods have been found for the isolation of cells from the embryo, such that they divide, but do not differentiate. The cells may be derived from the early embryo or primordial germ cells (Matsui *et al.*, 1992). In some cases, if stem cells are aggregated with another embryo they retain the ability to colonize all of the tissues of the resulting offspring, including the germline. By site-specific recombination in the cells it is possible to introduce genes or to modify existing genes in a predictable manner, with little risk of inadvertent mutation to other genes (Hooper, 1992). As it is possible to confirm precise genetic changes in the cells before aggregation with the second embryo these cells have provided a very efficient means for the introduction of genetic changes in that species. While this approach offers the opportunity for precise modification it has the disadvantage that an intermediate chimeric generation is required before the effect of the modification can be assessed and that this would require up to 2 years in livestock species.

Potential value of nuclear transfer

The development of methods for nuclear transfer from embryonic cells encouraged the thought that it might become possible to devise methods for nuclear transfer from cells after precise genetic changes had been made in the cells. In livestock, nuclear transfer from inner cell mass cells has become routine (Bondioli *et al.*, 1990) and mouse ES cells retain some similarities to inner cell mass cells (Beddington and Robertson, 1989). These two observations suggested that when ES cells from livestock species become available then nuclear transfer from such cells would provide an ideal means for the introduction of genetic change. Hypothetically, in such a system, it would be possible to exploit the unusually high efficiency of recombination in ES cells before nuclear transfer to produce groups of genetically identical offspring, with the genetic modification. In addition, use of nuclear transfer would avoid the chimeric generation and ensure germline transmission from the offspring. Despite an extensive research effort over a period of 10 years there is still no confirmation that ES cells have been isolated from any species other than the mouse, although cells which resemble mouse ES cells can be seen for a number of passages during culture of pig, sheep and cattle cells (this conference). However, nuclear transfer has proved to be more powerful than was generally anticipated and genetic modification by this route now seems attainable through the use of primary cell cultures.

Nuclear Transfer

Method of nuclear transfer

Nuclear transfer in mammals is achieved by the fusion of a donor cell to an unfertilized egg or early embryo which has been enucleated (shown schematically in Fig. 5.1). Details of the procedures have been described previously (Campbell *et al.*, 1993, 1996b). Development after nuclear transfer is influenced by a great many factors which have been discussed elsewhere (Campbell *et al.*, 1996b), including a requirement that normal ploidy be maintained in the reconstructed embryo. When a nucleus is transferred from a cell that has begun to differentiate, the pattern of gene expression must be 'reprogrammed' from that of the differentiated phenotype to that required for early development. The experiments in which cell cycle stage has been varied also suggest that the efficiency of this process is influenced by both donor and recipient cell cycle stage. The role of cell cycle in development after nuclear transfer will be reviewed before a consideration of the new opportunities that are becoming available.

Cell cycle in nuclear transfer

Although the cell cycle stage of both donor and recipient cells influence when DNA replication occurs in the reconstructed embryo, the recipient cell may have a dominant role because of the influence of meiosis (maturation or mitosis) promoting factor (MPF) in the cytoplasm (Barnes *et al.*, 1993; Campbell *et al.*, 1993). During meiosis, MPF activity increases at the time of formation of the meiotic spindles, and at metaphase II remains high until fertilization or parthenogenetic activation. Regardless of the cell cycle stage of the donor nucleus, transfer to a cytoplast with a high level of MPF is followed

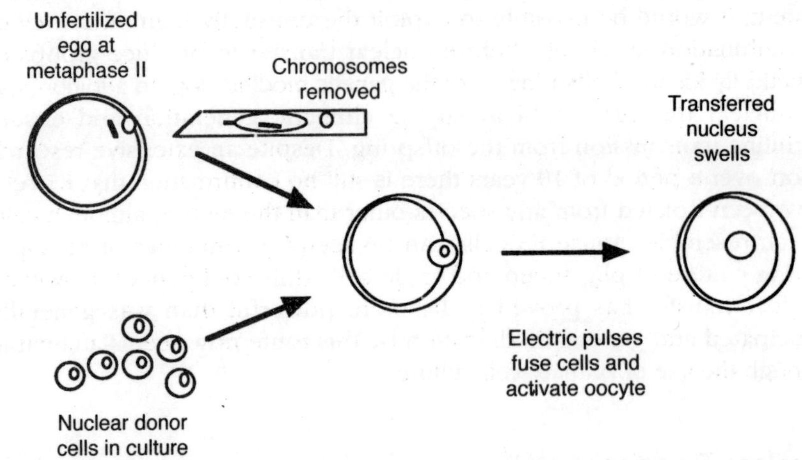

Fig. 5.1. Schematic representation of the method of nuclear transfer used in these studies. There were some variations in the time of oocyte activation in relation to cell fusion which are not shown. These details of the procedures are given in the referenced publications.

by nuclear membrane breakdown, chromosome condensation and, following reformation of the nuclear membrane, DNA replication. By contrast, following transfer to an oocyte with a low level of MPF activity, the nucleus determines whether DNA replication occurs. These observations suggested two different approaches to the maintenance of normal ploidy following nuclear transfer. If the cytoplast has a high level of activity then a diploid nucleus is appropriate, whereas pre-activation of the cytoplast creates a universal recipient able to maintain normal ploidy after transfer of a nucleus at any stage of the cell cycle (Campbell *et al.*, 1993, 1994). These different approaches have both been exploited effectively, however, the few comparisons of different recipient oocytes which have been made suggest that there are big differences in embryo development, even between those treatment combinations which were expected to retain normal ploidy (Table 5.1).

By contrast, there are several observations to show that development is influenced by donor nucleus stage, even among those groups in which normal ploidy is expected (Table 5.1). This general conclusion is interpreted to indicate two different effects: (i) a greater opportunity for reprogramming of gene expression during specific phases of the cell cycle; and (ii) a benefit from transfer to similar phases of the cell cycle.

Development after nuclear transfer is more likely to occur if the donor cells are taken from the 'window of opportunity' which includes G2, mitosis G1 and G0 (see Table 5.1). Cells in G0, known as quiescence, have exited the growth cycle and become relatively inactive. Stem cell populations

Table 5.1. Effect of donor and recipient cell cycle stage upon development of mouse and sheep embryos reconstructed by nuclear transfer to the morula or blastocyst stage of development. Cell cycle combinations in which normal ploidy is expected are indicated in bold. (Summary taken from Campbell *et al.*, 1996a.)

		Donor cell cycle stage				
Recipient cell cycle stage	Donor cell type	G0	G1	Early S	Late S/G2	G2
MII	Mouse two-cell		**77.8**	0.0		20.8
MII	Mouse four-cell		**43.0**	0.0	0.0	
G1/S-phase			**60.0**	**14.0**	**0.0**	
Late S/G2			**0.0**	**0.0**	**12.0**	
MII	Mouse eight-cell		**27.0**	0.0		
Post-activated MII	Sheep embryo culture	**12.8**				
MII		**16.3**				
S-phase		**11.7**				

remain in this stage until they receive the stimulus to re-enter the growth cycle. The impression that such a window exists is gained from experiments in several species with different cell types. The response to nuclear transfer varies between species (see below). There is a need for a systematic comparison between these donor cell cycle phases in several different cell types in more than one species.

Comparisons with mouse embryo blastomere donor cells show advantages in using donor cells in G2, mitosis or G1 over other stages of the growth cell cycle. The effect of cell cycle was also studied in sheep. Donor nuclei cells were induced to become quiescent as a means of obtaining a stable, diploid population. Live lambs were born following transfer of nuclei from differentiated cells derived from sheep embryos, fetus and adult mammary gland tissue (Wilmut *et al.*, 1997). The cells were maintained in culture for prolonged periods before being induced to leave the growth cycle and become quiescent.

Normal development of embryos produced by nuclear transfer is assumed to depend upon a pattern of gene expression that closely resembles that observed after fertilization. Immediately after fertilization, development depends upon proteins and RNA produced in the oocyte before ovulation (see Thompson, 1996). At a species-specific stage of development, transcription from the embryonic genome begins and the maternal messages are destroyed. Then there are characteristic changes in gene expression as the cells divide and differentiate. When a nucleus is transferred from a cell that is transcriptionally active, transcription must first cease before it is initiated at the appropriate stage of development. Abnormal development might reflect the presence of unusual transcripts or the lack of those that are required. It has been suggested that if a donor nucleus is in the 'window', factors which influence chromatin structure and gene expression are removed from the DNA helix. As a result, there is greater access to the DNA for those factors in the oocyte cytoplasm which reprogramme gene expression. Although this hypothesis is compatible with all published information, there have been very few studies of transcription after nuclear transfer or of chromatin structure in donor cells.

Response to nuclear transfer in different species

There is an apparent association between the stage at which transcription from the embryonic genome begins and the efficiency of nuclear transfer. There are few direct comparisons, but it seems that nuclear transfer is relatively ineffective in those species, such as the mouse, in which initiation of transcription from the embryonic genome occurs earliest. By comparison, it has been most effective in amphibians in which the transition occurs very late in development. This observation is interpreted to indicate that further reprogramming of gene expression occurs during each mitosis and that, while there is only one cell division before the transition in the mouse, there are three in sheep and cattle and 12 in amphibians. This association leads to the prediction that the response in rabbits and cattle to the new approach to nuclear transfer will be similar to that seen in sheep, but that there may be differences in pigs and mice, in which the major transition occurs at the four- and two-cell stage respectively (see discussion by Thompson, 1996).

Limitations to the present procedures

There are several limitations to the present methods. Only a small proportion of reconstructed embryos develop to become live offspring, varying between 0.04% with adult cells to 1.7% and 1.0% for fetal- and embryo-derived cells respectively (Wilmut *et al.*, 1997). Even when considering only the proportion of embryos that became live lambs after they had developed to morulae or blastocysts in culture before transfer to recipients these same proportions are 3.4%, 7.5% and 4.6%, respectively. If all offspring were to carry a desired genetic change, see below, this would represent a significant gain in efficiency on direct injection, but there is still a major cost in obtaining each lamb.

Secondly, there may be complications at the birth of the lambs. A number of the lambs derived by nuclear transfer died soon after birth. In most cases this has been because of congenital abnormalities in the cardiovascular or urinogenital systems. In addition, some lambs were unusually large, although the extent of the increase cannot be estimated because of the geneticists habit of transferring embryos between breeds as a means of an instantaneous confirmation that the offspring are indeed derived in the manner described. An increase in birth weight has been associated with several different treatments, including the culture of *in vivo*-produced sheep zygotes for as little as 3 days in medium containing serum (see Walker *et al.*, 1996), nuclear transfer (Willadsen *et al.*, 1991; Wilson *et al.*, 1995), alterations to the relationship between the stage of the developing embryo and maternal endocrine environment (Wilmut and Sales, 1981; Kleeman *et al.*, 1994) and extreme levels of non-protein nitrogen in the diet (McEvoy *et al.*, 1997). As embryos produced by nuclear transfer must also be cultured to a stage at which they can be transferred into the uterus of a recipient female, it is not clear if the increase in size associated with nuclear transfer is due to the nuclear transfer, the culture or to both. Neither the environmental factors causing the increase in size, nor the mechanisms by which the fetus increases in size are known. However, it is clear that there are profound differences in the progress of parturition. Gestation is typically extended by several days and the onset of labour often slow. Despite the prolongation of gestation, lung development in the lambs is often immature.

While it may be acceptable to use nuclear transfer for biotechnology with the present limitations, large-scale agricultural applications must surely depend upon their elimination. In that case the breeder will expect to have a conception rate which approaches that after normal breeding, with little increase in perinatal mortality.

Applications

Several advantages will arise from the opportunity to introduce precise genetic changes in livestock by nuclear transfer. Gene addition will be more efficient and require fewer animals. This offers welfare and economic benefits. The expression of added genes may also be expected to be more predictable once experience has been gained with specific sites of insertion. As the site of the modification will be selected and then confirmed before the offspring are produced there should be no unintended mutation at that site, although the possibility of mutation elsewhere cannot be absolutely excluded. Above all, there will be for the first time the chance to change existing genes. This may be used to modify production of selected proteins or to study the role of the gene product or the regulation of expression of that gene.

The first uses of this new technique will be in biotechnology, rather than agriculture. Although nuclear transfer already offers several advantages over direct injection it is still an expensive procedure and the costs can more readily be justified when the product is a high-value therapeutic protein or the greater understanding of a disease. Furthermore, a greater number of candidate genes has been identified in humans. However, in the longer term it seems probable that as the genome mapping projects identify interesting new loci so gene targeting through nuclear transfer will offer the means to study either the role of a gene product or the mechanisms which regulate gene expression. In the following sections several different fields of use in biotechnology will be considered in turn, before a brief review of agricultural applications.

Pharmaceuticals

A number of therapeutic proteins are already being produced in the milk of farm animals, including α_1-antitrypsin (Wright *et al.*, 1991; Carver *et al.*, 1993). The transgenic animals were produced by direct injection. Gene targeting will allow a more efficient and precise means for the production of animals with additional genes. This advantage is perhaps greatest in cattle because a female born as a twin to a male calf is almost always an infertile 'freemartin'. In cattle, as in other species, a majority of embryos developing after direct injection die during early development. Whereas it is possible in other species to transfer several such embryos to each recipient, the risk of infertile freemartins limits the value of this strategy in cattle. By contrast, if genes are added before nuclear transfer all of the offspring will be of the same sex, which can be selected as that which is most appropriate for the particular application.

The greatest benefit to arise from use of nuclear transfer is that a specific change may be made. Experience will be required to confirm the most reliable strategy which may be via the insertion of coding sequences into a milk protein gene. This could both direct secretion of the desired protein and stop the production of the milk protein. Alternatively, a fusion gene could be inserted into a site known to allow high levels of expression of such genes. The construct would include both the coding elements for the human protein and the regulatory elements for the milk protein gene being used to direct synthesis to the mammary gland. Experience in this field will assist later applications to modify milk composition.

Xenotransplantation

At present many patients requiring organ transplantation die before a suitable donor becomes available. There is an ever-increasing waiting list for such organs. There is currently considerable interest in the possibility of using pigs as the source of organs for transplantation into humans. One major problem is the hyperacute rejection response, which destroys very rapidly an organ transplanted between species. This rejection is due to the presence of naturally occurring antibodies in human circulation that immediately recognize certain foreign (xeno-) antigens on transplanted tissue. These antibodies initiate a very rapid complement-mediated response that causes the lysis of the transplanted tissue. Present transgenic strategies include introducing (human) genes whose products suppress the complement-mediated lysis or modify the xeno-antigen (White, 1996). In addition to facilitating such transgenic modification of pigs, gene targeting in conjunction with nuclear transfer would enable quite different strategies to be adopted. Gene targeting could prevent expression of the genes responsible for the xeno-antigen e.g. α_1-3-galactosyl-transferase. Alternatively, strategies using cell-based germline modification could include modifying the pig major histocompatibility complex (MHC) genes so that the longer term MHC-mediated tissue rejection was suppressed, thus reducing/eliminating the requirement for chronic immunosuppression of recipients with drugs. The research necessary for this application will, for the first time, establish routine methods of nuclear transfer in pigs.

Disease models

Medical research benefits from models of human diseases in animals for detailed study of the disease and the development of new therapies. Gene targeting in mice, through the use of ES cells, has been used to provide models of human genetic diseases, such as cystic fibrosis. However, many of the mouse models have significant limitations because of the physiological difference between mice and humans. Furthermore, there are severe practical limitations to routine drug administration and physiological sampling, because of the small size of the animals. The opportunity to target precise genetic changes in livestock could offer more useful animal models.

Sheep have been used extensively for the study of many human respiratory diseases because there are considerable similarities between human and sheep lungs. Sheep may thus be an ideal candidate in which to generate a model for human cystic fibrosis by targeting mutations to the cystic fibrosis gene. Such animals could be used to study the physiological changes in detail, to evaluate new small-molecule-based therapies and for the development of gene therapy strategies.

Disease models in farm animals will be important in understanding genetic diseases in these species. Studies of scrapie in sheep and bovine spongiform encephalopathies in cattle will provide much needed understanding of these spongiform encephalopathies in these species, and indirectly of Creutzfeldt–Jakob disease in humans.

Susceptibility models

Similar benefits may flow from the use of animals to investigate genetic differences in susceptibility to disease. There is a genetic component to most of the major causes of adult death, including cancer, heart attack and stroke. As the human genome projects identify regions associated with susceptibility to specific diseases then it will be possible to use gene targeting to study the role of candidate genes in the aetiology of the disease. In some cases this may stimulate the development of drugs able to overcome the susceptibility.

Physiological studies

There is every reason to expect that the opportunity to target precise genetic changes in livestock will offer advantages to all of the existing transgenic projects described in this book. Gene targeting has the potential to change any aspect of gene function and to modify the gene product or prevent its production completely. By changes to the regulatory region, the site and level of expression may be changed. Alternatively, the mechanisms controlling gene expression may be analysed by modifications to the regulatory region, leading in turn to the opportunity to modify expression either in livestock or human patients. In just the same way that studies of undesirable gene expression may help in the treatment of disease, similar studies of gene expression in health or exceptional agricultural productivity will confer different benefits. An understanding of the role of specific genes in agricultural production will assist accurate genetic selection or the introduction of targeted changes in livestock.

Conclusions

A new era in biotechnology, biomedicine, farm animal breeding and research is being heralded by the development of methods for precise genetic change. Specifically, nuclear transfer from cultured cells after the introduction of genetic changes will allow entirely new approaches to the study of genes in disease and health. In the past, physiological studies have identified the role of secretions from each organ. Classically, the role of an organ was defined by the removal of that organ, the purification of a product from the tissue and confirmation that normal health was restored by administration of the product. As the gene mapping projects identify new genes in humans and livestock an analogous series of studies can be expected to define the role of specific gene products. In the past such studies were limited to mice, but the new nuclear transfer techniques will provide the same powerful tool in other species. The value of this new opportunity cannot yet be estimated, but it will certainly be large.

Acknowledgements

We gratefully acknowledge the assistance of our colleagues in the conduct of the experiments described in this review. The research was supported by the Ministry of Agriculture Fisheries and Food, Roslin Institute and PPL Therapeutics.

References

Barnes, F.L., Collas, P., Powell, R., King, W.A., Westhusin, M. and Shepherd, D. (1993) Influence of recipient oocyte cell cycle stage on DNA synthesis, nuclear envelope breakdown, chromosome constitution, and development in nuclear transplant bovine embryos. *Molecular Reproduction and Development* 36, 33–41.

Beddington, R.S.P. and Robertson, E.J. (1989) An assessment of the developmental potential of embryonic stem cells in the midgestation mouse embryo. *Development* 105, 733–737.

Bondioli, K.R., Westhusin, M.E. and Looney, C.R. (1990) Production of identical bovine offspring by nuclear transfer. *Theriogenology* 33, 165–174.

Campbell, K.H.S., Ritchie, W.A.R. and Wilmut, I. (1993) Nuclear–cytoplasmic interactions during the first cell cycle of nuclear transfer reconstructed bovine embryos: implications for DNA replication and development. *Biology of Reproduction* 49, 933–942.

Campbell, K.H.S., Loi, P., Cappai, P. and Wilmut, I. (1994) Improved development of ovine nuclear transfer embryos reconstructed during the presumptive S-phase of enucleated activated oocytes. *Biology of Reproduction* 50, 1385–1393.

Campbell, K.H.S., Loi, P., Otaegui, P.J. and Wilmut, I. (1996a) Cell cycle co-ordination in embryo cloning by nuclear transfer. *Reviews in Reproduction* 1, 40–46.

Campbell, K.H.S., McWhir, J., Ritchie, W.A.R. and Wilmut, I. (1996b) Sheep cloned by nuclear transfer from a cultured cell line. *Nature* 385, 810–813.

Carver, A.S., Dalrymple, M.A., Wright, G., Cottom, D.S., Reeves, D.B., Gibson, Y.H., Keenan, J.L., Barrass, J.D., Scott, A.R., Colman, A. and Garner, I. (1993) Transgenic livestock as bioreactors: stable expression of human alpha-1-antitrypsin by a flock of sheep. *Bio/Technology* 11, 1263–1270.

Hooper, M.L. (1992) In: Evans, H.J. (ed.) *Embryonal Stem Cells: Introducing Planned Changes into the Germline.* Harwood Academic Publishers, Chur, Switzerland.

Kleemann, D.O., Walker, S.K. and Seamark, R.F. (1994) Enhanced fetal growth in sheep administered progesterone during the first three days of pregnancy. *Journal of Reproduction and Fertility* 102, 411–417.

Matsui, Y., Zsebo, K. and Hogan, B.L.M. (1992) Derivation of pluripotential embryonic stem cells from murine primordial germ cells in culture. *Cell* 70, 841–847.

McEvoy, T.G., Robinson, J.J., Aitken, R.P. and Robertson, I.S. (1997) Dietery excesses of urea influence the viability and metabolism of preimplantation sheep embryos and may affect fetal growth among survivors. *Animal Reproductive Science* 47, 71–90.

Pursel, V.G., Pinkert, C.A., Miller, K.F., Bolt, D.J., Campbell, R.G., Palmiter, R.D., Binster, R.L. and Hammer, R.E. (1989) Genetic engineering of livestock. *Science* 244, 1281–1288.

Thompson, E.M. (1996) Chromatin structure and gene expression in the preimplantation mammalian embryo. *Reproduction, Nutrition and Development* 36, 619–635.

Walker, S.K., Hartwich, K.M. and Seamark, R.F. (1996) The production of unusually large offspring following embryo manipulation: concepts and challenges. *Theriogenology* 45, 111–120.

White, D. (1996) Alteration of complement activity: a strategy for xenotransplantation. *Trends in Biotechnology* 14, 3–5.

Willadsen, S.M., Janzen, R.E., McAlister, R.J., Shea, B.F., Hamilton, G. and McDermand, D. (1991) The viability of late morulae and blastocysts produced by nuclear transplantation in cattle. *Theriogenology* 35, 161–170.

Wilmut, I. and Clark, A.J. (1990) Basic techniques for transgenesis. *Journal of Reproduction and Fertility* 43 (Suppl.), 265–275.

Wilmut, I. and Sales, D.I. (1981) Effect of an asynchronous environment on embryonic development in sheep. *Journal of Reproduction and Fertility* 61, 179–184.

Wilmut, I., Schnieke, A.E., McWhir, J., Kind, A.J. and Campbell, K.H.S. (1997) Viable offspring from fetal and adult mammalian cells. *Nature* 385, 810–813.

Wilson, J.M., Williams, J.D., Bondioli, K.R., Looney, C.R., Westhusin, M.E. and McCalla, D.F. (1995) Comparison of birth weight and growth characteristics of bovine calves produced by nuclear transfer (cloning), embryo transfer and natural mating. *Animal Reproduction Science* 38, 73–83.

Wright, G., Carver, A., Cottom, D., Reeves, D., Scott, A., Simons, P., Wilmut, I., Garner, I. and Colman, A. (1991) High-level expression of active human alpha-1-antitrypsin in the milk of transgenic sheep. *Bio/Technology* 9, 830–834.

6

Embryonic Stem Cell Chimeras and Somatic Cell Nuclear Transplantation for Production of Transgenic Cattle

J.M. Robl[2], J.B. Cibelli[1,2], P.G. Golueke[1], J.J. Kane[1], C. Blackwell[1], J. Jerry[2], E.S. Dickenson[1], F.A. Ponce de Leon[2] and S.L. Stice[1]

[1]*Advanced Cell Technology, Inc.* and [2]*Department of Veterinary and Animal Sciences, Paige Laboratory, University of Massachusetts, Amherst, Massachusetts, USA*

Introduction

Genetic modification of cattle could be useful in increasing the efficiency of meat and milk production. Small numbers of transgenic cattle have been made using traditional pronuclear microinjection (Pursel and Rexroad, 1993). However, no transgenic lines have yet been commercialized for agriculture. At least part of the reason for this slow progress is the limitations and inefficiency of the transgenic production technology.

One of the limitations of pronuclear microinjection is that the gene insertion site is random. This typically results in variations in expression levels and several transgenic lines must be produced to obtain one line with appropriate levels of expression to be useful. Because integration is random, it is advantageous that a line of transgenic animals be started from one founder animal, to avoid difficulties in monitoring zygosity and potential difficulties that might occur with interactions among multiple insertion sites (Cundiff *et al.*, 1993). Furthermore, if inbreeding is to be avoided, starting a transgenic line from one hemizygous animal with a random insert would require breeding several generations and significant time for introgression of the transgene into the population before breeding and testing homozygotes (Cundiff *et al.*, 1993). Even without concern for inbreeding, it would take 6.5 years before reproduction could be tested in homozygous animals (Seidel, 1993). Finally, the quality of the genetics of a homozygous transgenic line would lag behind that of the general population because of the reduced population within which to select future generations of transgenic animals and the difficulty of bringing new genetics into a population in which the transgene is fixed.

A second limitation of the pronuclear microinjection procedure is its efficiency, which ranges from 0.34 to 2.63% of the gene-injected embryos developing into transgenic animals and a fraction of these appropriately expressing the gene (Pursel and Rexroad, 1993). This inefficiency results in a high cost of producing transgenic cattle because of the large number of recipients needed and, more importantly, unpredictability in the genetic background into which the gene is inserted because of the large number of embryos needed for microinjection. For agricultural purposes a high-quality genetic background is essential; therefore, long-term backcrossing strategies must be used with pronuclear microinjection.

An ideal system for producing transgenic animals for agricultural applications should be highly efficient and use small numbers of recipient animals to produce transgenics. It should allow the insertion of a transgene into a specific genotype. The insertion would preferably be into a predetermined site that would confer high expression and not affect general viability and productivity of the animal. Furthermore, the identification of a locus for insertion should allow multiple lines to be produced and crossed to produce homozygotes, and new genetic background could easily be added to the transgenic line by the production of new transgenics at any time. Therefore, the ideal system would likely require the transfection and selection of cells that could be easily grown in culture yet retain the potency to form germ cells and pass the gene to subsequent generations.

One such system for producing transgenic animals has been developed and widely used in the mouse. This approach involves the use of embryonic stem (ES) cells. Mouse ES cells are relatively easy to grow as colonies *in vitro*. The cells can be transfected by standard procedures and transgenic cells clonally selected by antibiotic resistance (Doetschman, 1994). Furthermore, the efficiency of this process is such that sufficient transgenic colonies (hundreds to thousands) can be produced to allow a second selection for homologous recombinants (Doetschman, 1994). Mouse ES cells can then be combined with a normal host embryo and, because they retain their potency, can develop into all the tissues in the resulting chimeric animal, including the germ cells. The transgenic modification can then be transmitted to subsequent generations. One of our objectives for the production of transgenic cattle has been to use a similar approach and develop pluripotential embryonic cell lines (PEC) from cattle that could be transfected and selected *in vitro* and, after association with a host embryo, contribute to various tissues, including the germline, in the resulting offspring.

Another system for producing transgenic animals has been demonstrated by the recent studies of Wilmut *et al.* (1997) in the sheep. In this approach somatic cells grown *in vitro* are reprogrammed by fusion to a recipient egg. The resulting embryo is then capable of developing into a full-term offspring. Because somatic cells such as fibroblasts can be easily grown, transfected and selected *in vitro*, this would be an ideal system for producing transgenic animals. Furthermore, because the entire animal would be derived from the transgenic donor cell it would not be necessary to breed one generation to produce a pure transgenic, as with the chimeric approach, and all the offspring would be genetically identical. Cloning would also be an advantage in that the offspring could be of a predetermined sex so sex-limited traits, such as milk production, could be tested in females produced in the first generation. A second objective of our work, therefore, has been to develop a system for using somatic cell nuclear transplantation to produce transgenic cattle.

Production of Transgenic Cattle PECs

The defining requirements we used for designating cells as ES cells were: (i) the cells should be derived from the inner cell mass (ICM) of a blastocyst stage embryo; (ii) they should be capable of dividing in culture without showing signs of morphological differentiation; and (iii) they should contribute to cells of the germline and endodermal, mesodermal and ectodermal tissues when combined with a host embryo to form a chimera. In addition, cells were evaluated in relation to mouse ES cells for morphology, several cytoplasmic markers and growth characteristics.

Morphologically, the colonies that were established from bovine ICMs maintained distinct margins, had high nuclear to cytoplasmic ratios, generally maintained a high density of lipid granules (Fig. 6.1) and were cytokeratin- and vimentin-negative as in the mouse but, contrary to the mouse, the cells grew as a single layer with individual cells being visible and were not positive for alkaline phosphatase. Another difference between mouse ES cells and bovine PECs was that bovine PECs were much slower growing than mouse ES cells, indicating a much longer cell cycle (estimated to be about 40 h).

Two methods were used to establish PEC colonies from day-7 *in vitro*-produced bovine blastocysts. The method involved isolating the ICM immunosurgically. Antisera were developed against bovine spleen cells in mice. The zona pellucida was removed using 0.5% pronase until the zona thinned and could be removed by pipetting. The blastocysts were exposed to a 1:100 dilution of anti-bovine mouse serum for 45 min then washed and treated with guinea pig complement. The lysed trophectodermal cells were removed by pipetting. For the second method the ICM was isolated mechanically using two 26-gauge needles. The needles were crossed and brought down on the zona-intact blastocysts, which were cut using a scissor action. Some of the trophectodermal cells remained with the ICM

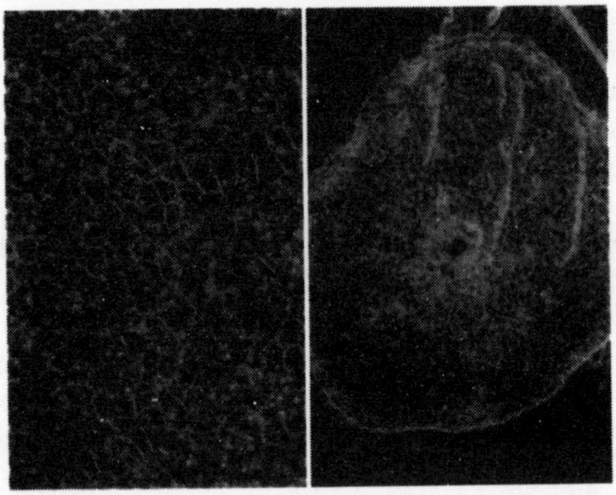

Fig. 6.1. Bovine PEC colony showing distinct margins with cells containing a high nuclear to cytoplasmic ratio and a high density of lipid granules.

and inevitably disappeared following plating and passaging. A PEC colony was considered as established after the third passage without signs of differentiation. For the immunosurgically isolated ICMs 5/9 (55%) formed PEC colonies and for the mechanically isolated ICMs 6/12 (50%) formed colonies. Because no difference was detected between these methods, the mechanical method was adopted for the advantage of simplicity.

Establishment of PEC colonies and maintenance of the undifferentiated state depend on an intimate contact between the ICM and the mouse fibroblast feeder layer. In an attempt to increase the contact during the initial establishment, day 7 *in vitro*-produced ICMs were placed either beneath or on top of mouse fetal fibroblast feeder layers. As above, a PEC colony was considered as established after the third passage without signs of differentiation. In agreement with previous results, 5/9 (55%) ICMs plated on top of the feeder layer produced colonies but only 4/11 (36%) of those placed beneath the feeder layer provided less appropriate interaction to inhibit differentiation of the ICMs.

Several methods of passaging bovine PEC colonies were attempted. Because it is beneficial to clonally propagate PECs following transfection and is necessary for homologous recombination, many attempts were made to trypsinize colonies to produce single cells and establish new colonies from these cells. To summarize, all attempts at clonally propagating bovine PECs were unsuccessful. Therefore, the routine method of passage that was established was to mechanically cut the colony into pieces that contained at least 50 cells and plate the clumps of cells on new feeder layers.

Following the development of methods of establishing and passaging bovine PECs and the identification of limitations in clonally propagating the cells, we turned to pursuing methods of transfecting and selecting for transgenic cells. The construct that was used contained a human cytomegalovirus promoter and β-galactosidase/neomycin resistance fusion gene (Friedrich and Soriano, 1991; β-GEO). Selection was based on treatment with Geneticin (G418, Sigma, St Louis, Missouri) to kill non-expressing cells. The β-GEO gene was used to verify incorporation and expression.

Prior to transfecting cells, it was necessary to determine the sensitivity of non-transgenic cells to G418. Colonies from three different embryos were challenged with 0, 50, 100 and 150 $\mu g\ ml^{-1}$ G418. A colony was considered dead when it completely lifted from the feeder layer. Survival varied among lines of cells with the first line surviving an average of 9 days at 100 $\mu g\ ml^{-1}$ and 7 days at 150 $\mu g\ ml^{-1}$. The second survived 12, 10 and 7 days at 50, 100 and 150 $\mu g\ ml^{-1}$, respectively and the third line survived 8, 7 and 5 days at 50, 100 and 150 $\mu g\ ml^{-1}$, respectively. To ensure death of all non-transgenic colonies, 150 $\mu g\ ml^{-1}$ G418 was chosen as the dose for subsequent transfection experiments.

Because it was not possible to trypsinize and produce a cell suspension of bovine PECs, the method of transfection was limited to either microinjection or lipofection. Various lipofection protocols were tested and found to be effective on fibroblast and Comma D cell cultures but were not effective on bovine PECs. Therefore, microinjection was used. PECs from three different lines were microinjected into the nucleus with a linearized version of the construct described above. At 1 day following microinjection the colonies were treated with 150 $\mu g\ ml^{-1}$ G418 continuously for 30 days (four or five passages). For the three lines 3753, 3508 and 3502 cells were injected and five, two and zero colonies, respectively, survived selection in G418. Some cells within each of these colonies expressed β-galactosidase activity and samples of cells were positive for the transgene when amplified by PCR (35 cycles) and analysed by Southern blot hybridization to the amplified product. Because the colonies essentially disappeared during selection it is likely that the transgenic lines were of clonal origin, although this was not confirmed. Variation in expression in cells within a colony was probably due to cell-to-cell variation in factors such as cell cycle stage, position effects and others.

Potency of the cells was tested by producing chimeras with host embryos. Prior to evaluating the incorporation of PECs into embryos, the relationship between the number of PECs injected into morulae and the rate of development to the blastocyst stage was investigated. As shown in Table 6.1 either four, eight or 12 cells were injected following isolation by trypsinization. Rate of development to the blastocyst stage decreased with increasing number of ES cells used. As an injection control, fibroblasts (either four, eight or 12 cells) were injected into morulae, and as a non-injection control, a group of non-treated embryos were cultured to the blastocyst stage. There were no differences in development rate due to the number of cells injected, but manipulation, or the injection of cells, did appear to have a detrimental effect on development. Although it was found that increasing the number of PECs injected decreased the rate of development it was also

Table 6.1. Effect of cell injection on development of bovine morulae to the blastocyst stage.

Type of cell	Number of cells injected	Number of morulae injected	Number of blastocysts (%)
PEC	4	62	15 (24)^{a,b}
PEC	8	65	10 (15)^{a}
PEC	12	67	9 (13)^{a}
Fibroblast	4	54	16 (30)^{a,b}
Fibroblast	8	58	11 (19)^{a,b}
Fibroblast	12	36	10 (28)^{a,b}
Control	0	46	19 (41)^{b}

^{a,b} Percentages with different superscripts are significantly different (P<0.05).

believed that decreasing the number of cells would decrease the level of chimerism in the embryos. A compromise position of injecting eight cells was chosen for further experiments.

Incorporation of PECs into bovine blastocysts was evaluated to determine if the PECs could interact with the host embryo and be incorporated into the inner cell mass of the blastocyst. PECs were labelled with 100 μg ml^{-1} of the fluorescent carbocyanine dye, DiI (Sigma, St Louis, Missouri), then injected into morula stage embryos. Four days later the resulting blastocysts were observed under the fluorescent microscope. Incorporation of labelled PECs into both the ICM and the trophectoderm was detected in all blastocysts. To further verify that the cells had been incorporated into the ICM, the trophectoderm was removed by immunosurgery and the isolated ICM was observed. In all cases labelled cells were detected in the ICM. This indicated that the PECs could be incorporated into the compacted morula and ICM and form the early precursors of the fetus.

The next step in examining the potency of the PECs was to test chimerism in fetuses recovered at 40 days of gestation. Eighteen day-7 blastocysts, injected with eight to ten PECs were transferred into six recipient cows. Forty days after transfer the fetuses were recovered by Caesarean section. The total number of fetuses recovered was 12, with six being normally developing and six dead and in the process of being resorbed. Of the six normal fetuses, the β-GEO transgene was detected, by PCR (35 cycles) and Southern analysis of the PCR product, in some tissues in all of them (Table 6.2). Of the abnormal fetuses it was possible to analyse some tissues in one and it, too, was transgenic. In addition to analysing somatic tissues, preparations enriched in PGCs were isolated and analysed in the normal fetuses and two showed evidence of having transgenic cells. The results of this experiment indicated that the PECs did have the capacity to differentiate into many different kinds of tissues and survive at least 40 days *in vivo*. Further work is in progress on alternative methods of generating transgenic PECs and evaluating survival of PECs to term in chimeric animals.

Table 6.2. Contribution of transgenic PECs to various tissues in 40-day bovine fetuses.

Tissue	Fetus number[a]					
	1	2	3	4	5	6
Heart	+	+	−	+	+	+
Muscle	+	+	n.d.[b]	−	n.d.[b]	+
Brain	−	+	+	−	+	+
Liver	n.d.[b]	−	+	−	+	+
Gonads	−	+	+	+	+	+
PGC[c]	+	−	+	−	−	−

[a] Fetal tissues were analysed by 35 cycles of PCR followed by Southern hybridization of the amplified product.
[b] Not determined.
[c] PGCs were isolated from a trypsin digest of the genital ridge, then individually

These results are promising for the development of a highly efficient method of producing PECs in the bovine. Bovine PECs may be very useful as a source of *in vitro*-produced cells for transplantation into humans. However, the use of bovine PECs for gene targeting does not appear to be practical. Generally, it is expected that between 100 and 1000 transgenic colonies need to be produced and screened to find one or more colonies with the correct insert (Doetschman, 1994). Our highest level of efficiency in transgene incorporation was five colonies from 3753 cells injected. To produce 100 transgenic colonies would require injecting about 75,000 cells. This would require 25 days of injection if one person injected 3000 cells per day, which is possible with experience. Producing more than 100 transgenic colonies is probably unrealistic without significant increases in efficiency.

Production of Transgenic Cattle Somatic Cell Nuclear Transplant Embryos

Fibroblasts were chosen as the donor cell because of their ease of isolation, growth and transfection. Bovine fetal fibroblasts were produced from 30–100 mm crown rump length (approximately 40–80 days of gestation) fetuses obtained from the slaughterhouse. Fetuses were shipped by overnight express mail on ice. In some cases, when a 2-day shipment was used, healthy fibroblast lines could still be produced. After propagation for three passages, fibroblasts were transfected by electroporation with a closed circular construct of β-GEO. Following electroporation, transfected cells were selected on 400 μg ml^{-1} of G418. After 3 weeks of selection, single colonies were isolated, propagated and used for nuclear transfer experiments.

Nuclear transplant blastocysts and fetuses were produced from fibroblasts using standard procedures. Basically, *in vitro*-matured oocytes were obtained from Trans Ova Genetics, Inc. by overnight express mail. Oocytes were enucleated following fluorescent labelling of the DNA to verify enucleation. Trypsinized fibroblast cells were transferred to the perivitelline space and fused to the oocyte cytoplast by electroporation. Activation was induced by a combination of calcium ionophore and 6-dimethylaminopurine. The rate of development to the blastocyst stage was about 10% (353/3625) for nuclear transfer embryos and 14% (106/758) for activated controls. Some blastocysts were shipped to Ultimate Genetics, Inc. for transfer into recipient cows. Two blastocysts were transferred into each recipient. Fetuses recovered at day 40 were morphologically normal and fibroblast cells recovered from these fetuses expressed β-galactosidase at a high level. Development to term is in progress.

The results indicate that fibroblast nuclear transplantation may be an ideal method of producing transgenic cattle. Transfection, selection and clonal propagation are relatively easy in primary fibroblasts. The CMV promoter, along with several other constitutive promoters, drive gene expression at a high rate in fibroblasts allowing for routine antibiotic selection. These factors have allowed us to produce a number of transgenic lines with high-expressing random gene inserts. Our preliminary results also indicate that fibroblasts can be grown for a sufficient number of passages *in vitro*, without becoming senescent, to allow a second round of selection for a targeted insert. The fibroblast nuclear transplant system may be a method that will finally allow the commercial production of transgenic livestock for improved agricultural production.

References

Cundiff, L.V., Bishop, M.D. and Johnson, R.K. (1993) Challenges and opportunities for integrating genetically modified animals into traditional animal breeding plans. *Journal of Animal Science* 71 (Suppl. 3), 20–25.

Doetschman, T. (1994) Gene transfer in embryonic stem cells. In: Pinkert, C. (ed.) *Transgenic Animal Technology: a Laboratory Handbook*. Academic Press, New York, pp. 115–146.

Friedrich, G. and Soriano, P. (1991) Promoter traps in embryonic stem cells: a genetic screen to identify and mutate developmental genes in mice. *Genes and Development* 5, 1513–1523.

Purcel, V.G. and Rexroad, C.E., Jr (1993) Status of research with transgenic farm animals. *Journal of Animal Science* 71 (Suppl. 3), 10–19.

Seidel, G.E., Jr (1993) Resource requirements for transgenic livestock research. *Journal of Animal Science* 71 (Suppl. 3), 26–33.

Wilmut, I., Schnieke, A.E., McWhir, J., Kind, A.J. and Campbell, K.H.S. (1997) Viable offspring derived from fetal and adult mammalian cells. *Nature* 385, 810–813.

7

Status of Sperm-mediated Delivery Methods for Gene Transfer

E.J. Squires

Department of Animal & Poultry Science, University of Guelph, Guelph, Canada

The production of transgenic animals by sperm cell-mediated gene transfer continues to be a fascinating and intriguing possibility due to the simplicity and potential use of the technique for all animals. Some success in producing transgenic fish, lower vertebrates and invertebrates using this method has been reported by a number of laboratories. However, except for the group that first reported the production of transgenic mice, no one has produced transgenic birds or mammals by this method. Many examples of successful experiments where DNA has been 'bound' to sperm cells have been reported. Methods for transfer of DNA to sperm include simply incubating the sperm and DNA together, use of liposomes containing DNA or electroporation of sperm with the DNA. Treatment with DNase or extensive washing does not entirely remove this DNA. Recent work has suggested that some of the DNA is internalized into the sperm nucleus and becomes incorporated into sperm chromatin. The incorporation of DNA into the sperm genome may be a necessary prerequisite for producing transgenic animals, since unincorporated DNA may be degraded both within the sperm cell and after fertilization of the egg. An interesting methodology for the transfection of male germ cells by injection of liposome/DNA complexes into seminiferous tubules has been described which may overcome this difficulty. Other methods to increase the integration of the intact transgene in the sperm cell genome may allow stable expression and replication of the DNA in the fertilized egg.

Introduction

The potential use of sperm cells as vectors for gene transfer was suggested by Brackett *et al.* (1971) who demonstrated that SV40 virus binds to sperm. Interest in this area was renewed by reports that transgenic mice (Lavitrano *et al.*, 1989) and pigs (Gandolfi *et al.*, 1989) were produced after *in vitro*

© CAB INTERNATIONAL 1999. *Transgenic Animals in Agriculture* (eds J.D. Murray, G.B. Anderson, A.M. Oberbauer and M.M. McGloughlin)

fertilization of oocytes with sperm that had been incubated with naked DNA. However, other laboratories were unable to repeat these experiments (Brinster *et al.*, 1989; Gavora *et al.*, 1991). Since then, work has continued on using sperm cells to transfer DNA, due to the simplicity and potential use of the technique for all animals. Work with pigs (Gandolfi *et al.*, 1996), cattle (Schellander *et al.*, 1995) and chickens (Martinez *et al.*, 1992; Nakanishi and Iritani, 1993) has been largely unsuccessful in producing transgenics, although limited success has been claimed by some groups (Rottmann *et al.*, 1992; Squires and Drake, 1994, 1997; Sperandio *et al.*, 1996).

There have been many examples of successful gene transfer; with fish (Tsai *et al.*, 1995; Patil and Khoo, 1996), Japanese abalone (Tsai *et al.*, 1997), sea urchin (Arezzo, 1989) and *Xenopus laevis* (Habrova *et al.*, 1996). Species differences in producing transgenics may be due to differences in the stability of the foreign DNA, both in the sperm and after transfer to the egg. In this regard, it is interesting that microinjection of foreign DNA into the cytoplasm of fertilized fish eggs can produce transgenics (Chourrout *et al.*, 1986), while in other species the DNA is microinjected into the male pronucleus of the fertilized egg. This may reflect the presence of nucleases in the sperm or egg of higher vertebrates that degrade or rearrange the foreign DNA.

DNA Uptake by Sperm

Work using sperm cells as vectors for gene transfer has largely focused on methods to maximize the efficiency of transfer of DNA to sperm, while maintaining the viability of the sperm after transfer. Many examples of successful experiments where DNA has been 'bound' to sperm cells have been reported (Castro *et al.*, 1990; Atkinson *et al.*, 1991). Treatment with DNase or extensive washing does not remove all of this DNA. DNA binding apparently occurs with particular proteins on the sperm and is reported to be antagonized by glycoproteins present in seminal fluid (Zani *et al.*, 1995). Others (Rottman *et al.*, 1992) have reported that chicken sperm cells do not need to be washed before being treated with liposomes containing DNA. Recent work has suggested that 15–20% of the DNA is internalized into the sperm nucleus and becomes incorporated into sperm chromatin (Zoraqi and Spadafora, 1997). Although some DNA binding occurs after simply incubating the sperm and DNA together, the efficiency of binding can be dramatically increased by the use of liposomes containing DNA or electroporation of sperm with the DNA.

Use of liposomes

We have systematically investigated the use of various liposome preparations for efficacy in encapsulating DNA and transfer of DNA to chicken sperm cells while still maintaining the fertility of the sperm (Squires and Drake, 1993). Liposomes containing DNA in Beltsville Poultry Semen Extender (Sexton and Fewlass, 1978) were prepared by reverse phase evaporation. The trapping efficiency of DNA into the liposomes was estimated by using ^{32}P-labelled DNA and determining the percentage of the total amount of radioactivity that was recovered in the liposome pellet after centrifugation. The liposomes containing the labelled DNA (0.1 ml) were mixed with sperm cells (0.1 ml) for 10 min at 40°C, and the liposomes were separated from the sperm cells by centrifugation through 0.25 M sucrose or silicone oil.

The trapping efficiency of the DNA into the liposomes was increased with increasing concentration of lipid, but this reduced the subsequent transfer of DNA to the sperm. Increasing the positive charge by including stearylamine in the lipid mixture improved the trapping efficiency of the DNA into liposomes and the transfer of DNA to sperm cells. Including lysophosphatidylcholine in the lipid mixture promotes fusion of the liposomes with sperm cells and the combination of lysophosphatidylcholine and stearylamine gave the highest transfer efficiency to sperm. The fertility of the sperm cells was dramatically reduced by exposure to liposomes made from dimyristoylphosphatidyl choline, dilauroylphosphatidyl choline or lipids extracted from sperm cell membranes. Increasing sperm cell numbers reduced the total transfer of DNA from liposomes to sperm cells. However, the number of sperm cells should not be lowered below 5×10^7 sperm because of reduced fertility.

We found the optimum conditions for DNA transfer to chicken sperm cells while maintaining fertility are using liposomes comprised of 10 μmol ml^{-1} dipalmitoylphosphatidyl choline, 5 mol% stearylamine and 20 mol% lysophosphatidylcholine with 2.5×10^8 sperm. Other experiments with lipofectin reagent at 0.006 or 0.06 μmol ml^{-1} were also effective in transferring DNA to sperm, but lower numbers of sperm cells (5×10^7 sperm) must be used, which decreased fertility somewhat. Nakanishi and Iritani (1993) reported that 51.6% of sperm retained exogenous DNA after treatment with lipofectin, but the experimental conditions were not given. We have found that the optimum conditions for loading sperm with DNA varies with different species. In particular, fish sperm are quite different from other species since they are activated by water and do not have an acrosome. This requires that salmonid sperm be diluted in buffer (20 mM Tris, pH 9, containing 80 mM NaCl, 40 mM KCl and 1 mM CaCl) to prevent activation.

Use of electroporation

There have been reports of DNA transfer by electroporation to bovine (Gagne *et al.*, 1991) and chicken (Nakanishi and Iritani, 1993) sperm as well as sperm from several fish species including zebrafish (Patil and Khoo,

1996) and loach (Tsai *et al.*, 1995). Electroporation reduced the fertility of chicken sperm by causing premature breakdown of the acrosome (Nakanishi and Iritani, 1993). Similar reductions in the fertility of bovine sperm were seen after electroporation (Gagne *et al.*, 1991), while the fertility of fish sperm was largely unaffected by electroporation (Tsai *et al.*, 1995). In using electroporation, the electric field strength and DNA concentration must be optimized.

Fertilization Methods

A potential problem of intravaginal insemination is that 'sorting' of sperm could occur, preventing the sperm containing the DNA from fertilizing the egg. In some experiments (Squires and Drake, 1994), we compared the efficiency of intravaginal insemination versus insemination using intramagnal catheters (Lakshmanan *et al.*, 1990). In our experience, we found little difference in the frequency of transgenesis by either method. Trefil *et al.* (1996) used intramagnal insemination of hens to improve the fertilizing ability of spermatozoa treated with lipofectin. The efficiency of sperm-mediated gene transfer could potentially be increased by sorting the sperm after transfection to select sperm containing the foreign DNA, perhaps using fluorescence-enhanced flow cytometry (Nakanishi and Iritani, 1993). In conditions where it is not possible to maintain sperm fertility while loading sperm cells with foreign DNA, the sperm cells could possibly be injected into the egg by intracytoplasmic sperm injection (ICSI).

Detection of the Transgene

A variety of methods have been used to determine if the foreign DNA was present in the developing animal. PCR or Southern blotting of genomic DNA can measure the presence of foreign DNA in the blastocyst, embryo or young animal. The expression of the transgene is usually estimated by measuring the enzyme activity coded by the transgene (CAT, chloramphenicol acetyl transferase, β-galactosidase, green fluorescent protein, etc.) or by Western analysis or enzyme-linked immunosorbent assay (ELISA) of the gene products.

Nakanishi and Iritani (1993) checked for the presence of DNA in freshly laid chicken eggs by PCR. They reported that the incidence of eggs containing foreign DNA was 67% after fertilization with sperm treated with lipofectin–DNA, 47% after fertilization with sperm treated with naked DNA and 23% after fertilization with sperm electroporated with DNA. However, there was no evidence of genomic integration of the foreign DNA. Similar results were previously reported by Rottman *et al.* (1992). They reported that, while naked DNA was not taken up by chicken sperm, liposome-encapsulated

DNA was. The foreign DNA was present in 26% of 11- to 13-day-old embryos but was not incorporated into the genomic DNA. Episomal replication of the foreign DNA had apparently occurred.

In order for stable integration of the transgene to be demonstrated, it is necessary to show that the transgene is transferred to the offspring in a Mendelian fashion. In our work (Squires and Drake, 1994) we were able to demonstrate by PCR analysis of genomic DNA that the transgene was present in the offspring from a backcross to control birds, but the occurrence was less than the expected 50% of progeny. In the first generation backcross, this may have been due to mosaicism, as has been reported for transgenics produced by other methods (Overbeek *et al.*, 1991; Love *et al.*, 1994). However, similar results were also obtained in the next generation, suggesting that the transgene is partially eliminated during cell division. We were also unable to detect significant expression of the transgene in any of the birds that were positive by PCR analysis.

Problems with the Method

The main obstacles to the use of sperm cells as vectors for DNA transfer to the egg are the rearrangement/degradation of the foreign DNA and lack of incorporation into the genome. Tsai *et al.* (1995) reported success in producing transgenic loach using sperm electroporated with a growth hormone gene construct from chinook salmon. While the frequency of transgenesis was about 50%, a number of different size bands were seen on Southern blots of digests of genomic DNA, suggesting that rearrangement of the transgene had occurred before integration. In a later report (Tsai *et al.*, 1997) where they produced Japanese abalone transgenics with a CAT construct, the size of the transgene fragment on Southern blots was larger than expected. Recent work (Zoraqi and Spadafora, 1997) has shown that plasmid DNA internalized into sperm is associated with the nuclear scaffold and it has been suggested that topoisomerase II may play a role in the non-homologous recombination of foreign DNA into specific sites in the sperm genome. The plasmid DNA also becomes extensively rearranged. Preliminary reports suggest that nuclease activities are stimulated by the presence of high amounts of foreign DNA and this can be inhibited by pretreating the sperm with the apoptotic inhibitor, aurintricarboxylic acid (Spadafora *et al.*, 1997).

Problems with the integration efficiency and expression of transgenes are also found with other methods of gene transfer. The expression of transgenes can be affected by cotransfection with other actively transcribed genes (Thorey *et al.*, 1993). Improvements in efficiency may also occur if the DNA is targeted to the nucleus. In this regard, a nuclear DNA attachment element has been identified which confers integration site-independent expression of the lysozyme gene (Stief *et al.*, 1989). Another problem in common with gene transfer by microinjection is multiple copy insertion.

This can be reduced by treating the transgene with alkaline phosphatase to avoid ligation of linear DNA and the formation of head-to-tail arrays. In addition, mosaicism of expression in which the transgene is not expressed in all tissues has been reported (Tsai *et al.*, 1995).

New Approaches

Studies by Brinster and Zimmerman (1994) and Brinster and Avarbock (1994) have shown that spermatagonial cells can be injected into the seminiferous tubules of sterile recipient mice. The testicular stem cells seeded the recipient seminiferous tubules and populated the recipient testes with spermatozoa that were capable of fertilizing eggs in female mice and producing normal live offspring. The use of this technique to produce transgenic animals requires that undifferentiated spermatogonia be cultured *in vitro* so they can be transfected with the transgene.

Recently, Kim *et al.* (1997) have further investigated this idea by transfecting male stem cells with DNA *in vivo*. They injected lipofectin–DNA complexes into seminiferous tubules of mice and at random sites in the testes of pigs. In all animals, the developing male germ cells were first destroyed with busulphan, an alkylating agent. After the remaining male stem cells differentiated, 7–13% of mouse spermatozoa contained the transgene and 15–25% of the pig seminiferous tubules contained sperm with the transgene. However, the busulphan treatment was extremely toxic, killing a large percentage of the animals and reducing the numbers of spermatozoa in the survivors. Earlier work by Sato *et al.* (1994) indicated that DNA injected into mouse testis appeared in the sperm up to 7 days later, but had disappeared by 28 days.

Kim *et al.* (1997) also conducted experiments in which epididymal spermatozoa were treated with lipofectin–DNA complexes. While the exogenous DNA was bound to the sperm and was resistant to DNase treatment, male pronuclei removed from oocytes fertilized by the transfected spermatozoa did not contain the exogenous DNA. However, exogenous DNA was occasionally detected in the cytoplasm. It is thus apparent that in order to transfer foreign DNA using sperm as a vector, the exogenous DNA must be incorporated into the sperm genome. The DNA in fully differentiated sperm cells is highly condensed into a small volume by protamine and it may be extremely difficult to integrate foreign DNA into the sperm genome at this stage.

Conclusions

The successful generation of transgenic animals using sperm cells as carriers for the foreign DNA is still an attractive possibility. DNA can be efficiently loaded into the sperm cells using liposomes or by electroporation and successfully transferred to the egg. The main obstacles are the degradation and rearrangement of the transgene and the lack of integration of the transgene into the developing embryo. Methods that stabilize the foreign DNA and increase the efficiency of integration of DNA into the sperm chromatin would help to make possible the use of sperm cells for generating transgenic animals.

References

Arezzo, F. (1989) Sea urchin sperm as a vector of foreign genetic information. *Cell Biology International Reports* 13, 391–404.

Atkinson, P.W., Hines, E.R., Beaton, S., Matthaei, K.I., Reed, K.C. and Bradley, M.P. (1991) Association of exogenous DNA with cattle and insect spermatazoa *in vitro*. *Molecular Reproduction and Development* 29, 1–5.

Brackett, B.G., Baranska, W., Sawicki, W. and Koprowski, H. (1971) Uptake of heterologous genome by mammalian spermatozoa and its transfer to ova through fertilization. *Proceedings of the National Academy of Sciences USA* 68, 353–357.

Brinster, R.L. and Avarbock, M.R. (1994) Germline transmission of donor haplotype following spermatogonial transplantation. *Proceedings of the National Academy of Sciences USA* 91, 11303–11307.

Brinster, R.L. and Zimmermann, J.W. (1994) Spermatogenesis following male germ-cell transplantation. *Proceedings of the National Academy of Sciences USA* 91, 11298–11302.

Brinster, R.L., Sangren, E.P., Behringer, R.R. and Palmiter, R.D. (1989) No simple solution for making transgenic mice. *Cell* 59, 239–241.

Castro, F.O., Hernandez, O., Uliver, C., Solano, R., Milanes, C., Aguilar, A., Perez, A., de Armas, R., Herrera, L. and de la Fuente, J. (1990) Introduction of foreign DNA into the spermatazoa of farm animals. *Theriogenology* 34, 1099–1110.

Chourrout, D., Guyomard, R. and Houdebine, L.M. (1986) High efficiency gene transfer in rainbow trout (*Salmo gairdneri*) by microinjection into egg cytoplasm. *Aquaculture* 51, 143–150.

Gagne, M.B., Pothier, F. and Sirard, M.A. (1991) Electroporation of bovine spermatozoa to carry foreign DNA in oocytes. *Molecular Reproduction and Development* 29, 6–15.

Gandolfi, F., Lavitrano, M., Camaioni, A., Spadafora, C., Siracusa, G. and Lauria, A. (1989) The use of sperm-mediated gene transfer for the generation of transgenic pigs. *Journal of Reproduction and Fertility Abstract Series* 4, 10.

Gandolfi, F., Terqui, M., Modina, S., Brevini, T.A.L., Ajmone-Marsan, P., Foulon-Gauze, F. and Courot, M. (1996) Failure to produce transgenic offspring by intra-tubal insemination of gilts with DNA-treated sperm. *Reproduction, Fertility and Development* 8, 1055–1060.

Gavora, J.S., Benkel, B., Sasada, H., Cantwell, W.J., Fiser, P., Teather, R.M., Nagai, J. and Sabour, M.P. (1991) An attempt at sperm-mediated gene transfer in mice and chickens. *Canadian Journal of Animal Science* 71, 287–291.

Habrova, V., Takac, M., Navratil, J., Macha, J., Ceskova, N. and Jonak, J. (1996) Association of rous sarcoma virus DNA with *Xenopus laevis* spermatozoa and its transfer to ova through fertilization. *Molecular Reproduction and Development* 44, 332–342.

Kim, J.-H., Jung-Ha, H.-S., Lee, H.-T. and Chung, K.-S. (1997) Development of a positive method for male stem cell-mediated gene transfer in mouse and pig. *Molecular Reproduction and Development* 46, 515–526.

Lakshmanan, N., Duby, R.T. and Smyth J.R. Jr (1990) Intramagnal catheterization: a novel method for intramagnal insemination of chickens. *Poultry Science* 69 (Suppl. 1), 77.

Lavitrano, M., Camaioni, A., Fazio, V.M., Doici, S., Farace, M.G. and Spadafora, C. (1989) Sperm cells as vectors for introducing foreign DNA into eggs: genetic transformation of mice. *Cell* 57, 717–723.

Love, J., Cribbin, C., Mather, C. and Sang, H. (1994) Transgenic birds by DNA microinjection. *Biotechnology* 12, 60–63.

Martinez, R., Perez, A., Castro, F.O., Lleonart, R., Castro, O., Garcia, R., Aguilar, A., Herrera, L. and De La Fuente, J. (1992) Conditions for Southern blot analysis for the detection of single copy genes: application to the screening for transgenic chickens. *Biotecnologia Aplicada* 9, 83–86.

Nakanishi, A. and Iritani, A. (1993) Gene transfer in the chicken by sperm-mediated methods. *Molecular Reproduction and Development* 36, 258–261.

Overbeek, P.A., Aguilar-Cordova, E., Hanten, G., Schaffner, D., Patel, P., Lebovitz, R.M. and Lieberman, M.W. (1991) Coinjection strategy for visual identification of transgenic mice. *Transgenic Research* 1, 31–37.

Patil, J.G. and Khoo, H.W. (1996) Nuclear internalization of foreign DNA by zebrafish spermatozoa and its enhancement by electroporation. *Journal of Experimental Zoology* 274, 121–129.

Rottmann, O.J., Antes, R., Höfer, P. and Maierhofer, G. (1992) Liposome-mediated gene transfer via spermatazoa into avian eggs. *Journal of Animal Breeding and Genetics* 109, 64–70.

Sato, M., Iwase, R., Kasai, K. and Tada, N. (1994) Direct injection of foreign DNA into mouse testis as a possible alternative to sperm-mediated gene transfer. *Animal Biotechnology* 5, 19–31.

Schellander, K., Peli, J., Schmoll, G. and Brem, G. (1995) Artificial insemination in cattle with DNA-treated sperm. *Animal Biotechnology* 6, 41–50.

Sexton, T.J. and Fewlass, T.A. (1978) A new poultry semen extender 2. Effect of the diluent components on the fertilizing capacity of chicken semen stored at 5°C. *Poultry Science* 57, 277–284.

Spadafora, C., Maione, B., Zoraqi, G., Zaccagnini, G., Pittoggi, C. and Lorenzini, R. (1997) Interaction between sperm cells and foreign DNA: fate of nuclear internalized exogenous DNA. In: *Proceedings of Keystone Symposia 'Germ Cell Differentiation', Frisco, Colorado*, March 1997.

Sperandio, S., Lulli, V., Bacci, M.L., Forni, M., Maione, B., Spadafora, C. and Lavitrano, M. (1996) Sperm-mediated DNA transfer in bovine and swine species. *Animal Biotechnology* 7, 59–77.

Squires, E.J. and Drake, D. (1993) Liposome-mediated DNA transfer to chicken sperm cells. *Animal Biotechnology* 4, 71–88.

Squires, E.J. and Drake, D. (1994) Transgenic chickens by liposome-sperm-mediated gene transfer. In: Smith, C., Gavora, J.S., Benkel, B., Chesnais, J., Fairfull, W., Gibson, J.P., Kennedy, B.W. and Burnside, E.B. (eds) *Proceedings of the 5th World Congress on Genetics Applied to Livestock Production*, Vol. 21. University of Guelph, Ontario, pp. 350–353.

Squires, E.J. and Drake, D. (1997) Transgenic chickens by liposome-sperm-mediated gene transfer. In: Houdebine, L.M. (ed.) *Transgenic Animals: Generation and Use.* Harwood Academic Publishers, Amsterdam, pp. 95–99.

Stief, A., Winter, D.M., Straetling, W.F.H. and Sippel, A.E. (1989) A nuclear DNA attachment element mediates elevated and position-independent gene activity. *Nature (London)* 341, 343–345.

Thorey, I.S., Cecena, G., Reynolds, W. and Oshima, R.G. (1993) *Alu* sequence involvement in transcriptional insulation of the keratin 18 gene in transgenic mice. *Molecular and Cellular Biology* 13, 6742–6751.

Trefil, P., Thoraval, P., Mika, J., Coudert, F. and Cambrine, G. (1996) Intramagnal insemination of hens can eliminate negative influence of lipofectin on fertilising ability of spermatozoa. *British Poultry Science* 37, 661–664.

Tsai, H.-J., Tseng, F.S. and Liao, I.C. (1995) Electroporation of sperm to introduce foreign DNA into the genome of loach (*Misgurnus anguillicaudatus*). *Canadian Journal of Fisheries and Aquatic Science* 52, 776–787.

Tsai, H.-J., Lai, C.-H. and Yang, H.-S. (1997) Sperm as a carrier to introduce an exogenous DNA fragment into the oocyte of Japanese abalone (*Haliotis divorsicolor suportexta*). *Transgenic Research* 6, 85–95.

Zani, M., Lavitrano, M., French, D., Lulli, V., Maione, B., Sperandio, S. and Spadafora C. (1995) The mechanism of binding of exogenous DNA to sperm cells: factors controlling DNA uptake. *Experimental Cellular Research* 217, 57–64.

Zoraqi, G. and Spadafora, C. (1997) Integration of foreign DNA sequences into mouse sperm genome. *DNA and Cell Biology* 16, 291–300.

8

Understanding the Origin of Avian Primordial Germ Cells: Implications for Germ Cell Culture and Transgenesis in Poultry

J.N. Petitte, S. D'Costa and L. Karagenç

Department of Poultry Science, North Carolina State University, Raleigh, North Carolina, USA

Introduction

The production of transgenic poultry is highly desirable for a number of reasons. During the last two decades, considerable advances have been made in our primary understanding of the molecular basis of many genetic traits important to the poultry industry. The application of this knowledge to the global industry through transgenic technology holds considerable promise for improving the profitability and quality of commercial poultry stocks and for the development of novel uses for poultry. Moreover, since the domestic fowl and other species, such as Japanese quail, continue to be model organisms in the basic biological sciences, a routine means of producing transgenic birds will no doubt provide a tool to further enhance our understanding of avian biology.

Among the various approaches toward the development of transgenic poultry, the use of embryonic chimeras has received considerable attention in the last 10 years. Current methods using embryonic chimeras employ early blastodermal cells or primordial germ cells (PGCs) as outlined in Fig. 8.1. This scheme is obviously an adaptation of procedures used for mammalian embryos, particularly the mouse, while taking advantage of the unique features provided by the avian embryo. Briefly, PGCs or precursors to PGCs in the form of blastodermal cells are isolated, cultured, transfected and returned to a recipient embryo to produce a germline chimera, which hopefully contains transgenic gametes that would yield fully transgenic offspring. For the purposes of this discussion, the emphasis will be on the use of primordial germ cells and the current attempts to culture these cells for use in producing transgenic poultry.

© CAB INTERNATIONAL 1999. *Transgenic Animals in Agriculture* (eds J.D. Murray, G.B. Anderson, A.M. Oberbauer and M.M. McGloughlin)

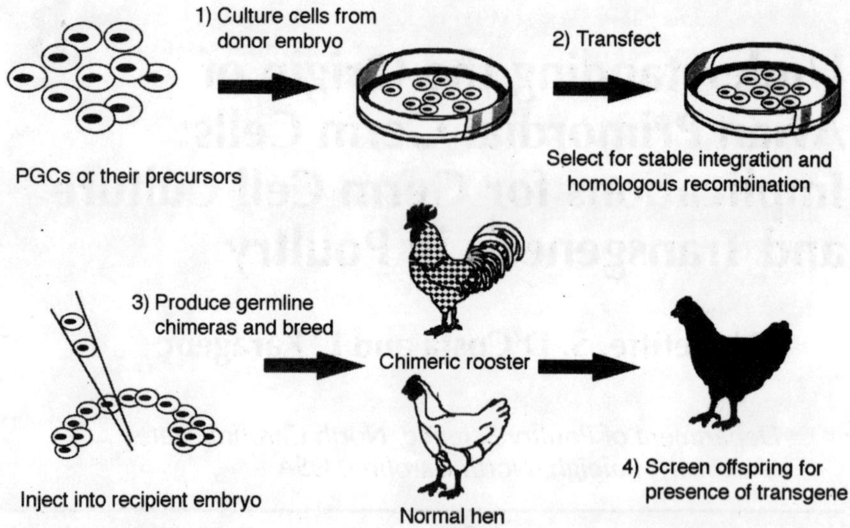

Fig. 8.1. A schematic representation of the steps in the production of transgenic poultry using primordial germ cells (PGCs) or their precursors. 1, PGCs can be obtained from cultured stage X embryo cells, germinal crescent, blood and the gonad (see Fig. 8.2). 2, During culture, the PGCs are transfected with the appropriate DNA construct and selected for stable integration and, if appropriate, evaluated for homologous recombination. 3, The PGCs are then returned to the embryo at the appropriate stage of development to generate a germline chimera. 4, Offspring from the chimera are then screened for the presence of the transgene, and these individuals will become the founding population for a line of transgenic poultry.

Early Development of the Avian Embryo

The two most economically important species of poultry are the domestic fowl (*Gallus gallus domesticus*) and turkey (*Meleagris gallopavo*). Fortunately, domestic fowl have been used in the study of vertebrate development for almost a century and a considerable body of work has amassed regarding the origin of primordial germ cells immediately before and after incubation. Unlike mammals, the main reproductive strategy of the bird entails the packaging of a large vitellus in a calcified eggshell containing the complete nutrient requirement for the development and hatching of a viable chick. Upon ovulation the ovum is fertilized in the distal end of the oviduct, the infundibulum, a process that is clearly polyspermic (Perry, 1987). However, the first cleavage divisions do not begin until about 5–6 h later, after the ovum has received its investment of albumen and the shell membranes and moves into the shell gland for the deposition of the shell (Eyal-Giladi and Kochav, 1976; Perry, 1987). During the approximately 22 h process required for eggshell formation, a considerable amount of cell division takes place. For several decades, early embryonic development in the shell gland was considered rather amorphous and the staging of embryos began with the onset of incubation (Hamburger and Hamilton, 1951). Subsequently, Eyal-Giladi and Kochav (1976) examined the pre-incubation period of development of the chick. They described and documented a series of well-defined stages from fertilization to the first few hours of incubation based upon significant changes in morphology, which have been shown subsequently to correlate with the early organization of the embryo (Eyal-Giladi, 1991). The staging system of Eyal-Giladi and Kochav (1976) is designated with Roman numerals while that of Hamburger and Hamilton (1951) is designated with Arabic numbers to avoid confusion. Recently, Gupta and Bakst (1993) reported that the early development of the turkey embryo has diverged enough from that of the chick to require a different staging system (for a direct comparison between chick and turkey see Bakst *et al.*, 1997).

For the current discussion it is important to understand that by the time the egg is laid the stage X chick embryo consists of a central, translucent cellular area, area pellucida, suspended over a non-yolky cushion of fluid (Fig. 8.2). This region is bordered by a peripheral ring of cells, the area opaca, which is in direct contact with the yolk. Upon incubation, the area pellucida differentiates into two distinct layers, an upper epiblast that will give rise to the embryo proper and a thin hypoblast lying immediately beneath the epiblast (Fig. 8.2). The hypoblast gives rise to portions of the extraembryonic membranes and does not contribute to the embryo proper. The process of hypoblast formation takes place upon incubation between stages XI–XIII. At stage XIV/stage 2, a thickening appears in the posterior end of the embryo and is the first sign of the formation of the primitive streak and the establishment of the three primary germ layers. As the streak elongates the hypoblast is displaced anteriorly by the emerging endoderm (Fig. 8.2). It is the hypoblast that forms the lower layer of the so-called germinal crescent, a region were chick primordial germ cells were first identified (Swift, 1914). This arrangement is important in understanding the origin of the avian germline prior to and after formation of the primitive streak.

Origin of Avian PGCs

For the purpose of producing transgenic poultry, primordial germ cells are the cell type of significance since they will give rise to ova or spermatozoa. The origin of the germline in vertebrates has been a topic of interest for biologists for a very long time. The current scenario for the development of avian germ cells is outlined in Fig. 8.3. PGC development follows a rather circuitous trek beginning with pre-primitive streak development, migration to the germinal crescent, passive migration in the embryonic vasculature, active migration to the presumptive gonad and final residency in the germinal ridge

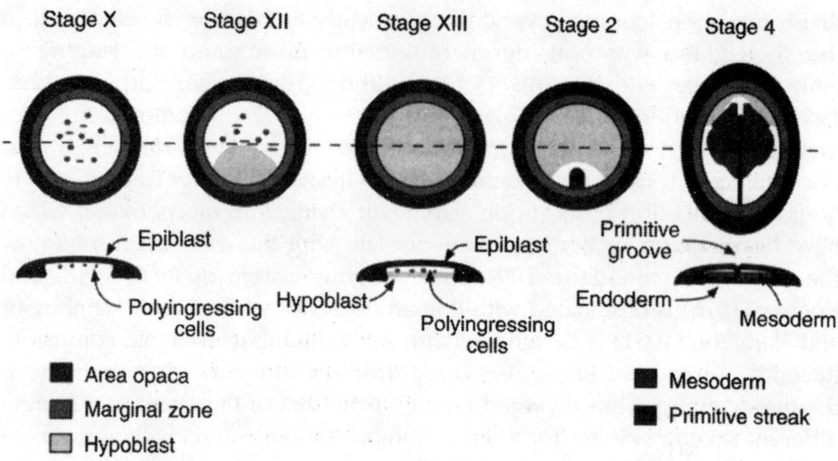

Fig. 8.2. A diagrammatic representation of the structure of the chick embryo from stage X (oviposition) to the formation of the primitive streak at stage 4 (about 12 h of incubation). The stage X embryo is characterized by the area opaca, which is in direct contact with the yolk, and the area pellucida, composed of the marginal zone and the central disk, suspended above the yolk. At this stage, the area pellucida epiblast contains clusters of polyingressing cells on the ventral surface that have moved from the dorsal surface of the epiblast. From stages XI–XIII, the hypoblast forms in a posterior–anterior direction. Some of the polyingressing cells become associated with the hypoblast. During the formation of the primitive streak, beginning at stage 2, the endoderm pushes the hypoblast anteriorly and laterally, taking on a crescent shape.

(Fig. 8.3; see Nieuwkoop and Sutasurya, 1979; Petitte *et al.*, 1997). PGCs are ordinarily identified through a combination of morphological characteristics, e.g. large size and eccentric nucleus, coupled with either a histochemical marker such as periodic acid–Schiff (PAS), which stains for glycogen (Meyer, 1960), or with immunohistochemical markers such as EMA-1 and SSEA-1 (stage-specific embryonic antigen 1), which recognize cell-surface carbohydrate epitopes (Urven *et al.*, 1988; Loveless *et al.*, 1990; Karagenç *et al.*, 1996). For the most part, avian PGCs have been identified in primitive streak and older embryos using PAS and in experimental situations where embryo fragments were cultured until PAS staining could be used.

PGCs are derived from the epiblast (Eyal-Giladi *et al.*, 1981), a process that begins at stage X (Karagenç *et al.*, 1996) before their migration to the germinal crescent (Ginsburg and Eyal-Giladi, 1986), a region in the stage 4–10 embryo located in an anterior, extraembryonic segment bordering the area opaca and area pellucida (Swift, 1914) (Fig. 8.3). Using a series of *in vivo* and *in vitro* techniques, Karagenç *et al.* (1996) utilized antibodies SSEA-1 and EMA-1 to identify a population of about 20 EMA-1/SSEA-1-positive cells that first appear in the stage X epiblast in the area pellucida and that translocate

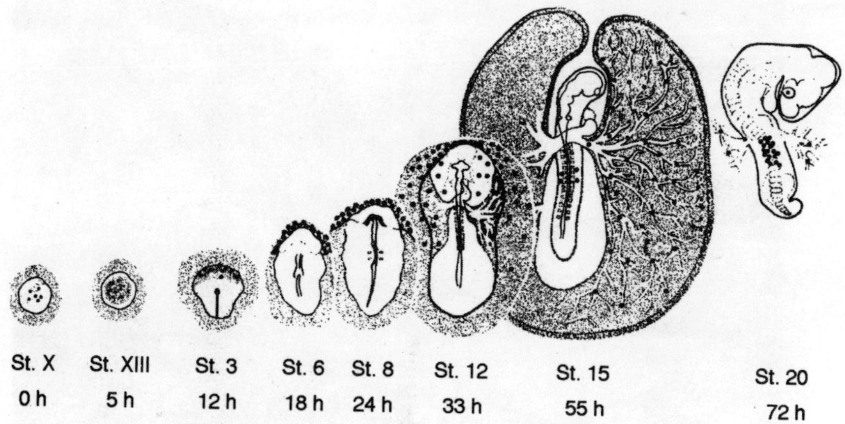

Fig. 8.3. An outline of the origin of primordial germ cells (PGCs) (dark spots in diagram) of the chick embryo from early stages of development through their final residence in the gonadal anlage. The germ cell lineage is first identified by the presence of about 20 SSEA-1 and EMA-1-positive cells in the stage X epiblast at the time of oviposition (Karagenç *et al.*, 1996). Upon incubation, these cells translocate to the hypoblast between stages XI to XIV. With the formation of the primitive streak (stage 3–4) the hypoblast moves in anteriorly to form the germinal crescent (Swift, 1914). From this time onward PGCs can be identified using periodic acid–Schiff (PAS) staining. While in the crescent (stages 6, 8, 12), the germ cells associate with the forming blood islands and then move into the embryonic circulation. Subsequently the germ cells actively leave the blood vessels (stage 15) and migrate along the dorsal mesentery to the gonadal anlage. Most of the germ cells have reached the presumptive gonad by stage 20. The entire process appears to involve passive and active phases of migration, chemotaxis, location of vascular elements and extracellular matrix components. Approximate hours of incubation are shown. (Redrawn after Nieuwkoop and Sutasurya, 1979, with modifications.)

to the hypoblast during stages XII–XIV (Fig. 8.4A, B). Experimentally, it was shown that the SSEA-1-positive cells on the hypoblast have the potential to give rise to PGCs *in vitro* and *in vivo* (Karagenç *et al.*, 1996, and unpublished observations). Additionally, the formation of the area pellucida appears to be a prerequisite for PGC development (Karagenç *et al.*, 1996), although germ cells arise spontaneously in cultured fragments of the central epiblast without normal axial development (Ginsburg and Eyal-Giladi, 1987). Recently, Kagami *et al.* (1997) reported that removing as few as 700 of the 10,000–15,000 cells of the central epiblast in unincubated embryos leads to germ cell-depleted chickens. It is tempting to conclude from this observation that avian germ cell allocation is a one-time event occurring in the geometric centre of the epiblast; however, this hypothesis does not explain the observed gradual allocation of germ cells from the epiblast that occurs from stage XII through to germinal crescent stages (Ginsburg and Eyal-Giladi, 1986;

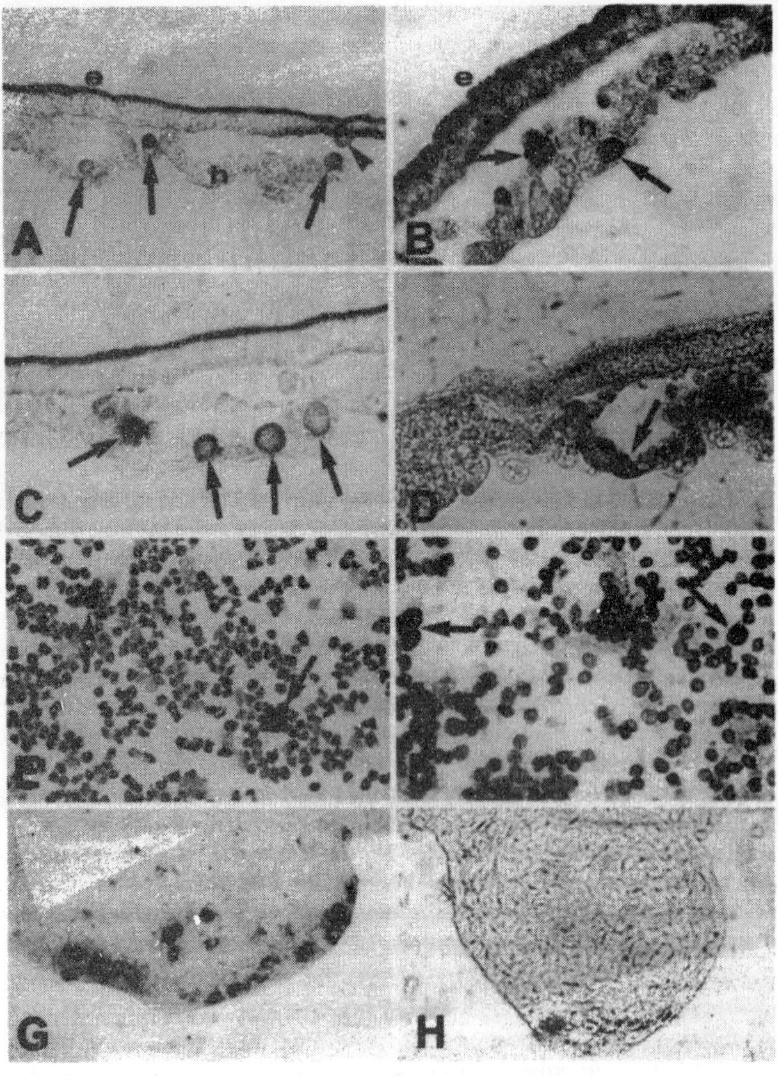

Fig. 8.4. Immunohistochemical staining of primordial germ cells (PGCs) using SSEA-1 in the chick embryo (A, C, E, G) and in the turkey embryo (B, D, F, H). A and B: cross-sections of chicken and turkey embryos, respectively, prior to formation of the primitive streak. Both species contain SSEA-1 labelled cells (arrows) in the hypoblast (h). PGCs originate from the epiblast (e) and translocate to the hypoblast. Arrowhead shows chick PGC emerging from the epiblast. C and D: sections through the germinal crescent of chicken and turkey embryos. Arrows indicate PGCs. E and F: blood smears from chicken and turkey embryos during the period of passive migration of germ cells (arrows) through the blood vessels. G and H: sections through the 5-day chick embryonic gonad and 8-day turkey embryonic gonad (equivalent developmental stages). Clusters of gonadal PGCs can only be identified with SSEA-1 in the chick.

Karagenç *et al.*, 1996) nor the body of data showing that PGCs can arise from anterior–posterior or transverse fragments of the stage X embryo and from cells of dispersed stage X embryos (Ginsburg and Eyal-Giladi, 1987, 1989; Karagenç *et al.*, 1996). Certainly additional experiments are needed to reconcile these observations.

In any case, once primordial germ cells reach the germinal crescent, they are committed to the germline, and excision or ultra-violet irradiation of the crescent depletes germ cell numbers (Reynaud, 1976, 1977; McCarrey and Abbott, 1978, 1982). While in the germinal crescent, the PGCs associate with the incipient blood islands and enter the embryonic circulation for a period of passive migration (Meyer, 1964; Fujimoto *et al.*, 1976a). Subsequently, the PGCs begin an active period of migration and exit the blood vessels to migrate along the dorsal mesentery and collect at the germinal ridges (Fujimoto *et al.*, 1976b). By 72 h of incubation most of the germ cells have reached the gonad. This process of passive and active migration is undoubtedly mediated by chemo-attraction, extracellular matrix components and the organization of the vascular system (Dubois and Croisille, 1970; Kuwana *et al.*, 1986; Nakamura *et al.*, 1988; Urven *et al.*, 1989).

All of these periods of PGC development appear to be conserved among various species of birds. However, the successful use of histological markers to identify avian germ cells varies with species (see Table 8.1). For example, while PAS staining is useful for identification of chick PGCs, quail PGCs are PAS-negative. Hence, it cannot be assumed that various markers in one species will be useful in another. Given the importance of the domestic turkey to the poultry industry, we compared the use of SSEA-1 and PAS in the turkey versus the chicken embryo for the identification of PGCs (S. D'Costa and J.N. Petitte, unpublished observations). Figure 8.4 shows that the turkey embryo at stage XI (Gupta and Bakst, 1993), stage 4, and stage 13 is SSEA-1-positive, while the gonad at day 8–9 of incubation is SSEA-1 negative. On the other hand, PGCs of the chick embryo are positive at all the corresponding stages. The role of SSEA-1 in germ cell function is unknown but may be related to migration and cell–cell adhesion, as suggested for mammalian germ cells. If this is the case in birds, the species difference in SSEA-1 expression may prevent turkey germ cells from completing their migration to the gonad. Such a feature would explain, in part, the results obtained by Reynaud (1976) who produced turkey/chick germline chimeras but could not clearly demonstrate functional gametes in the adult. Again, more research in this area is needed to clarify whether interspecies turkey/chick germline chimeras are possible.

Development of Germline Chimeras Using PGCs

The foregoing discussion of germ cell origins and development immediately suggests various intervention points to make germline chimeras using PGCs.

Table 8.1. A comparison of the major histochemical and immunological markers of primordial germ cells (PGCs) among chicken, quail, turkey and the mouse at various stages of germ cell development. Question marks indicate unknown use of the marker and NS indicates non-specific. In practice, the use of these markers must be coupled with the morphological characteristics of PGCs and their location for proper identification. Pertinent references are given.

		Stage of germ cell development					
Species	Marker	Pre-streak embryo	Stage 4–10, germinal crescent	Stage 11–17, blood	Stage 18–22, dorsal mesentery	Stage 23+, gonadal ridge	References
Chicken	AP	NS	NS	NS	+	+	Chiquoine and Rothenberg (1957)
	PAS	NS	+	+	+	+	Meyer(1964), Fujimoto *et al.* (1976b), Swartz (1982)
	EMA-1	+	+	+	+	+	Urven *et al.* (1988), Karagenç *et al.* (1996)
	SSEA-1	+	+	+	+	+	Karagenç *et al.* (1996)
Quail	PAS	NS	–	–	–	–	Nakamura *et al.* (1992)
	QH-1	+	+	?	?	+	Pardanuad *et al.* (1987)
	SSEA-1	+	+	?	?	?	Petitte and Karagenç, unpublished observations
Turkey	PAS	NS	+	+	+	+	Reynaud (1967, 1969)
	SSEA-1	+	+	+	–	–	See text
		6.5 dpc, egg cylinder	8.5 dpc, allantois	No equivalent stage	9–12 dpc, hindgut/dorsal mesentery	12.5–13 dpc, gonadal ridge	
Mouse	AP	+	+	–	+	+	Chiquoine (1954), Mintz and Russel (1957), Ozdzenski (1967), Ginsburg *et al.* (1990)
	EMA-1	?	+	–	+	+	Hahnel and Eddy (1986)
	SSEA-1	?	?	–	+	+	Fox *et al.* (1981)

NS, non-specific; PAS, periodic acid–Schiff; AP, alkaline phosphatase; SSEA-1, stage-specific embryonic antigen-1.

In fact, several groups have used PGCs from the germinal ridge, blood and gonads for this purpose. Reynaud (1969) definitively demonstrated the extragonadal origin of PGCs by injecting germinal crescent cells, which presumably contained PGCs, into the extraembryonic vasculature of sterilized recipient embryos. Subsequently, the donor PGCs populated the host gonad. This observation led to initial attempts to produce germline chimeric chicks or quail through germinal crescent cell transfer or using PGCs from blood (Shuman, 1981; Gonzales, 1989; Wentworth *et al.*, 1989; Simkiss *et al.*, 1989; Petitte *et al.*, 1991). Since that time, the results of other experiments have demonstrated that PGCs from these two sources can be used to produce embryonic germline chimeras as well as adult birds that produce gametes derived from the donor germ cells (Watanabe *et al.*, 1992; Yasuda *et al.*, 1992; Tajima *et al.*, 1993; Vick *et al.*, 1993; Naito *et al.*, 1994b; Ono *et al.*, 1996). In most cases, the depletion of endogenous germ cells improves the frequency of chimerism. Recently, Chang *et al.* (1995b) have shown that PGCs obtained from the germinal ridge can be used to produce germline chimeras when injected into the blood of recipient embryos during earlier stages of development. This is an important observation, since it was assumed that once the PGCs reached the gonadal anlage, their ability to migrate and repopulate the host gonad would be lost. It would be of interest to examine how long gonadal germ cells retain this capability.

Transgenesis Using PGCs

Several methods of inserting DNA into PGCs are available, and the use of each has its advantages and disadvantages. For the most part, only retroviral infection of germinal crescent PGCs has been successful in producing transgenic chickens (Vick *et al.*, 1993). This achievement was due mainly to the high integration efficiency of retroviral vectors. DNA complexed with liposomes offers a convenient method both *in situ* and *in vitro* (Brazolot *et al.*, 1991; Han *et al.*, 1994, 1996; Watanabe *et al.*, 1994; Hong *et al.*, 1998), and bombardment of the germinal crescent with DNA-coated particles can place DNA within the cells (Li *et al.*, 1995). Electroporation of germinal crescent or blood PGCs has yet to be tried because of the large numbers of cells required by most procedures. Recently, electroporation of dispersed gonadal tissue followed by separation of PGCs was tested as a means of transfecting gonadal PGCs (Hong *et al.*, 1998). This approach yielded levels comparable to liposome-mediated transfection. Nevertheless, stable integration of exogenous DNA in the PGC genome is normally a rare event using these methods, and the possibility of episomal inheritance of plasmid DNA calls for careful experimentation (Chapters 7 and 9, this volume). The only manner in which to guarantee stable integration of DNA into the germ cells is through the culture of PGCs with the application of some type of selectable marker. Traditionally, antibiotic resistance has been used for this purpose; however, constructs using jellyfish green fluorescent protein might serve as another means to identify cells where integration events have occurred.

Culture of PGCs

Whatever the source of PGCs and the means of transfection, the production of transgenic poultry would be facilitated immensely by the ability to establish long-term cultures of germ cells or their precursors. In addition to applications in transgenics, the ability to expand primordial germ cell populations could have a major impact on the conservation and preservation of avian genetic resources and endangered wild avian species (Delany and Pisenti, 1998). In this regard, it is noteworthy that cryopreserved chick primordial germ cells have been successfully used to produce chicks though the production of germline chimeras (Naito *et al.*, 1994a; Tajima *et al.*, 1996). For these applications to achieve their full potential, it is important to have a basic understanding of the complex biology of PGCs at the molecular level. Unfortunately, knowledge in this area is still rudimentary and is often based upon work with mammalian germ cells. Despite this limitation, there have been attempts to culture germ cells from gonadal PGCs for relatively short periods of time, usually less than a week. The first effort to culture chick PGCs was reported by Fritts-Williams and Meyer (1972). In this case, stage 25–26 embryonic gonads were used as the starting material since sexual differentiation was not morphologically apparent. Germ cells were enriched by applying a single-cell suspension of gonadal tissue to a Ficoll gradient. The fraction containing the PGCs was cultured for 7–14 days in hanging drops or roller tubes, but no proliferation was observed. Subsequently, Allioli *et al.* (1994) cultured gonadal PGCs long enough for infection with a retroviral vector and obtained at least 50% of the PGCs expressing *LacZ*. Wentworth *et al.* (1996) mentioned that attempts to culture gonadal PGCs co-cultured with gonadal somatic cells were dependent upon the initial concentration of PGCs, and reported that PGCs could not be regularly subcultured. Likewise, Chang *et al.* (1995a,b, 1997) were able to culture gonadal PGCs on a stromal cell feeder layer for 5 days and then used the cells to produce germline chimeras that transmitted the donor genotype to 1.3–3.5% of the progeny. Such results are quite promising, and testing the effect of longer periods of culture on the ability to produce germline chimeras is an obvious next step.

More culture work has been done using gonadal PGCs than any other source of germ cells because of the relative ease of obtaining sufficient numbers of cells to use for experimental purposes. However, there are some reports of short-term culture of PGCs at other stages. Lee *et al.* (1983) removed the hypoblast from the germinal crescent area of stage 4–5 embryos and observed the migration of PGCs over the hypoblast feeder layer. Likewise, Kuwana and Fujimoto (1984) have observed the movement of PGCs from the blood *in vitro* on a feeder layer of dorsal mesentery. While these experimental procedures were designed to examine the process of germ cell locomotion, they do suggest that it is possible to culture germ cells from these sources if provided with suitable culture conditions.

Unfortunately, our lack of understanding of what makes avian germ cells survive and proliferate *in vivo* is a genuine problem associated with the culture of avian PGCs. In the mouse, several factors are implicated in the survival and proliferation of PGCs. These include, leukaemia inhibitory factor (LIF), stem cell factor (SCF)/c-kit, transforming growth factor-b, fibroblast growth factor-2 (FGF-2), among others. Unfortunately, Chang *et al.* (1995a) found no observable benefit to adding mammalian SCF or LIF to the culture of gonadal PGCs. In the course of examining the origin of PGCs in the pre-streak embryo, it was observed that the culture of dispersed cells from the area pellucida of stage IX embryos could be induced to give rise to primordial germ cells only when plated on a feeder layer of STO cells (Karagenç *et al.*, 1996; Fig. 8.5). Not only could this culture system yield PGCs, but the numbers obtained were roughly equivalent to those at germinal crescent stages, suggesting that the factors responsible for emergence, proliferation or a combination of the two were being provided by the feeder cells (Fig. 8.6). This observation immediately suggested that the culture of stage X embryo cells on an STO feeder layer could be used to examine the role of various growth factors involved in avian germ cell development.

Among the obvious candidate growth factors is SCF, also called steel factor and mast cell growth factor. SCF plays a pleiotropic role in haematopoiesis, melanocyte development and germ cell development (Bernstein *et al.*, 1990) and is the ligand of the receptor tryosine kinase, c-kit. In mice, mutations in either of these proteins result in impaired germ cell production or migration (Chabot *et al.*, 1988; Geissler *et al.*, 1988;

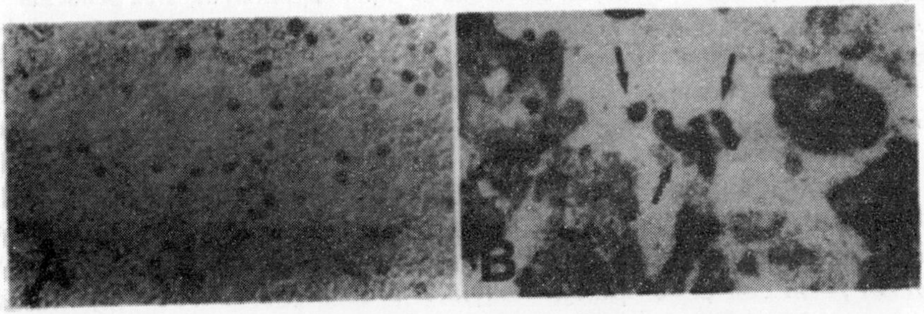

Fig. 8.5. The emergence of PGCs cultured on a feeder layer of STO cells. A, freshly seeded dispersed stage X embryo cells on a feeder layer of mouse STO fibroblasts; B, growth of the culture after 48 h and staining with SSEA-1. The culture consists of SSEA-1-positive and negative patches with PGCs often found in clusters lying above the feeder cells (arrows). Methods according to Karagenç *et al.* (1996).

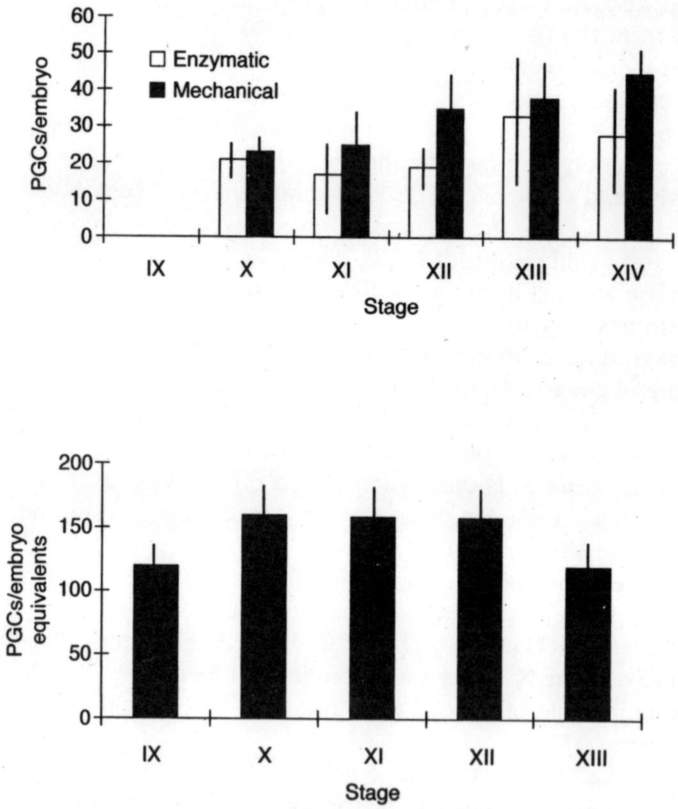

Fig. 8.6. The number of primordial germ cells (PGCs) obtained after 48 h of culture using dispersed chick embryo area pellucida cells without (upper panel) and with (lower panel) an STO feeder layer. Upper panel: whole dispersed embryos were cultured and stained with periodic acid–Schiff (Meyer, 1960). No germ cells were obtained using stage IX embryos. Lower panel: 5000 cells from the area pellucida were seeded on to an STO feeder layer in 96-well plates and stained for primordial germ cells using SSEA-1. Data are presented as number of PGCs per embryo. The number of PGCs obtained was considerably higher than for cultures without a feeder layer at any stage. (Data after Karagenç *et al.*, 1996.)

Copeland *et al.*, 1990). As in mammals, SCF in birds is involved in haematopoiesis and melanogenesis, but little is known about its role in avian germ cell development (Hayman *et al.*, 1993; Lahav *et al.*, 1994; Lecoin *et al.*, 1995). In chicken and quail, the cDNA has been cloned and, like the mammalian homologues, it encodes a transmembrane protein that can exist in two isoforms, a membrane-anchored form and a soluble form (Zhou *et al.*, 1993; Petitte and Kulik, 1996). Homologies between avian and mammalian forms of SCF are about 52%, and this explains the lack of biological activity of mammalian SCF on avian cells. During the prestreak period of germ cell development, transcripts for both forms of SCF and its receptor c-kit can be detected in whole embryo preparations using reverse transcriptase (RT)-PCR (Petitte and Karagenç, 1996, and unpublished observations). Likewise, preliminary work indicates that c-kit is expressed by avian PGCs at stage 28 when they have reached the germinal ridge. At the same time, the germinal ridge epithelium expresses SCF (J.N. Petitte and L. Karagenç, unpublished observations). All of these observations point to SCF/c-kit participating in avian germ cell development.

Another candidate growth factor that may be involved in germ cell survival and development is FGF-2. Resnick *et al.* (1992) and Matsui *et al.* (1992) reported that FGF-2 stimulates the proliferation of mouse PGCs on a variety of feeder layers. In addition, when combined with a SCF-expressing feeder layer and LIF, FGF-2 is an essential component for the long-term culture of mouse PGCs that resemble embryonic stem cells (Matsui *et al.*, 1992; Resnick *et al.*, 1992; Labosky *et al.*, 1994; Stewart *et al.*, 1994). In the chick embryo, FGF-2 transcripts can be detected in the stage XIII embryo (Mitrani *et al.*, 1990), and the application of FGF-2 changes the behaviour of epiblast cells *in vitro* and inhibits axial development of stage XIII embryos (Cooke and Wong, 1991). Finally, Dono and Zeller (1994) and Riese *et al.* (1995) reported that prior to primitive-streak formation, immunohistochemical staining of FGF-2 was located in the nuclei of all hypoblast cells and in some epiblast cells. At primitive streak stages, FGF-2 then appears in the cytoplasm.

Given the association of SCF/c-kit and FGF-2 expression with early and later stages of germ cell development, it is worthwhile examining their role in the culture of avian PGCs. Therefore, we examined the effect of SCF and FGF-2 in the development of PGCs *in vitro* using dispersed stage X cells. After 48 h of culture, STO cells supported the growth of avian PGCs far better than CV-1 cells, a green monkey kidney fibroblast cell line (Fig. 8.7). However, when STO-conditioned medium was added to the culture using CV-1 cells, the response was better than that observed with CV-1 cells alone, suggesting that soluble factors acting on the embryonic cells were involved. However, the addition of FGF-2 to the culture on CV-1 cells did not yield more PGCs than that with CV-1 cells alone. Nonetheless, when CV-1 cells that express the membrane-anchored form of avian SCF are used to culture dispersed stage X embryo cells, the number of PGCs is higher than that observed using CV-1 cells alone (Fig. 8.7). The addition of FGF-2 to the SCF-expressing CV-1 cells did not potentiate the response (Fig. 8.7). These data suggest that SCF may be involved in the emergence and survival of primordial germ cells early in development. Indeed, avian SCF transcripts and c-kit transcripts can be detected by RT-PCR in cultures of stage X cells on STO feeder layers (J.N. Petitte and L. Karagenç, unpublished observations). FGF-2, on the other hand, does not appear to support the survival and proliferation of PGCs from the stage X embryo. This is somewhat

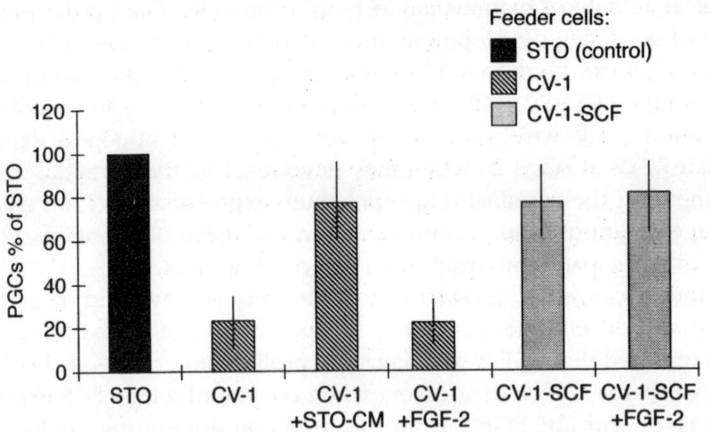

Fig. 8.7. Primordial germ cell culture of dispersed chick embryo cells from the area pellucida on feeder layers of STO, CV-1 and CV-1 cells expressing the membrane-anchored form of stem cell factor (CV-1-SCF). Data are represented as percentage of the response on STO cells alone. CV-1 cells normally do not support the development of PGCs; however, when STO-conditioned medium (STO-CM) is added, the number of germ cells increases relative to culture with CV-1 cells alone. The addition of 100 ng ml^{-1} FGF-2 had no effect on the response. When a CV-1 cell line expressing stem cell factor was used (CV-1-SCF), the percentage of germ cells was equivalent to that observed with STO-CM. Again, the addition of 100 ng ml^{-1} FGF-2 to the culture using CV-1-SCF did not yield more PGCs relative to STO cells alone. Data represent the average of at least three experiments.

unexpected, given the pivotal role of FGF-2 in mammalian PGC culture. However, all of the mammalian culture systems use PGCs from relatively late stages of germ cell development, and the need for FGF-2 in avian germ cell culture may be required for blood or gonadal PGCs. In any case, this culture system should provide a convenient means of examining the effect of other peptide growth factors and cytokines on the proliferation of chick PGCs *in vitro*. Such information could refine current culture methods so that PGCs could then be used for the routine production of transgenic poultry.

Conclusions

The development of robust methods of producing transgenic poultry would greatly facilitate the application of biotechnology to the poultry industry. Despite obstacles, in the last 10 years considerable progress has been made toward using PGCs for the production of transgenic poultry. Several choices are available in starting material and transfection methods. However, the ability to culture avian PGCs for periods long enough to allow selection and multiplication of cells containing integrated gene constructs requires a better understanding of the molecular requirements of the avian germline. Basic information on the process of germ cell development from their segregation from the somatic lineages to their final residence in the primitive gonad will help in attempts to improve the culture of avian PGCs. The data reported here and in other laboratories offer appreciable optimism for future refinements of avian primordial germ cell culture. This will allow academic and industrial institutions to take full advantage of the usefulness provided by the domestic fowl as a research model and as an agricultural commodity for the next century.

References

Allioli, N., Thomas, J.L., Chebloune, Y., Nigon, V.M., Verdier, G. and Legras, C. (1994) Use of retroviral vectors to introduce and express the beta-galactosidase marker gene in cultured chicken primordial germ cells. *Developmental Biology* 165, 30–37.

Bakst, M.R., Gupta, S.K. and Akuffo, V. (1997) Comparative development of the turkey and chicken embryo from cleavage through hypoblast formation. *Poultry Science* 76, 83–90.

Bernstein A., Chabot, B., Dubreuil, P., Reith, A., Mocka, K., Majumder, S., Ray, P. and Besner, P. (1990) The mouse W/c-kit locus. *Ciba Foundation Symposia* 148, 158–166.

Brazolot, C.L., Petitte, J.N., Etches, R.J. and Verrinder Gibbins, A.M. (1991) Efficient transfection of chicken cells by lipofection, and introduction of transfected blastodermal cells into the embryo. *Molecular Reproduction and Development* 30, 304–312.

Chabot, B., Stephenson, D.A., Chapman, V.M., Besmer, P. and Bernstein, A. (1988) The proto-oncogene c-kit encoding a transmembrane tyrosine kinase receptor maps to the mouse *W* locus. *Nature* 335, 88–89.

Chang, I., Tajima, A., Chikamune, T. and Ohno, T. (1995a) Proliferation of chick primordial germ cells cultured on stroma cells from the germinal ridge. *Cell Biology International* 19, 143–149.

Chang, I., Yoshiki, A., Kusakabe, M., Tajima, A., Chikamune, T., Natio, M. and Ohno, T. (1995b) Germline chimeras produced by transfer of cultured chick primordial germ cells. *Cell Biology International* 19, 569–576.

Chang, I., Jeong, D.K., Hong, Y.H., Park, T.S., Moon, Y.K., Ohno, T. and Han, J.Y. (1997) Production of germline chimeric chickens by transfer of cultured primordial germ cells. *Cell Biology International* 21, 495–499.

Chiquoine, A.D. (1954) The identification, origin and migration of the primordial germ cells in the mouse embryo. *Anatomical Record* 118, 135–146.

Chiquoine, A.D. and Rothenberg, E.J. (1957) A note on alkaline phosphatase activity of germ cells in *Amblystoma* and chick embryos. *Anatomical Record* 127, 31–35.

Cooke, J. and Wong, A. (1991) Growth-factor-related proteins that are inducers in early amphibian development may mediate similar steps in amniote (bird) embryogenesis. *Development* 111, 197–212.

Copeland, N.G., Gilbert, D.J., Cho, B.C., Donovan, P.J., Jenkins, N.A., Cosman, D., Anderson, D., Lyman, S.D. and Williams, D.E. (1990) Mast cell growth factor maps near the steel locus on mouse chromosome 10 and is deleted in a number of steel alleles. *Cell* 63, 175–183.

Delany, M.E. and Pisenti, J.M. (1998) Conservation of poultry genetic resources: consideration of the past, present and future. *Poultry and Avian Biology Reviews*, in press.

Dono, R. and Zeller, R. (1994) Cell type specific nuclear translocation of fibroblast growth factor-2 isoforms during chicken kidney and limb morphogenesis. *Developmental Biology* 163, 316–330.

Dubois, R. and Croisille, Y. (1970) Germ-cell line and sexual differentiation in birds. *Philosophical Transactions of the Royal Society, Series B* 259, 73–89.

Eyal-Giladi, H. (1991) The early embryonic development of the chick, as an epigenetic process. *Critical Reviews in Poultry Biology* 3, 143–166.

Eyal-Giladi, H. and Kochav, S. (1976) From cleavage to primitive streak formation: a complementary normal table and a new look at the first stages of the development of the chick. *Developmental Biology* 49, 321–327.

Eyal-Giladi, H., Ginsburg, M. and Farbarow, A. (1981) Avian primordial germ cells are of epiblastic origin. *Journal of Embryology and Experimental Morphology* 65, 139–147.

Fox, N., Damjanov, I., Martinez-Hernandez, A., Knowles, B.B. and Solter, D. (1981) Immunohistochemical localization of the early embryonic antigen (SSEA-1) in postimplantation mouse embryos and fetal and adult tissues. *Developmental Biology* 83, 391–398.

Fritts-Williams, M.L. and Meyer, D.B. (1972) Isolation and culture of homogenous populations of primordial germ cells in the chick. *Experimental Cell Research* 75, 512–514.

Fujimoto, T., Ninomiya, T. and Ukeshima, A. (1976a) Observations of the primordial germ cells in blood samples from the chick embryo. *Developmental Biology* 49, 278–282.

Fujimoto, T., Ukeshima, A. and Kiyofuji, R. (1976b) The origin, migration and morphology of the primordial germ cells in the chick embryo. *Anatomical Record* 185, 139–154.

Geissler, E.N., Ryan, M.A. and Housman, D.E. (1988) The dominant-white spotting (*W*) locus of the mouse encodes the *c-kit* proto-oncogene. *Cell* 55, 185–192.

Ginsburg, M. and Eyal-Giladi, H. (1986) Temporal and spatial aspects of the gradual migration of primordial germ cells from the epiblast into the germinal crescent in the avian embryo. *Journal of Embryology and Experimental Morphology* 95, 53–71.

Ginsburg, M. and Eyal-Giladi, H. (1987) Primordial germ cells of the young chick blastoderm originate from the central zone of the A. pellucida irrespective of the embryo-forming process. *Development* 101, 209–211.

Ginsburg, M. and Eyal-Giladi, H. (1989) Primordial germ cell development in cultures of dispersed central discs of Stage X chick blastoderms. *Gamete Research* 23, 421–428.

Ginsburg, M., Snow, M.H. and McLaren, A. (1990) Primordial germ cells in the mouse embryo during gastrulation. *Development* 110, 521–528.

Gonzales, D.S. (1989) Gonadal germ cell chimeras produced by microinjectiing primordial germ cells in to the vasculature of early chicken embryos. M.S. Thesis, University of Wisconsin, Madison, Wisconsin.

Gupta, S.K. and Bakst, M.R. (1993) Turkey embryo staging form cleavage through hypoblast formation. *Journal of Morphology* 217, 313–325.

Hahnel, A.C. and Eddy, E.M. (1986) Cell surface markers of mouse primordial germ cells defined by two monoclonal antibodies. *Gamete Research* 15, 25–34.

Hamburger, V. and Hamilton, H.L. (1951) A series of normal stages in development of the chick. *Journal of Morphology* 88, 49–92.

Han, J.Y., Shoffner, R.N. and Guise, K.S. (1994) Gene transfer by manipulation of primordial germ cells in the chicken. *Asian Journal of Animal Science* 7, 427–434.

Han, J.Y., Seo, D.S., Hong, Y.H., Choi, D.K. and Shin, Y.S. (1996) Expression of Lac Z gene in young chick gonad by the transfected primordial germ cell injection. *Korean Journal of Poultry Science* 23, 61–69.

Hayman, M.J., Meyer, S., Martin, F., Steinlein, P. and Beug, H. (1993) Self-renewal and differentiation of normal avian erythroid progenitor cells: regulatory roles of the TGFβ/c-ErbB and SCF/c-kit receptors. *Cell* 74, 157–169.

Hong, Y.H., Moon, Y.K., Jeong, D.K. and Han, J.Y. (1998) Improved transfection efficiency of chicken gonadal primordial germ cells for the production of transgenic poultry. *Transgenic Research*, in press.

Kagami, H., Tagami, T., Matsubara, Y., Harumi, T., Hanada, H., Marutama, K., Sakurai M., Kuwana, T. and Naito, M. (1997) The developmental origin of primordial germ cells and the transmission of donor-derived gametes in mixed-sex germline chimeras to the offspring in the chicken. *Molecular Reproduction and Development* 48, 501–510.

Karagenç, L., Cinnamon, Y., Ginsburg, M. and Petitte, J.N. (1996) Origin of primordial germ cells in the prestreak chick embryo. *Developmental Genetics* 19, 290–301.

Kuwana, T. and Fugimoto, T. (1984) Locomotion and scanning electron microscopic observations of primordial germ cells from the early embryonic chick blood *in vitro*. *Anatomical Record* 209, 337.

Kuwana, T., Maeda-Suga, H. and Fujimoto, T. (1986) Attraction of chick primordial germ cells by gonadal anlage *in vitro*. *Anatomical Record* 215, 403–406.

Labosky, P.A., Barlow, D.P. and Hogan, B.L. (1994) Mouse embryonic germ (EG) cell lines: transmission through the germline and differences in the methylation imprint of insulin-like growth factor 2 receptor (Igf2r) gene compared with embryonic stem (ES) cell lines. *Development* 120, 3197–3204.

Lahav, R., Lecoin, L., Ziller, C., Nataf, V., Carnahan, J.F., Martin, F.H. and Le Douarin, N.M. (1994) Effect of the *Steel* gene product on melanogenesis in avian neural crest cell cultures. *Differentiation* 58, 133–139.

Lecoin, C., Lahav, R., Martin, F.H., Teillet, M.A. and Le Douarin, N.M. (1995) Steel and c-kit in the development of avian melanocytes: a study of normally pigmented birds and of the hyperpigmented mutant silky fowl. *Developmental Dynamics* 203, 106–118.

Lee, H.Y., Schumann, J.B. and Nagele, R.G. (1983) Studies on the migration of primordial germ cells in early chick embryos: cultivation of hypoblast explants from the germinal crescent area. *In vitro* 19, 111–116.

Li, Y., Behnam, J. and Simkiss, K. (1995) Ballistic transfection of avian primordial germ cell *in ovo*. *Transgenic Research* 4, 26–29.

Loveless, W., Bellairs, R., Thorpe, S.J., Page, M. and Feiz, T. (1990) Developmental patterning of carbohydrate antigen FC10.2 during early embryogenesis in the chick. *Development* 108, 97–106.

McCarrey, J.R. and Abbott, U.K. (1978) Chick gonad differentiation following excision of primordial germ cells. *Developmental Biology* 66, 356.

McCarrey, J.R. and Abbott, U.K. (1982) Functional differentiation of chick gonads following depletion of primordial germ cells. *Journal of Embryology and Experimental Morphology* 68, 161.

Matsui, Y., Zsebo, K. and Hogan, B.L.M. (1992) Derivation of pluripotential embryonic stem cells from murine primordial germ cells in culture. *Cell* 70, 841–847.

Meyer, D.B. (1960) Application of period acid–Schiff technique to whole chick embryos. *Stain Technology* 35, 83–89.

Meyer, D.B. (1964) The migration of primordial germ cells in the chick embryo. *Developmental Biology* 10, 154–190.

Mintz, B. and Russell, E.S. (1957) Gene-induced embryological modifications of primordial germ cells in the mouse. *Journal of Experimental Zoology* 34, 207–237.

Mitrani, E., Gruenbaum, Y., Shohat, H. and Ziv, T. (1990) Fibroblast growth factor during mesoderm induction in the early chick embryo. *Development* 109, 387–393.

Naito, M., Tajima, T., Tagami, T., Yasuda, Y. and Kuwana, T. (1994a) Preservation of chick primordial germ cells in liquid nitrogen and subsequent production of viable offspring. *Journal of Reproduction and Fertility* 102, 321–325.

Naito, M., Tajima, T., Yasuda, Y. and Kuwana, T. (1994b) Production of germline chimeric chickens with high transmission rate of donor-derived gametes, produced by transfer of primordial germ cells. *Molecular Reproduction and Development* 3, 153.

Nakamura, M., Kuwana, T., Miyayama, Y. and Fujimoto, T. (1988) Extragonadal distribution of primordial germ cells in the early chick embryo. *Anatomical Record* 222, 90–94.

Nakamura, M., Yoshinaga, K. and Fujimoto, T. (1992) Histochemical identification and behaviour of quail primordial germ cells injected into chick embryos by the intravascular route. *Journal of Experimental Zoology* 261, 479–483.

Nieuwkoop, P.D. and Sutasurya, L.A. (1979) The migration of primordial germ cells. In: *Primordial Germ Cells in the Chordates*. Cambridge University Press, Cambridge, Massachusetts, pp. 113–127.

Ono, T., Muto, S., Matsumoto, T., Mochii, M. and Eguchi, M. (1995) Gene transfer into circulating primordial germ cells of quail embryos. *Experimental Animal* 4, 275–278.

Ozdzenski, W. (1967) Observations on the origin of primordial germ cells in the mouse. *Zoologica Poloniae* 17, 367–381.

Pardanaud, L., Buck, C. and Dieterlen-Lievre, F. (1987) Early germ cell segregation and distribution in the quail blastodisc. *Cell Differentiation* 22, 47–59.

Perry, M.M. (1987) Nuclear events from fertilisation to the early cleavage stages in the domestic fowl (*Gallus domesticus*). *Journal of Anatomy* 15, 99–109.

Petitte J.N. and Karagenç, L. (1996) Growth factors during early events in avian embryo development. *Poultry and Avian Biological Reviews* 7, 75–85.

Petitte J.N. and Kulik, M.J. (1996) Cloning and characterization of cDNAs encoding two forms of avian stem cell factor. *Biochimica et Biophysica Acta* 1307, 149–151.

Petitte, J.N., Clark, M.E. and Etches, R.J. (1991) Assessment of functional gametes in chickens after transfer of primordial germ cells. *Journal of Reproduction and Fertility* 92, 225–229.

Petitte, J.N., Karagenç, L. and Ginsburg, M. (1997) The origin of the avian germline and transgenesis in birds. *Poultry Science* 76, 1084–1092.

Resnick, J.L., Bixler, L.S., Cheng, L. and Donovan, P.J. (1992) Long-term proliferation of mouse primordial germ cells in culture. *Nature* 359, 550–551.

Reynaud, G. (1967) Mise en evidence des cellules germinales primordiales dans les jeunes blastodermes d'Oiseau par la technique de coloration P.A.S. *Comptes Rendus Hebdomadaires des Séances de l'Academie des Sciences, Série D, Sciences Naturelles* 265, 1636–1639.

Reynaud, G. (1969) Transfert de cellules germinales primordiales de Dindon a l'embryon de Poulet par injection intravasculaire. *Journal of Embryology and Experimental Morphology* 21, 485–507.

Reynaud, G. (1976) Capacities reproductrices et descendance de poulets ayant subi un transfert de PGC durant la vie embryonnaire. *Wilhelm Roux's Archives of Developmental Biology* 179, 85–110.

Reynaud, G. (1977) Peuplement en cellules gerinales d'embryons de Poulet de 5 jours et demi d'incubation apres irradiation aux rayons ultraviolet des jeunes blastodermes. *Comptes Rendus Hebdomadaires des Séances de l'Academie des Sciences, Série D, Sciences Naturelles* 284, 843–846.

Riese, J., Zeller, R. and Dono, R. (1995) Nucleo-cytoplasmic translocation and secretion of fibroblast growth factor-2 during avian gastrulation. *Mechanisms of Development* 49, 13–22.

Shuman, R.M. (1981) Primordial germ cell transfer in the chicken, *Gallus domesticus.* M.S. Thesis, University of Minnesota, St Paul, Minnesota.

Simkiss, K., Rowelett, K., Bumstead, N. and Freman, B.M. (1989) Transfer of primordial germ cell DNA between embryos. *Protoplasma* 151, 164–168.

Stewart, C., Gadi, I. and Bhatt, H. (1994) Stem cells from primordial germ cells can reenter the germline. *Developmental Biology* 161, 626–628.

Swartz, W.J. (1982) Acid and alkaline phosphatase activity in migrating primordial germ cells of the early chick embryo. *Anatomical Record* 202, 379–385.

Swift, C.H. (1914) Origin and early history of primordial germ cells in the chick. *American Journal of Anatomy* 15, 483–516.

Tajima, A., Naito, M., Yasuda, Y. and Kuwana, T. (1993) Production of germline chimera by transfer of primordial germ cells in the domestic chicken (*Gallus domesticus*). *Theriogenology* 40, 509–519.

Tajima, A., Naito, M., Yasuda, Y. and Kuwana, T. (1996) Production of germline chimeras by transfer of cryopreserved gonadal primordial germ cells (gPGCs) in chicken. In: *XX World's Poultry Congress Proceedings*, Vol. 1, pp. 385–388.

Urven, I.E., Erickson, C.A., Abbott, U.K. and McCarrey, J.R. (1988) Analysis of germline development in the chick using anti-mouse EC cell antibody. *Development* 103, 299–304.

Urven, I.E., Abbott, U.K. and Erickson, C.A. (1989) Distribution of extracellular matrix in the migratory pathway of avian primordial germ cells. *Anatomical Record* 224, 14–21.

Vick, L., Li, Y. and Simkiss, K. (1993) Transgenic birds from transformed primordial germ cells. *Proceedings of the Royal Society of London* 251, 179–182.

Watanabe, M., Kinutani, M., Naito, M., Ochio, O. and Takashima, Y. (1992) Distribution analysis of transferred donor cells in avian blastodermal chimeras. *Development* 114, 331–336.

Watanabe, M., Naito, M., Sasaki, E., Sakurai, M., Kuwana, T. and Oishi, T. (1994) Liposome-mediated DNA transfer into chicken primordial germ cells *in vivo*. *Molecular Reproduction and Development* 38, 268–274.

Wentworth, B.C., Tsai, H Hallet, J.H., Gonzales, D.S. and Rajcic, S.G. (1989) Manipulation of avian primordial germ cells and gonadal differentiation. *Poultry Science* 68, 999.

Wentworth, B., Tsai, H., Wentworth, A., Wong, E., Proudman, J. and El Halawani, M. (1996) Primordial germ cells for genetic modification of poultry. In: Miller, R.H., Pursel, V.G. and Norman, H.D. (eds) *Beltsville Symposia in Agricultural Research XX: Biotechnology's Role in the Genetic Improvement of Farm Animals*, pp. 202–227.

Yasuda, Y., Tajima, A., Fujimoto, T. and Kuwana, T. (1992) A method to obtain avian germline chimaeras using isolated primordial germ cells. *Journal of Reproduction and Fertility* 96, 521–528.

Zhou, J., Ohtaki, M. and Sakurai, M. (1993) Sequence of a cDNA encoding chicken stem cell factor. *Gene* 127, 269–270.

9

Generation of Transgenic Poultry by Transfection of Primordial Germ Cells

E.A. Wong[1], A.L. Wentworth[2], B.C. Wentworth[2], J.A. Proudman[3] and M.E. El Halawani[4]

[1] *Virginia Polytechnic Institute and State University, Blacksburg, Virginia, USA;* [2] *University of Wisconsin, Madison, Wisconsin, USA;* [3] *USDA-ARS, Beltsville, Maryland, USA;* [4] *University of Minnesota, St Paul, Minnesota, USA*

Avian primordial germ cells (PGCs) migrate via the extraembryonic blood from the germinal crescent to the developing gonads where they differentiate into spermatogonia and oogonia. PGCs thus represent a potential target cell for the introduction of foreign DNA into the avian germline. Retroviral transduction and DNA transfection of PGCs have been successfully utilized as gene transfer methods to produce transgenic offspring. The advantage of PGC methodology is the relative simplicity of the technique. A limitation of this approach is the need for second generation progeny to obtain birds carrying the transgene in all cells because the founder animal is a transgenic chimera only in gonadal tissue.

Avian Transgenesis

The development of powerful methodologies for the precise manipulation of the mammalian genome, for example knock-out mice, has intensified the effort to develop efficient methods for generating transgenic poultry. The difference in the reproductive system of birds, however, does not allow direct application of mammalian transgenic technology. Therefore, a number of different approaches have been developed for avian transgenesis (reviewed in Sang, 1994) including the use of replication-competent and replication-defective retroviral vectors, direct injection of DNA into the cytoplasm of the germinal discs, and transfection of blastodermal cells or primordial germ cells (PGCs).

© CAB INTERNATIONAL 1999. *Transgenic Animals in Agriculture*
(eds J.D. Murray, G.B. Anderson, A.M. Oberbauer and M.M. McGloughlin)

Foreign genes were first successfully introduced into the germline of chickens by using a replication-competent avian leukosis virus (ALV) (Salter *et al.*, 1986, 1987; Crittenden and Salter, 1990). Infection of embryos with wild-type or recombinant ALV produced viremic founder lines, which passed proviral DNA to F1 progeny. Expression of the subgroup A virus *env* gene from the recombinant ALV in transgenic chickens conferred resistance to ALV subgroup A infection (Crittenden and Salter, 1990). This result demonstrated the feasibility of using gene transfer as a method for developing disease resistant strains of poultry. The need, however, to develop virus-free transgenic poultry led to the development of replication-defective retroviral vectors. Microinjection of a replication-defective virus vector beneath the blastoderm of the unincubated fertile chicken egg produced male birds that contained vector DNA in their semen (Bosselman *et al.*, 1989a; Thoraval *et al.*, 1995). The vector DNA was passed on to 2–3% of the F1 offspring, demonstrating that the transgenic birds were germline chimeras. Expression of the foreign gene from the proviral DNA was detected in embryonic tissues (Bosselman *et al.*, 1989b), second-generation progeny (Briskin *et al.*, 1991), and chick embryo fibroblasts derived from transgenic embryos of F2 progeny (Thoraval *et al.*, 1995). Bosselman *et al.* (1989b) further demonstrated that embryonic expression of a transduced chicken growth hormone gene resulted in elevated levels of serum growth hormone.

Although the use of retroviral vectors has proven to be successful, this method is not preferred due to low public acceptance and the potential risk of generating replication-competent, recombinant retroviruses. Therefore, the development of non-viral methods for producing transgenic poultry has been actively pursued. Sperm-mediated gene transfer has been successfully used to introduce DNA into chickens (Nakanishi and Iritani, 1993; Squires and Drake, 1994). The direct injection of DNA into the cytoplasm of germinal discs before the first cleavage division has produced transgenic embryos (Naito *et al.*, 1991; Inada *et al.*, 1997) and adult chickens (Love *et al.*, 1994; Naito *et al.*, 1994a). Transfection or retroviral infection of blastodermal cells has also been successfully used to transfer foreign DNA into poultry (Brazolot *et al.*, 1991; Savva *et al.*, 1991; Allioli *et al.*, 1994; Bresler *et al.*, 1994; Watanabe *et al.*, 1994).

Genetic manipulation of PGCs is an alternative strategy for avian gene transfer. The advantage of this system is the simplicity of the technique relative to injection into the germinal disc or transfection and manipulation of blastodermal cells. A limitation is that the founder animal is a transgenic chimera only in gonadal tissue, thus requiring second generation progeny to obtain transgenic birds. This paper will review the methodology and successes to date using PGCs from the germinal crescent, blood or gonads as targets for gene transfer.

Origin and Characteristics of Primordial Germ Cells

Primordial germ cells originate in the epiblast of the stage X blastoderm (Eyal-Giladi *et al.*, 1981). During gastrulation, the PGCs gradually translocate from the epiblast to the anteriorly migrating hypoblast and become localized in an extra-embryonic area that lies anterior and anteriolateral to the primitive streak, known as the germinal crescent (Swift, 1914; Ginsburg and Eyal-Giladi, 1986; Muniesa and Dominguez, 1990; Ginsburg, 1997). These PGCs become incorporated into the developing blood vascular system (Fujimoto *et al.*, 1976b) during stages 13–16 (Hamburger and Hamilton, 1951). Circulating PGCs extravasate the blood vascular system in a region posterior to the vitelline artery and enter the neighbouring thickened epithelium of the splanchnopleure, the developing gonadal ridge (Ando and Fujimoto, 1983; Ukeshima *et al.*, 1987; Hong *et al.*, 1995). PGCs then migrate to the developing gonadal tissue and differentiate into spermatogonia in testes or oogonia in the ovary (Urven *et al.*, 1988).

Primordial germ cells are characterized by their large size (12–20 μm in diameter), large eccentrically placed nuclei and prominent, often fragmented, nucleoli (Fujimoto *et al.*, 1976a). Chicken PGCs also have high glycogen content and thus are readily identified by periodic acid–Schiff base (PAS) staining (Meyer, 1960) from the time of their appearance in the germinal crescent to colonization of the gonads (Urven *et al.*, 1988). A number of antibodies that recognize mammalian PGCs (EMA-1, FC10.2, SSEA-1) have been used to identify chicken PGCs (Urven *et al.*, 1988; Loveless *et al.*, 1990, Karagenç *et al.*, 1996). Quail PGCs do not stain with PAS (Nakamura *et al.*, 1991) but can be identified with the QH-1 or QCR1 antibodies (Pardanaud *et al.*, 1987; Ono *et al.*, 1996). Turkey PGCs have also been shown to stain with PAS and the EMA-1 antibody (B.C. Wentworth, unpublished observations).

General Methodology for Producing Germline Chimeras by Transfer of PGCs

To generate germline chimeras, PGCs can be harvested from developing chick embryos at three different stages: (i) from the germinal crescent of a stage 5–7 chick embryo prior to migration; (ii) from the blood of a stage 13–17 chick embryo during migration; or (iii) from the developing gonad of a stage 25–27 chick embryo after migration. One hundred to 250 PGCs are detectable in the germinal crescent (Fujimoto *et al.*, 1976b; Vick *et al.*, 1993b), whereas the reported number of PGCs in the blood range from one to 45 PGCs per 1.0 μl of blood (Singh and Meyer, 1967; Al-Thani and Simkiss, 1991; Vick *et al.*, 1993b; Naito *et al.*, 1994c). Thus injection of 100–200 PGCs should be sufficient to allow effective competition with endogenous circulating PGCs. Isolated PGCs are microinjected into the blood vascular system of an appropriately staged host embryo (stage 13–17) through a small 'window' cut in the shell and shell membrane. The 'window' is sealed by a variety of methods such as surgical tape, Scotch Magic tape or a coverslip and wax. The donor PGCs circulate in the blood vascular system, from which they later exit, colonize the host gonad and differentiate into either oocytes or spermatozoa.

Clearly, success with this method is dependent upon the ability of isolated PGCs to successfully migrate to the gonadal ridge and compete with endogenous PGCs for colonization of the gonad. The founder animal should be a germline chimera, i.e. contains a mixture of germ cells derived from donor and host PGCs. Partial sterilization of the host by treatment with the drug busulphan improves the efficiency of PGC colonization of host gonadal tissue (Vick *et al.*, 1993a; Bresler *et al.*, 1994). Mating of the germline chimera to a non-transgenic animal produces non-transgenic progeny as well as transgenic progeny with the foreign DNA in all cells. Thus, second generation transgenic progeny can be readily identified by examining DNA extracted from a blood sample. The ratio of transgenic to non-transgenic second generation progeny reflects the efficiency with which donor PGCs have colonized the host gonad.

Germline Chimeras and Transgenic Poultry from Manipulated Germinal Crescent PGCs

Successful transfer of PGCs isolated from the germinal crescent to produce germline chimeras was first reported in quail (Wentworth *et al.*, 1989). This early success opened the possibility that PGCs from the germinal crescent could be genetically modified to produce transgenic poultry. Savva *et al.* (1991) first reported the introduction of foreign DNA into germinal crescent PGCs using a replication-defective reticuloendothelial virus containing the *lacZ* gene. Integration of the retroviral DNA was detected in the gonad of an 18-day embryo using a sensitive PCR–dot blot procedure. In a further study from Simkiss' group (Vick *et al.*, 1993b), germinal crescent PGCs were transduced with a defective avian leukosis virus containing the *lacZ* gene. Twenty-three per cent of the gonads (5/22) from 18-day embryos were positive by Southern blot. One male with vector DNA in his sperm was selected for breeding experiments. One out of 24 progeny sired by this male were positive for the vector DNA extracted from blood. In these studies germinal crescent PGCs were injected into the blood vascular system of recipient embryos. Injection of PGCs into the host germinal crescent was also an effective strategy. Han *et al.* (1994) reported that germinal crescent PGCs, which had been transfected with a *lacZ* containing plasmid, were still capable of migrating to the host gonads and expressing β-galactosidase activity. These results demonstrate that germinal crescent PGCs can be isolated and genetically modified while still retaining the ability to migrate and colonize the gonad of a recipient embryo.

Employing a novel strategy, Li *et al.* (1995) introduced foreign DNA into germinal crescent PGCs *in ovo* using ballistic transfection. Hatchlings produced by this method were raised to sexual maturity and shown to contain vector DNA in their sperm. Ten out of 45 G1 progeny were positive for the vector DNA. The vector DNA in these G1 progeny gradually disappeared as the birds matured, suggesting that the vector DNA was transmitted episomally.

Germline Chimeras and Transgenic Poultry from Manipulated Blood PGCs

Embryonic blood PGCs are a convenient source of PGCs because of the ease with which blood PGCs can be obtained and the fact that blood PGCs already exist as a single-cell suspension. In the blood of a stage 13–14 chick embryo, 0.048% of the cells are PGCs (Yasuda *et al.*, 1992). Studies that utilized transfer of retrovirally marked PGCs or interspecies transfer of PGCs have clearly demonstrated that exogenous blood PGCs can successfully target themselves to the host gonads (Simkiss *et al.*, 1989; Nakamura *et al.*, 1991; Naito *et al.*, 1994c; Ono *et al.*, 1996). In the study by Nakamura *et al.* (1991), timing of the injection was critical. When PGCs from stage 13–14 chicks were injected into stage 15–16 quail embryos, 90% of the PGCs appeared in the gonads of the host embryo. The remaining 10% were located ectopically in the head, trunk and limbs. Injection of blood PGCs into host embryos at progressively later stages of development resulted in a decreasing percentage of PGCs migrating to the gonad. When blood PGCs were injected into stage 20 host embryos only 6% of the PGCs migrated to the gonads. These results demonstrate that only a narrow window of time exists during which PGCs are competent to migrate to the developing gonads.

An early report indicated that the use of blood PGCs was not likely to produce germline chimeras. Petitte *et al.* (1991) transferred blood PGCs between marked strains of chickens. They analysed the progeny using a feather colour marker to assess the ability of donor PGCs to form functional gametes. Donor PGCs from dwarf white leghorn (DWL) embryos were injected into barred Plymouth rock (BPR) embryos. Fifty-nine male and female chicks that hatched were raised to sexual maturity and test mated with BPR fowl. All of the 3117 offspring examined showed the typical BPR phenotype and none showed the phenotype expected for a DWL × BPR.

More recent studies have shown that transfer of blood PGCs can produce germline chimeras. To concentrate PGCs from a blood sample, Yasuda *et al.* (1992) separated PGCs from blood cells using Ficoll gradient centrifugation. They reported an 80-fold enrichment of PGCs, from 0.048% to 3.9% of the cells. Using this method, Tajima *et al.* (1993) successfully generated germline chimeras following injection of 100 blood PGCs. Based on feather colour, up to 12% of the chicks that hatched were derived from the donor PGCs. Vick *et al.* (1993b) showed that 3% of the embryos that received blood PGCs marked with a replication-defective retrovirus tested positive for the foreign DNA. One male bird was raised to sexual maturity and test mated to determine the percentage of his gametes that were derived from donor PGCs. Approximately 2% (1/56) of the progeny were derived from the donor PGCs.

Naito *et al.* (1994c) combined concentration of blood PGCs by Ficoll gradient centrifugation with depletion of endogenous PGCs by withdrawal of 4–10 μl of blood from the recipient embryo to generate germline chimeras at high frequency. Ninety-five per cent (19/20) of the birds examined by progeny testing were germline chimeras. Interestingly, when PGCs from white leghorns (WL) were transferred into BPR recipients, the average frequency of donor-derived offspring was 81% for three male chimeras and 96% for one female chimera. In contrast, the average frequency of donor-derived offspring was 23% for six male chimeras and 6% for five female chimeras when BPR PGCs were transferred into WL. Depletion of endogenous PGCs by withdrawal of blood from the host increased the frequency of donor-derived offspring from 14% to 23% in male chimeras generated by transfer of BPR PGCs into WL.

Blood PGCs have been successfully cultured on a feeder layer of stromal cells derived from the germinal ridge (Chang *et al.*, 1995a). PGCs were first labelled with a fluorescent dye to distinguish them from endogenous PGCs. A 4.8-fold increase in the number of labelled PGCs was reported along with a 3.9-fold increase in endogenous PGCs after 4 days in culture. This increase was only observed when the PGCs were grown on stromal cells derived from 5-day (stage 27) embryos and not with chick embryo fibroblasts. PGCs isolated from the blood of 2-day-old chick embryos have also been successfully stored in liquid nitrogen for 4–5 months (Naito *et al.*, 1994b). Frozen–thawed PGCs retained 94% viability and were still capable of producing viable offspring. These results indicate the potential for cryopreservation of PGCs as a possible approach to the conservation of genetic material in avian species.

Using a simpler approach, Watanabe *et al.* (1994) transfected blood PGCs *in vivo*. A cationic liposome–DNA complex containing the *lacZ* gene was injected into a stage 10–15 chick embryo. Twenty-four hours after injection, β-galactosidase activity was detected in blood cells, endothelial cells and endocardium cells of the heart in all embryos examined. Examination of other embryos 2–3 days after injection of the DNA–liposome complex revealed β-galactosidase activity in 4% of the embryos injected with a RSV–*lacZ* construct and 40% of the embryos injected with a β-actin–*lacZ* construct. The mean number of β-galactosidase-positive PGCs per embryo was 0.2 and 2.1, respectively. These results indicate that endogenous PGCs were successfully transfected *in vivo* and that the β-actin promoter is more active in PGCs than the RSV promoter. We have similarly found that injection of a DNA–liposome construct into stage 15 turkey embryos transfected host embryonic tissues and blood PGCs. Vector DNA could be detected by PCR in 43% (10/23) of the heart samples and 21% (4/19) of the gonadal samples (B.C. Wentworth *et al.*, unpublished observations). These encouraging results demonstrate that *in vivo* transfection of blood PGCs is a potentially efficacious method of generating germline chimeras. This method would be advantageous because of its simplicity. The efficiency, however, is a critical issue since all progeny of the founder animal need to be screened for the foreign DNA. In contrast, in transfection studies using donor PGCs from marked strains, only donor-derived progeny are analysed for the presence of foreign DNA because they can be identified phenotypically.

Germline Chimeras and Transgenic Poultry from Manipulated Gonadal PGCs

In the 5-day (stage 25–27) chick embryo most of the PGCs have completed their migration to the gonads and have initiated a stage of rapid proliferation. One advantage to culturing PGCs derived from the gonad is the elimination of the need to prepare a separate feeder layer of cells for the PGCs. The somatic gonadal cells typically form a layer of stromal cells on which the PGCs proliferate. Allioli *et al.* (1994) cultured PGCs isolated from the gonads of 5-day-old chick embryos and transfected the cultured PGCs with a replication-defective retroviral vector containing the β-galactosidase gene. The number of PGCs increased with days in culture and more than half of the PGCs were positive for β-galactosidase activity. These results demonstrated that gonadal PGCs could proliferate *in vitro* and express a foreign gene.

Chang *et al.* (1995b) also showed that cultured chick PGCs isolated from a stage 27 chick embryo could still migrate and proliferate in a host embryonic gonad. The transplanted PGCs had undergone at least three to seven cell divisions. In a subsequent study, Chang *et al.* (1997) reported that transfer of gonadal PGCs resulted in 6/60 recipients that were germline chimeras. Both male and female germline chimeras produced progeny derived from donor PGCs at a frequency of 1.3–3.1%. Like blood PGCs, gonadal PGCs can also be stored in liquid nitrogen for up to 3 months without loss of the ability to form germline chimeras (Tajima *et al.*, 1996). These results demonstrate that PGCs from the gonads of stage 25–27 chicken embryos, which have long since passed beyond the normal migration stage *in vivo*, are still capable of migrating to and colonizing the embryonic gonad of a stage 15–17 recipient embryo.

Similarly, our earlier work indicated that PGCs from the indifferent gonads of equivalent stage 27 turkey embryos could be cultured and transfected *in vitro* to generate transgenic F1 offspring (Wentworth *et al.*, 1995). Analysis of subsequent generations of our transgenic turkeys show that the vector DNA has persisted into the F3 generation. Surprisingly, the vector DNA appears to be transmitted as an episome by pedigree analysis, Southern blot analysis, and plasmid rescue studies (E.A. Wong *et al.*, unpublished observations). Transmission of the vector DNA does not follow normal Mendelian inheritance as would be expected for DNA that has integrated into the genome (Fig. 9.1). In pedigree 168, an untested founder female was mated with semen pooled from 12 transgenic and non-transgenic F0 males. The pooled semen tested positive by PCR for the transgene. Six F1 progeny were tested, of which two, no. 204 and no. 206, were positive by PCR for the transgene in DNA extracted from blood. Mating of positive F1 hen no. 204 and negative hen no. 205 with PCR-positive, pooled semen from 54 F1 males generated F2 progeny, all of which were positive by PCR or genomic Southern blot analysis. Further mating of Southern blot-positive male no. 1340 to negative female no. 1292 produced all positive F3 progeny. This last mating clearly demonstrated that the foreign DNA could be transmitted through the male germline. In addition, the pattern of inheritance was consistent with episomal transmission and not stable integration of the foreign DNA. In support of this hypothesis, Southern blot analysis of genomic DNA digested with a restriction enzyme that cleaves the vector DNA once produced only a single band that co-migrated with linear vector DNA. No junction fragments diagnostic of an integration event were detectable. Finally, vector DNA could be rescued from undigested genomic DNA by transformation of *Escherichia coli*. The rescued plasmid was identical to the plasmid used for transfection by restriction mapping and partial DNA sequencing. Presently it is not known by what mechanism the vector DNA replicates episomally in the transgenic turkeys. It is possible that an avian origin of replication has been fortuitously included in the construct or a cryptic origin of replication is being used.

Evidence for episomal persistence of foreign DNA has been observed during the generation of transgenic birds after cytoplasmic microinjection of DNA into one-cell fertilized eggs (Sang and Perry, 1989), sperm-mediated gene transfer (Nakanishi and Iritani, 1993; Squires and Drake, 1994), and *in vivo* ballistic transfection (Li *et al.*, 1995). Perhaps the presence of numerous microchromosomes in avian species creates an environment that allows episomal replication of foreign DNA molecules.

Episomal transmission of transgenes has been observed in other transgenic animals. Extrachromosomal replication of DNA injected into the cytoplasm has been reported for the nematode (Stinchcomb *et al.*, 1985), sea urchin (McMahon *et al.*, 1985), *Drosophila* (Steller and Pirrotta, 1985), *Xenopus* (Etkin *et al.*, 1984) and fish (Guyomard *et al.*, 1989). In mice, inclusion of putative origins of replication from polyoma virus or the *c-myc* gene resulted in extrachromosomal replication of the transgene (Rassoulzadegan *et al.*, 1986; Sudo *et al.*, 1990). The episomal plasmid was shown to be transmitted in a non-Mendelian fashion to the F1 and F2 generations through both eggs and sperm (Rassoulzadegan *et al.*, 1986) or only eggs (Sudo *et al.*, 1990).

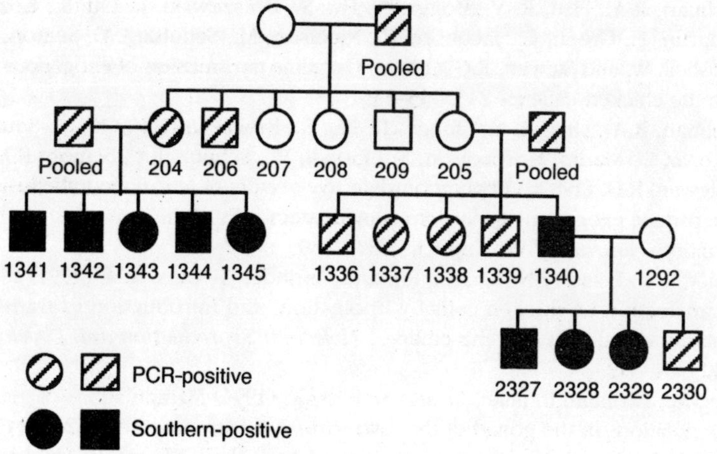

Fig. 9.1. Transmission of foreign DNA in pedigree 168. Females are represented by circles and males are represented by squares. Hatched or filled symbols indicate that the DNA samples were positive by PCR analysis only or by both PCR and genomic Southern blot analyses, respectively. Open symbols indicate that the DNA samples were negative by PCR analysis. The founder female was not tested.

The encouraging progress that has been made in the development of a number of techniques for transferring DNA into poultry opens the exciting possibility for manipulating the avian genome. Two potential applications of avian gene transfer technology involve the development of poultry that are disease resistant or that produce heterologous proteins in eggs. The ability to culture PGCs for days without loss of the capability to form a germline chimera will allow the application of powerful gene targeting methodologies that have been developed for murine embryonic stem cells. What remains to be determined is whether extended selection for PGCs with targeted integration events will adversely affect the ability of PGCs to migrate and colonize the host gonad. With all of these tools in hand, avian transgenesis is about to enter an exciting phase.

References

Allioli, N., Thomas, J.-L., Chebloune, Y., Nigon, V.-M., Verdier, G. and Legras, C. (1994) Use of retroviral vectors to introduce and express the β-galactosidase marker gene in cultured chicken primordial germ cells. *Developmental Biology* 165, 30–37.

Al-Thani, R. and Simkiss, K. (1991) Effects of an acute *in vivo* application of concanavalin A on the migration of avian primordial germ cells. *Protoplasma* 161, 52–57.

Ando, Y. and Fujimoto, T. (1983) Ultrastructural evidence that chick primordial germ cells leave the blood vascular system prior to migrating to the gonadal anlagen. *Development, Growth and Differentiation* 25, 345–352.

Bosselman, R.A., Hsu, R.-Y., Boggs, T., Hu, S., Bruszewski, J., Ou, S., Kozar, L., Martin, F., Green, C., Jacobsen, F., Nicolson, M., Schultz, J.A., Semon, K.M., Rishell, W. and Stewart, R.G. (1989a) Germline transmission of exogenous genes in the chicken. *Science* 243, 533–535.

Bosselman, R.A., Hsu, R.-Y., Boggs, T., Hu, S., Bruszewski, J., Ou, S., Souza, L., Kozar, L., Martin, F., Nicolson, M., Rishell, W., Schultz, J.A., Semon, K.M. and Stewart, R.G. (1989b) Replication-defective vectors of reticuloendotheliosis virus transduce exogenous genes into somatic stem cells of the unincubated chicken embryo. *Journal of Virology* 63, 2680–2689.

Brazolot, C.L., Petitte, J.N., Etches, R.J. and Verrinder-Gibbins, A.M. (1991) Efficient transfection of chicken cells by lipofection, and introduction of transfected blastodermal cells into the embryo. *Molecular Reproduction and Development* 30, 304–312.

Bresler, M., Behnam, J., Luke, G. and Simkiss, K. (1994) Manipulations of germ-cell populations in the gonad of the fowl. *British Poultry Science* 35, 241–247.

Briskin, M.J., Hsu, R.-Y., Boggs, T., Schultz, J.A., Rishell, W. and Bosselman, R.A. (1991) Heritable retroviral transgenes are highly expressed in chickens. *Proceedings of the National Academy of Sciences USA* 88, 1736–1740.

Chang, I.-K., Tajima, A., Chikamune, T. and Ohno, T. (1995a) Proliferation of chick primordial germ cells cultured on stroma cells from the germinal ridge. *Cell Biology International* 19, 143–149.

Chang, I.-K., Yoshiki, A., Kusakabe, M., Tajima, A., Chikamune, T., Naito, M. and Ohno, T. (1995b) Germline chimera produced by transfer of cultured chick primordial germ cells. *Cell Biology International* 19, 569–576.

Chang, I.-K., Jeong, D.K., Hong, Y.H., Park, T.S., Moon, Y.K. Ohno, T. and Han, J.Y. (1997) Production of germline chimeric chickens by transfer of cultured primordial germ cells. *Cell Biology International* 21, 495–499.

Crittenden, L.B. and Salter, D.W. (1990) Expression and mobility of retroviral inserts in the chicken germline. In: *Transgenic Models in Medicine and Agriculture.* Wiley Liss, New York, pp. 73–87.

Etkin, L.D., Pearman, B., Roberts, M. and Bektesh, S.L. (1984) Replication, integration and expression of exogenous DNA injected into fertilized eggs of *Xenopus laevis*. *Differentiation* 26, 194–202.

Eyal-Giladi, H., Ginsburg, M. and Farbarov, A. (1981) Avian primordial germ cells are of epiblastic origin. *Journal of Embryology and Experimental Morphology* 65, 139–147.

Fujimoto, T., Ninomiya, T. and Ukeshima, A. (1976a) Observations of the primordial germ cells in blood samples from the chick embryo. *Developmental Biology* 49, 278–282.

Fujimoto, T., Ukeshima, A. and Kiyofuji, R. (1976b) The origin, migration and morphology of the primordial germ cells in the chick embryo. *The Anatomical Record* 185, 139–154.

Ginsburg, M. (1997) Primordial germ cell development in avians. *Poultry Science* 76, 91–95.

Ginsburg, M. and Eyal-Giladi, H. (1986) Temporal and spatial aspects of the gradual migration of primordial germ cells from the epiblast into the germinal crescent in the avian embryo. *Journal of Embryology and Experimental Morphology* 95, 53–71.

Guyomard, R., Chourrout, D., Leroux, C., Houdebine, L.M. and Pourrain, F. (1989) Integration and germline transmission of foreign genes microinjected into fertilized trout eggs. *Biochimie* 71, 857–863.

Hamburger, V. and Hamilton, H.L. (1951) A series of normal stages in the development of the chick embryo. *Journal of Morphology* 88, 49–92.

Han, J.Y., Shoffner, R.N. and Guise, K.S. (1994) Gene transfer by manipulation of primordial germ cells (PGCs) in the chicken. *Asian–Australasian Journal of Animal Science* 7, 427–434.

Hong, Y.H., Seo, D.S., Jeong, D.K., Choi, K.D. and Han, J.Y. (1995) Migration of the primordial germ cells and gonad formation in the early chicken embryo. *Asian–Australasian Journal of Animal Science* 8, 557–562.

Inada, S., Hattori, M.A., Fujihara, N. and Morohashi, K. (1997) *In vivo* gene transfer into the blastoderm of early development stage of chicken. *Reproduction, Nutrition and Development* 37, 13–20.

Karagenç, L., Cinnamon, Y., Ginsburg, M. and Petitte, J.N. (1996) Origin of primordial germ cells in the prestreak chick embryo. *Developmental Genetics* 19, 290–301.

Li, Y., Behnam, J. and Simkiss K. (1995) Ballistic transfection of avian primordial germ cell *in ovo*. *Transgenic Research* 4, 26–29.

Love, J., Gribbin, C., Mather, C. and Sang H. (1994) Transgenic birds by DNA microinjection. *Bio/Technology* 12, 60–63.

Loveless, W., Bellairs, R., Thorpe, S.J., Page, M. and Feizi, T. (1990) Developmental patterning of the carbohydrate antigen FC10.2 during early embryogenesis in the chick. *Development* 108, 97–106.

McMahon, A.P., Flytzanis, C.N., Hough-Evans, B.R., Katula, K.S., Britten, R.J. and Davidson, E.H. (1985) Introduction of cloned DNA into sea urchin egg cytoplasm: replication and persistence during embryogenesis. *Developmental Biology* 108, 420–430.

Meyer, D.B. (1960) Application of the periodic acid–Schiff technique to whole chick embryos. *Stain Technology* 35, 83–89.

Muniesa,. P. and Dominguez, L. (1990) A morphological study of primordial germ cells at pregastrular stages in the chick embryo. *Cell Differentiation and Development* 31, 105–117.

Naito, M., Agata, K., Otsuka, K., Kino, K., Ohta, M., Hirose, K., Perry, M.M. and Eguchi, G. (1991) Embryonic expression of beta-actin–lacZ hybrid gene injected into the fertilised ovum of the domestic fowl. *International Journal of Developmental Biology* 35, 69–75.

Naito, M., Sasaki, E., Ohtaki, M. and Sakurai, M. (1994a) Introduction of exogenous DNA into somatic and germ cells of chickens by microinjection into the germinal disc of fertilized ova. *Molecular Reproduction and Development* 37, 167–171.

Naito, M., Tajima, A., Tagami, T., Yasuda, Y. and Kuwana, T. (1994b) Preservation of chick primordial germ cells in liquid nitrogen and subsequent production of viable offspring. *Journal of Reproduction and Fertility* 102, 321–325.

Naito, M., Tajima, A., Yasuda, Y. and Kuwana, T. (1994c) Production of germline chimeric chickens, with high transmission rate of donor-derived gametes, produced by transfer of primordial germ cells. *Molecular Reproduction and Development* 39, 153–161.

Nakamura, M., Maeda, H. and Fujimoto, T. (1991) Behaviour of chick primordial germ cells injected into the blood stream of quail embryos. *Okajimas Folia Anatomica Japan* 67, 473–478.

Nakanishi, A. and Iritani, A. (1993) Gene transfer in the chicken by sperm-mediated methods. *Molecular Reproduction and Development* 36, 258–261.

Ono, T., Yokoi, R. and Aoyama, H. (1996) Transfer of male or female primordial germ cells of quail into chick embryonic gonads. *Experimental Animals* 45, 347–352.

Pardanaud, L., Buck, C. and Dieterlen-Lievre, F. (1987) Early germ cell segregation and distribution in the quail blastodisc. *Cell Differentiation* 22, 47–60.

Petitte, J.N., Clark, M.E. and Etches, R.J. (1991) Assessment of functional gametes in chickens after transfer of primordial germ cells. *Journal of Reproduction and Fertility* 92, 225–229.

Rassoulzadegan, M., Leopold, P., Vailly, J. and Cuzin, F. (1986) Germline transmission of autonomous genetic elements in transgenic mouse strains. *Cell* 46, 513–519.

Salter, D.W., Smith, E.J., Hughes, S.H., Wright, S.E., Fadly, A.M., Witter, R.L. and Crittenden, L.B. (1986) Gene insertion into the chicken germline by retroviruses. *Poultry Science* 65, 1445–1458.

Salter, D.W., Smith, E.J., Hughes, S.H., Wright, S.E. and Crittenden, L.B. (1987) Transgenic chickens: insertion of retroviral genes into the chicken germline. *Virology* 157, 236–240.

Sang, H. (1994) Transgenic chickens – methods and potential applications. *Trends in Biotechnology* 12, 415–420.

Sang, H. and Perry, M.M. (1989) Episomal replication of cloned DNA injected into the fertilised ovum of the hen, *Gallus domesticus. Molecular Reproduction and Development* 1, 98–106.

Savva, D., Page, N., Vick, L. and Simkiss, K. (1991) Detection of foreign DNA in transgenic chicken embryo using the polymerase chain reaction. *Research in Veterinary Science* 50, 131–133.

Simkiss, K., Rowlett, K., Bumstead, N. and Freeman, B.M. (1989) Transfer of primordial germ cell DNA between embryos. *Protoplasma* 151, 164–166.

Singh, R.P. and Meyer, D.B. (1967) Primordial germ cells in blood smears from chick embryos. *Science* 156, 1503–1504.

Squires, E.J. and Drake, D. (1994) Transgenic chickens by liposome-sperm-mediated gene transfer. *Proceedings of the 5th World Congress on Genetics Applied to Livestock Production* 21, 350–353.

Steller, H. and Pirrotta, V. (1985) Fate of DNA injected into early *Drosophila* embryos. *Developmental Biology* 109, 54–62.

Stinchcomb, D.T., Shaw, J.E., Carr, S.E. and Hirsh, D. (1985) Extrachromosomal DNA transformation of *Caenorhabditis elegans. Molecular and Cellular Biology* 5, 3484–3496.

Sudo, K., Ogata, M., Sato, Y., Iguchi-Ariga, M.M. and Ariga, H. (1990) Cloned origin of DNA replication in c-myc gene can function and be transmitted in transgenic mice in an episomal state. *Nucleic Acids Research* 18, 5425–5432.

Swift, C.H. (1914) Origin and early history of the primordial germ-cells in the chick. *American Journal of Anatomy* 15, 483–516.

Tajima, A., Naito, M., Yasuda, Y. and Kuwana, T. (1993) Production of germline chimera by transfer of primordial germ cells in the domestic chicken (*Gallus domesticus*). *Theriogenology* 40, 509–519.

Tajima, A., Naito, M., Yasuda, Y. and Kuwana, T. (1996) Production of germline chimeras by transfer of cryopreserved gonadal primordial germ cells (gPGCs) in chicken. In: *Proceedings of the XX World's Poultry Congress*, New Delhi, India. World Poultry Science Association, India Branch, New Delhi, pp. 385–388.

Thoraval, P., Afanassieff, M., Cosset, F.-L., Lasserre, F., Verdier, G., Coudert, F. and Dambrine, G. (1995) Germline transmission of exogenous genes in chickens using helper-free ecotropic avian leukosis virus-based vectors. *Transgenic Research* 4, 369–376.

Ukeshima, A., Kudo, M. and Fujimoto, T. (1987) Relationship between genital ridge formation and settlement site of primordial germ cells in chick embryos. *Anatomical Record* 219, 311–314.

Urven, L.E., Erickson, C.A., Abbot, U.K. and McCarrey, J.R. (1988) Analysis of germline development in the chick embryo using an anti-mouse EC cell antibody. *Development* 103, 299–304.

Vick, L., Luke, G. and Simkiss, K. (1993a) Germline chimeras can produce both strains of fowl with high efficiency after partial sterilization. *Journal of Reproduction and Fertility* 98, 637–641.

Vick, L., Ying, L. and Simkiss, K. (1993b) Transgenic birds from transformed primordial germ cells. *Proceedings of the Royal Society of London* B 251, 179–182.

Watanabe, M., Naito, M., Sasaki, E., Sakurai, M., Kuwana, T. and Oshi, T. (1994) Liposome-mediated DNA transfer into chicken primordial germ cells *in vivo*. *Molecular Reproduction and Development* 38, 268–274.

Wentworth, B.C., Tsai, H., Hallett, J.H., Gonzales, D.S. and Rajcic-Spasojevic, G. (1989) Manipulation of avian primordial germ cells and gonadal differentiation. *Poultry Science* 68, 999–1010.

Wentworth, B., Tsai, H., Wentworth, A., Wong, E., Proudman, J. and El Halawani, M. (1995) Primordial germ cells for genetic modulation of poultry. In: Miller, R.H., Pursel, V.G. and Norman, H.D. (eds) *Beltsville Symposia in Agricultural Research, XX, Biotechnology's Role in the Genetic Improvement of Farm Animals*. American Society of Animal Science, Savoy, Illinois, pp. 202–227.

Yasuda, Y., Tajima, A., Fujimoto, T. and Kuwana, T. (1992) A method to obtain avian germline chimeras using isolated primordial germ cells. *Journal of Reproduction and Fertility* 96, 521–528.

10

Expression of Insulin-like Growth Factor-I in Skeletal Muscle of Transgenic Swine

V.G. Pursel[1], R.J. Wall[1], A.D. Mitchell[1], T.H. Elsasser[1], M.B. Solomon[1], M.E. Coleman[2], F. DeMayo[3] and R.J. Schwartz[3]

[1] *USDA-ARS, Beltsville, Maryland, USA;* [2] *GeneMedicine, Inc., The Woodlands, Texas, USA;* [3] *Baylor College of Medicine, Houston, Texas, USA*

Although growth hormone is considered the primary growth-promoting hormone in mammals, many of its effects are thought to be mediated by insulin-like growth factor-I (IGF-I), which is a potent mitogen that stimulates cell proliferation and synthesis of DNA and protein. The aim of this research was to determine whether directing expression of IGF-I specifically to striated muscle would enhance lean muscle growth in swine. Transgenic pigs were produced by microinjection of zygotes with a fusion gene consisting of the regulatory sequences of an avian skeletal α-actin gene and a cDNA encoding human IGF-I. All but one of 13 transgenic pigs expressed the IGF-I transgene. Muscle IGF-I concentrations varied from 20 to 1702 ng g^{-1} muscle in transgenic pigs compared with less than 10 ng g^{-1} muscle in control pigs. Muscle IGF-I concentrations were in general agreement with abundance of IGF-I mRNA on Northern blots. Serum IGF-I concentrations in transgenic pigs (160 ± 6.8 ng ml^{-1}) did not differ from that of littermate control pigs (143 ± 6.5 ng ml^{-1}). Daily weight gain from 20 to 60 kg body weight was similar for transgenic and littermate control pigs (865 ± 29.6 g vs. 876 ± 18.9 g day^{-1}). Body composition of eight transgenic and eight control pigs was estimated by X-ray absorptiometry (DXA) scanning at 60 kg body weight. The DXA results indicated transgenic female pigs had significantly less fat and more lean tissue than female littermate controls (P <0.05 for each). However, body composition of transgenic and control boars did not differ. Subsequently nine of ten founders transmitted their transgene to G1 progeny, which will be used for evaluation of growth rate, feed efficiency and carcass composition. Transgenic and control pigs did not differ in general appearance, and no gross abnormalities, pathologies or health-related problems were encountered. Based on these results we conclude that enhancing IGF-I specifically in skeletal muscle may have a positive effect on carcass composition of swine.

© CAB INTERNATIONAL 1999. *Transgenic Animals in Agriculture* (eds J.D. Murray, G.B. Anderson, A.M. Oberbauer and M.M. McGloughlin)

Introduction

Endocrine control of normal tissue development and growth in mammals is complex. A group of peptide hormones, i.e. growth hormone-releasing factor (GRF), somatostatin, growth hormone (GH), insulin, insulin-like growth factor-I (IGF-I), insulin-like growth factor-II (IGF-II), thyrotrophic hormone and gonadotrophic hormones work in concert to regulate and coordinate the metabolic pathways responsible for tissue formation and development. Even though correlations between growth and circulating levels of some of these peptide hormones have often produced conflicting results, the preponderance of data indicates that genetic capacity for growth is related to increased circulating levels of GH and IGF-I (Sejrson, 1986). Furthermore, injection of pigs, sheep and cattle with exogenous GH and stimulation of GH secretion in lambs by immunizing them against somatostatin have generally produced enhanced feed efficiency, increased growth rate, and reduced subcutaneous fat. These findings all suggest the possible usefulness of manipulating genes for the peptide hormones to modify the growth characteristics and carcass composition of farm animals.

Transfer of GH fusion genes has received the bulk effort in growth-related transgenic research with farm animals (see review by Pursel and Rexroad, 1993), however, IGF-I mediates many of the effects of GH, so enhanced expression of IGF-I may alter growth characteristics without such a dramatic affect on the systemic physiology of the animal. IGF-I is a mitogenic peptide that plays an important role in differentiation and postnatal development (Zapf *et al.*, 1984). IGF-I has the potential of acting as an endocrine agent when its synthesis and secretion by the liver is stimulated by GH. In addition, elevated GH stimulates IGF-I synthesis in numerous tissues throughout the body where it may act locally as a paracrine and/or autocrine agent. The relative importance of these two modes of action in regulation of growth is unknown.

Mathews *et al.* (1988) produced a single line of human (h) IGF-I transgenic mice in which a mouse metallothionein (mMT) promoter was used to increase serum IGF-I by 50–60% above that of littermate control mice. This IGF-I expression resulted in a 1.3-fold increase in body weight in comparison with littermate control mice without an increase in skeletal growth. The elevated plasma IGF-I was effective in feeding back on the hypothalamus to inhibit GH synthesis and secretion, thus depressing serum GH to non-detectable levels, and may have confounded effects of IGF-I on muscle mass. Subsequently, the same mMT–hIGF-I construct was transferred into four pigs, only one of which expressed the transgene. Unfortunately, this pig died at an early age before growth performance could be evaluated (Pursel *et al.*, 1989). Reiss *et al.* (1996) transferred hIGF-I under control of a rat α-myosin heavy chain promoter into mice to direct IGF-I expression specifically to the heart. This strategy elevated plasma IGF-I by 84%, and increased body weight and heart weight compared with littermate control mice by about 15% and 50%, respectively.

Coleman *et al.* (1995) constructed an α-skeletal actin–hIGF-I transgene to direct IGF-I expression specifically in skeletal muscle of transgenic mice. The rationale behind this approach was to direct sufficient expression of IGF-I in skeletal muscle to act as a paracrine mitogenic agent without altering plasma IGF-I concentration sufficiently to have a systemic endocrine effect and depress growth hormone secretion and release.

A line of IGF-I transgenic mice was evaluated that had a 47-fold higher concentration of IGF-I per gramme of muscle than that of control mice. Over expression of IGF-I elicited hypertrophy of all classes of myofibres and a shift in myofibre type toward more oxidative fibre types. Neither IGF-I concentrations in the serum nor body weight were significantly increased in transgenic mice compared with sibling control mice.

This α-skeletal actin–hIGF-I transgene has now been transferred into swine to investigate its potential for improving productivity. This communication describes the production and characterization of the founder IGF-I transgenic pigs and the successful transmission of the fusion gene to G1 progeny.

Material and Methods

Animals

Seventy-six sexually mature DK-43 hybrid gilts (Dekalb Swine Breeders, Inc., Dekalb, Illinois) that had previously displayed oestrus two or more times were used for the experiment. Gilts were penned and fed outdoors in groups of ten, and checked for oestrus once daily with a mature boar before assignment to experimental use. The gilts were confined to a gestation stall for individual feeding of altrenogest (Regu-Mate®, Hoechst Roussel Pharmaceuticals Inc., Somerville, New Jersey). Experimental protocols used in this research were approved by the Beltsville Area Institutional Animal Care and Use Committee.

Synchronization of oestrus and ovulation

Oestral cycles of all ovum donors and recipients were regulated by feeding 0.17 mg altrenogest per kg body weight daily for 5 or 6 days starting on day 11–15 of the oestral cycle (onset of oestrus = day 0).

Donors received 1000, 1200 or 1500 IU eCG (Diosynth Inc., Chicago, Illinois) by subcutaneous injection at 29 h and 750 IU hCG by intramuscular (i.m.) injection at 111 h after the last feeding of altrenogest. Donor gilts were bred by natural service or artificially inseminated twice between 7 and 33 h after the hCG injection. Twenty-eight donor gilts (designated D-R) were also used as recipients of microinjected zygotes that had previously been recovered from other donors. Micromanipulated ova were transferred into these donors immediately after their ova had been flushed from the oviducts.

Control recipient gilts (C-R) were injected with 750 IU hCG at 125 h after a last feeding of altrenogest (C-R gilts were last fed altrenogest 8 h before and were injected with hCG 6 h after donor gilts).

Ovum recovery, microinjection and transfer

Ova were recovered from donors between 50 and 55 h after the hCG injection. Anaesthesia was induced by administering the following per 100 kg body weight: 400 mg Ketamine HCl (Ketaset®, Aveco Co., Fort Dodge, Iowa); 200 mg xylazine (Rompun®, Haver Lockhart, Bayvet Division, Miles Laboratories, Shawnee, Kansas); 100 mg Telozol® (50 mg tiletamine HCl and 50 mg zolazepam HCl, Aveco Co., Fort Dodge, Iowa); 4 mg butorphanol tartrate (Torbugesic®, Fort Dodge Laboratories, Fort Dodge, Iowa); and 6 mg atropine sulphate (Butler Company, Columbus, Ohio). One-fifth of the anaesthetic dose was administered i.m. and the remaining dose was administered intravenously (i.v.) after 10 min. Oviducts were exteriorized by midventral laparotomy, and each oviduct was retrograde flushed with 20 ml of Beltsville embryo culture medium (without 5% CO_2), which was the only medium used for ovum culture from recovery to transfer (Pursel and Wall, 1996).

Ova were centrifuged at 15,000 $\times$ *g* for 10 min to permit visualization of the pronuclei (Wall *et al.*, 1985). One pronucleus of each zygote was microinjected with several hundred copies of the transgene. Within 2 h after microinjection, 12–35 injected ova were transferred into one or both oviduct of either D-R gilts or C-R gilts. Ova were recovered from four to six donors each day. Ova from early recoveries were transferred into subsequent donors (D-R), and ova from later recoveries were transferred into C-R gilts.

Gene construct

The fusion gene used for microinjection (Fig. 10.1) was composed of the avian skeletal α-actin (α-SkA) promoter −424 to +1, the natural capsite and the 5′-UTR, exon 1, first intron and portions of exon 2 up to the initiation ATG joined to human IGF-I class 1a cDNA and the avian skeletal α-actin 3′ UTR and contiguous 1.5 kb of non-coding sequences (Coleman *et al.*, 1995).

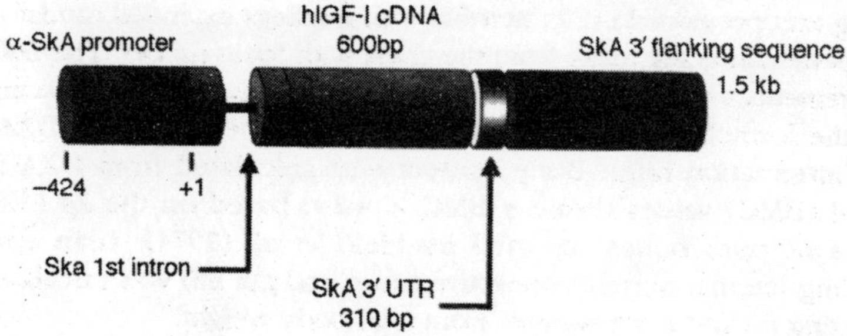

Fig. 10.1. Schematic diagram of the skeletal α-actin–IGF-I fusion gene. The construct was composed of the avian skeletal α-actin promoter −424 to +1, the natural capsite (+1), the 5′-UTR (exon 1, 60 bp), first intron (line, 123 bp), portions of exon 2 up to the initiation ATG (15 bp), human IGF-I cDNA (504 bp), skeletal α-actin 3′-UTR (310 bp) and contiguous 1.5 kb non-coding flanking sequences.

Farrowing

Recipients were brought into the farrowing barn on day 108 of pregnancy. One day after birth, the tip of each pig's tail was removed for DNA analysis by PCR. Pigs that were found to be positive for the transgene by PCR were subsequently confirmed by Southern blot hybridization analysis.

Animals and diets

At weaning (day 28) each transgenic was paired with a non-transgenic littermate of the same sex and housed in a nursery with free-choice feed and water. At about 20 kg body weight pigs were moved from the nursery and housed with two to four pigs per pen. Body weights were recorded at weekly intervals. The pigs were given *ad libitum* access to a pelleted corn–soybean meal diet containing 3.5 Mcal of digestible energy, 18% crude protein and 1.2% lysine until they reached 60 kg body weight, at which time they were fed at about 85% of *ad libitum*.

Carcass composition

At approximately 60 kg body weight, the mass of carcass fat, bone and lean of eight transgenic and eight littermate control pigs was estimated by dual-energy X-ray absorptiometry (DXA) analysis (Mitchell *et al.*, 1996). Pigs were fasted overnight and then anaesthetized as indicated above to prevent movement during the scanning procedure. Total body DXA scans were performed using a Lunar DPX-L instrument (Lunar, Madison, Wisconsin). The pig was positioned on its sternum with hindlegs extended caudally and forelegs held laterally away from the chest with foam rubber. The DXA fat measurements were adjusted to correct for previously reported inaccuracies using the formula % fat = 493 − [(348.8) (R_{st})], where R_{st} is the DXA soft tissue attenuation ratio. Bone content was calculated from DXA bone mineral (BMC) values (bone = BMC × 4.14) based on the 24.14% ash content of pork bones reported by Field *et al.* (1974). Lean content (including internal organs, connective tissue and gut fill) was calculated by subtracting fat and bone weights from total body weight.

Muscle and plasma IGF-I assay

A muscle biopsy was recovered from the longissimus muscle of each transgenic pig when they reached about 90 kg body weight using the biopsy device described by Schöberlein (1989). The tissue was frozen in liquid nitrogen immediately after biopsy and maintained at −70°C until protein and mRNA were extracted. Each muscle sample (100 mg wet weight) was lyophilized and powdered before homogenization in 1 ml 1 M acetic acid using a bead beater (Biospec Products Inc., Bartlesville, Oklahoma). The homogenized samples were held at −20°C for 1 h, centrifuged in a microfuge for 10 min, the supernatant transferred into a new tube and lyophilized to dryness. Each sample was solubilized in 0.1 × volume of 25 mM Tris-Cl (pH 8.0) and 1% (v/v) triton, and clarified by centrifugation before assay. The concentration of IGF-I was determined using 50 µl per assay by the non-competitive, two-site IRMA assay (kit DSL-5600, Diagnostic Systems Laboratories, Webster, Texas). Rat muscle spiked with rhIGF-I was used as a standard.

Total RNA was isolated from tissues by selective precipitation from phenol-extracted, guanidine thiocyanate homogenates. Northern blots were prepared as previously described by Coleman *et al.* (1995). Total RNA (10 µg) isolated from longissimus muscle biopsies was fractionated on agarose–formaldehyde gels and transferred to Gene Screen nylon membranes (Dupont NEN). Blots were hybridized with a ^{32}P-labelled hIGF-I probe and exposed overnight without intensifying screens.

Blood was collected from the anterior vena cava from each transgenic and control pig using a syringe that contained ethylenediamine tetraacetic acid (EDTA) to prevent coagulation. Plasma was separated by centrifugation and stored at −70°C until assayed. Plasma IGF-I was measured by double antibody RIA using rabbit anti-human IGF-I (UBK-487, NIDDK, Bethesda, Maryland) as the primary antiserum and recombinant human IGF-I (for standard and iodination tracer). Samples were prepared using the acid glycylglycine acidification technique (Elsasser *et al.*, 1988). The samples were all evaluated in a single assay. The intra-assay coefficient of variation was 6.7%.

Mating for G1 progeny

When transgenic gilts reached approximately 180 days of age they were housed in outside pens and oestrus was checked once daily with a mature boar. At the second or a subsequent oestrus transgenic gilts were mated to a non-transgenic boar (Orange Line hybrid, Dekalb Swine Breeders, Inc., Dekalb, Illinois).

When transgenic boars reached 8 months of age each boar was mated to two or three non-transgenic DK-43 hybrid gilts. When possible, each boar was mated twice per oestrus to each gilt.

Statistical analysis

Data were analysed using GLM procedures (SAS, 1988). The model used for analysis of birth weight, weaning weight, daily gain, body fat and plasma IGF-I concentrations consisted of the main effects of genotype, sex and resulting interactions with the pig considered to be the experimental unit. Means were compared by least significant differences. Frequency of transgene transmission to progeny was analysed using the chi-squared test.

Results and Discussion

Transgene integration

A total of 1207 zygotes were microinjected and transferred into 51 recipient gilts, 23 and 28 of which were C-R and D-R, respectively. Twenty-seven recipients became pregnant (11 C-R and 16 D-R) and produced 167 piglets (Table 10.1). The transgene integrated in 17 piglets, as determined by PCR and Southern blot hybridization, of which 14 and 13 were alive at birth and weaning, respectively. The 1.4% integration efficiency is well within the range of efficiencies reported by others (Ebert and Schindler, 1993).

Expression of IGF-I transgene

Expression of the transgene hIGF-I mRNA in longissimus muscle was detected in 9/12 transgenic pigs (Fig. 10.2). A subsequent Northern assay indicates that pig no. 50403, which was not included (Fig. 10.2), was expressing IGF-I at a level similar to pig no. 52006. In contrast, IGF-I concentrations in longissimus muscle were elevated more than twofold above control muscle in 11/12 transgenic pigs when expression was evaluated by the IRMA assay procedure after tissue extraction (Fig. 10.3). For the nine pigs in which IGF-I mRNA was detected, the quantity of hIGF-I

Table 10.1. Production of αACT-IGF-I transgenic founder pigs.

Item	Totals
No. of zygotes injected and transferred	1207
No. of recipients	51
No. of recipients farrowing	27
% of recipients farrowing	52.9
No. of pigs born (live)	167 (145)
No. of pigs per litter	6.2
% of injected zygotes developing (preg. gilts)	14 (26)
No. of transgenic pigs born (live)	17 (14)
% of pigs transgenic	10.2
% of injected zygotes	1.4

mRNA observed in the Northern blot was in general agreement with IGF-I concentrations found by the IRMA assay.

According to the assay results shown in Fig. 10.3, the range in levels of transgene expression among founders varied from about twofold to more than 170-fold above the IGF-I concentration detected in littermate control pigs (<10 ng IGF-I g^{-1} longissimus muscle). The wide range in expression

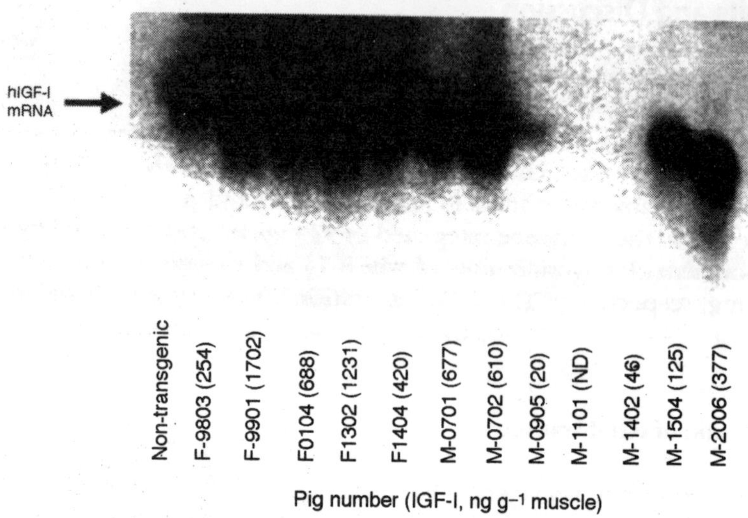

Fig. 10.2. Northern blot of total RNA (10 mg) isolated from longissimus muscle biopsies of five transgenic females (F), seven transgenic males (M), and a control pig (wild-type). The blot was hybridized with a ^{32}P-labelled hIGF-I probe that revealed a major transcript of approximately 1.0 kb IGF-I. For comparison with mRNA, IGF-I concentrations shown in Fig. 10.3 are indicated in parentheses for each animal.

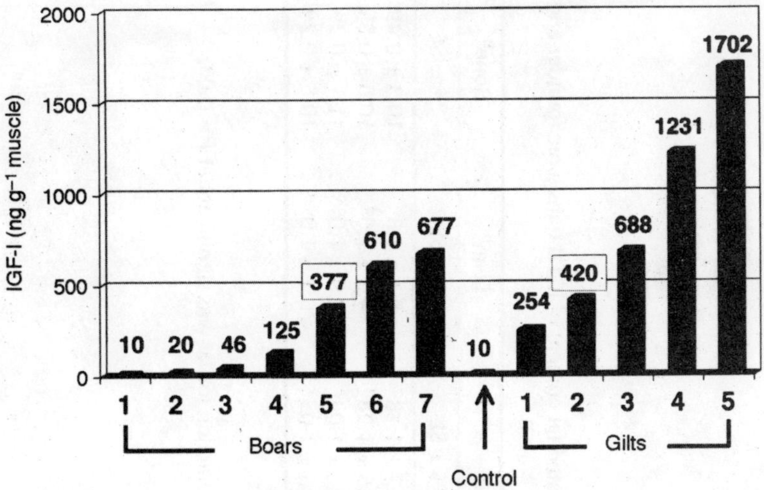

Fig. 10.3. IGF-I concentrations in longissimus muscle (ng IGF-I g^{-1} wet muscle) were determined by non-competitive, two-site IRMA assay.

among a group of transgenic founders is consistent with numerous other transgenes, and is thought to be the result of integration occurring in a wide variety of chromosomal locations, as has been shown by *in situ* hybridization analysis (Shamay *et al.*, 1991; Kuipers *et al.*, 1997), of mosaicism in some founders, and varying numbers of gene copies integrating (Palmiter and Brinster, 1986).

Plasma IGF-I concentrations of the transgenic founders (160 ± 6.8 ng ml^{-1}) did not differ from their littermates (143 ± 6.5 ng ml^{-1}; $P = 0.09$), but the trend for transgenics to be higher than controls was consistent with previous observations in the IGF-I transgenic mice (Coleman *et al.*, 1995).

Growth performance and body composition

The birth and weaning weights of the transgenic pigs did not differ from those of their littermates ($P > 0.10$ for each). The average daily gain from 20 kg to 60 kg body weight for the six transgenic boars and six transgenic gilts that harboured the IGF-I transgene was similar to that of littermate controls (Table 10.2; 865 ± 29.6 vs. 876 ± 18.9 g day^{-1}, $P > 0.10$).

Body composition of transgenic and control female pigs differed significantly at 60 kg body weight according to estimations based on DXA analyses (Table 10.2). Transgenic gilts contained less fat (15.5 vs. 22.2, $P = 0.015$) and more lean tissue (74.5 vs. 68.1, $P = 0.034$) than control gilts. In contrast, body composition of transgenic and control boars did not differ. Estimated bone content was similar for transgenic and control pigs for both

Table 10.2. Least squares mean plasma IGF-I concentrations, daily weight gain and estimated fat content of IGF-I transgenic founder and control pigs.

		Plasma IGF-I		Weight gain day^{-1a}		Fat[b]		Lean[b]	Bone[b]
Gene	Sex	*n*	ng ml^{-1} ± SE	*n*	grammes ± SE	*n*	% ± SE	% ± SE	% ± SE
Control	F	6	140 ± 8.7	6	890 ± 36.0	2	22.2 ± 1.78^{c}	68.1 ± 1.8^{c}	10.3 ± 0.49
Transgenic	F	5	142 ± 9.6	6	886 ± 36.0	2	15.5 ± 1.98^{d}	74.5 ± 2.0^{d}	10.8 ± 0.55
Control	M	5	147 ± 9.6	6	863 ± 36.0	6	19.7 ± 1.03^{d}	69.9 ± 1.0^{c,d}	11.1 ± 0.29
Transgenic	M	6	178 ± 9.6	6	843 ± 36.0	6	19.5 ± 1.05^{d}	70.1 ± 1.0^{c,d}	10.8 ± 0.29

[a] Daily gain from 20 to 60 kg body weight during free-choice feeding.

[b] Estimated by dual energy X-ray absorptiometry at 60 kg body weight.

[c,d] Means in column with differing superscript differ significantly $P < 0.05$; gene × sex interaction for fat was significant at $P = 0.03$.

sexes. While these differences in body composition are based on relatively few animals, preliminary results with progeny produced by these transgenic founders confirm these results (A.D. Mitchell and V.G. Pursel, unpublished observations).

General health

The IGF-I transgenic and control pigs did not differ in physical appearance, general behaviour, ability to tolerate summertime heat stress or reproductive capacity. These observations differ from those reported previously for transgenic pigs expressing bovine growth hormone, who were lethargic, did not tolerate elevated ambient temperatures, and had reproductive problems (Pursel *et al.*, 1989).

The general health of the transgenic pigs did not differ from their littermate controls. Of the 14 living transgenic pigs that were born, only one died before weaning; autopsy did not reveal the cause of death. Subsequently, one gilt (no. 1404) died suddenly at 6 months of age with no apparent cause, and another gilt (no. 9803) died just before the expected parturition. At necropsy the latter gilt was diagnosed with endocarditis of heart valves and cardiac haemorrhage. Whether these deaths were associated with expression of the IGF-I transgene in cardiac muscle is unknown. Based on results with this same fusion gene in mice we would expect IGF-I expression in cardiac muscle, but at a lower level than in skeletal muscle (Coleman *et al.*, 1995).

Reproduction and gene transmission

Transgenic gilts exhibited their first oestrus at the same age as the littermate controls (213.8 days and 214 days, respectively) and conceived on the first or second breeding. The farrowing results for three transgenic gilts and 17 non-transgenic gilts, which were mated by seven transgenic boars, are presented in Fig. 10.4. In addition, gilt no. 9903 was carrying 11 fully formed pigs at the time of death just before parturition, and gilt no. 1405 became ill after entering the farrowing barn and farrowed two dead pigs after a protracted parturition. The 13 pigs produced by these two gilts were not assayed for presence of the IGF-I transgene.

Nine of ten transgenic founders transmitted the IGF-I transgene to one or more progeny (Fig. 10.4). The rate of transmission by seven of the transgenic pigs did not differ significantly from the 50% expected on the basis of Mendelian inheritance. One transgenic boar (no. 1504) failed to transmit the transgene to his 25 progeny, and two boars (nos 0905 and 0701) transmitted the transgene to significantly less than half of their progeny ($P < 0.01$ and $P < 0.02$, respectively). The failure to transmit and a

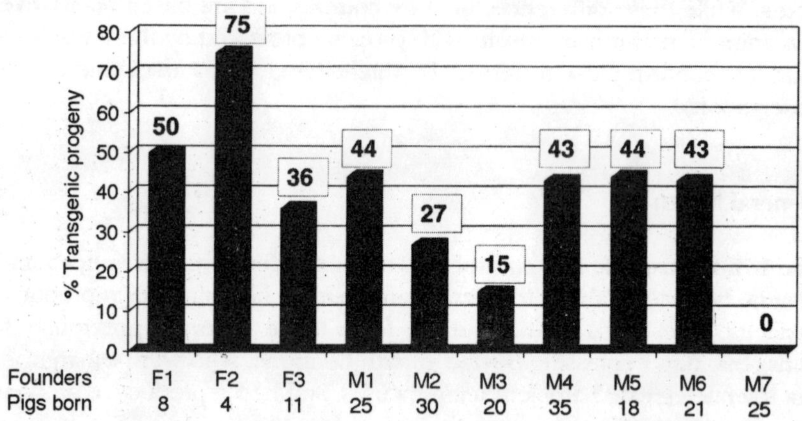

Fig. 10.4. The percentage of progeny that inherited the IGF-I transgene from the founder transgenic parent. The total number of progeny produced by the founder is indicated.

low rate of transmission probably results from transgene mosaicism in germ cells, as reported originally in mice (Palmiter *et al.*, 1984) and more recently in pigs (Pursel *et al.*, 1990).

Conclusions

The regulatory elements from avian skeletal α-actin gene and 5′ flanking sequences can provide a high level of IGF-I expression in skeletal muscle of transgenic pigs. Elevated IGF-I expression in skeletal muscle did not significantly increase serum IGF-I concentrations, alter the growth rate or adversely impact the general health of transgenic pigs in comparison with their littermate controls. Transgenic founder males and females appeared to have normal reproductive capacity, and 9/10 transgenic founders transmitted their transgene to progeny. These findings agree closely with those obtained when the same fusion gene was transferred into mice (Coleman et al., 1995). Further research is required to determine whether expression of this IGF-I transgene stimulates myofibre hypertrophy as it did in mice.

We believe the present findings, when combined with and compared with other IGF-I transgenic results, provide a strong indication as to whether circulating IGF-I (classical endocrine function) or locally produced IGF-I (paracrine function) is more important in regulation of animal growth. If one contrasts the results of Mathews *et al.* (1988) and Reiss *et al.* (1996), where transgene expression elevated circulating IGF-I (endocrine effect), with that of present results in pigs and in mice (Coleman *et al.*, 1995), where IGF-I expression is directed to skeletal muscle (paracrine effect), it appears that the endocrine effect has considerably more overall impact on growth and development.

References

Coleman, M.E., DeMayo, F., Yin, K.C., Lee, H.M., Geske, R., Montgomery, C. and Schwartz, R.J. (1995) Myogenic vector expression of insulin-like growth factor I stimulates muscle cell differentiation and myofibre hypertrophy in transgenic mice. *Journal of Biological Chemistry* 270, 12109–12116.

Ebert, K.M. and Schindler, J.E.S. (1993) Transgenic farm animals: progress report. *Theriogenology* 39, 121–136.

Elsasser, T.H., Rumsey, T.S., Hammond, A.C. and Fayer, R. (1988) Influence of parasitism in plasma concentrations of growth hormone, somatomedin-C and somatomedin-binding proteins in calves. *Journal of Endocrinology* 116, 191–200.

Field, R.A., Riley, M.L., Mello, F.C., Corbridge, M.H. and Kotula, A.W. (1974) Bone composition in cattle, pigs, sheep and poultry. *Journal of Animal Science* 39, 493–499.

Kuipers, H.W., Langford, G.A. and White, J.G. (1997) Analysis of transgene integration sites in transgenic pigs by fluorescence *in situ* hybridization. *Transgenic Research* 6, 253–259.

Mathews, L.S., Hammer, R.E., Behringer, R.R., D'Ercole, A.J., Bell, G.I., Brinster, R.L. and Palmiter, R.D. (1988) Growth enhancement of transgenic mice expressing human insulin-like growth factor-I. *Endocrinology* 123, 2827–2833.

Mitchell, A.D., Conway, J.M. and Scholz, A.M. (1996) Incremental changes in total and regional body composition of growing pigs measured by dual-energy X-ray absorptiometry. *Growth, Development and Aging* 60, 113–123.

Palmiter, R.D. and Brinster, R.L. (1986) Germline transformation of mice. *Annual Review of Genetics* 20, 465–499.

Palmiter, R.D., Wilkie, T.M., Chen, H.Y. and Brinster R.L. (1984) Transmission distortion and mosaicism in an unusual transgenic mouse pedigree. *Cell* 36, 869–877.

Pursel, V.G. and Rexroad, C.E. Jr (1993) Status of research with transgenic farm animals. *Journal of Animal Science* 71 (Suppl. 3), 10–19.

Pursel, V.G. and Wall, R.J. (1996) Effects of transferred ova per recipient and dual use of donors as recipients on production of transgenic swine. *Theriogenology* 46, 201–209.

Pursel, V.G., Pinkert, C.A., Miller, K.F., Bolt, D.J., Campbell, R.G., Palmiter, R.D. and Brinster, R.L. (1989) Genetic engineering of livestock. *Science* 244, 1281–1288.

Pursel, V.G., Hammer, R.E., Bolt, D.J., Palmiter, R.D. and Brinster R.L. (1990) Genetic engineering of swine: integration, expression and germline transmission of growth-related genes. *Journal of Reproduction and Fertility* 41 (Suppl.), 77–87.

Reiss, K., Cheng, W., Ferber, A., Kajstura, J., Li, P., Li, B.S., Olivetti, G., Homcy, C.J., Baserga, R. and Anversa, P. (1996) Overexpression of insulin-like growth factor-1 in the heart is coupled with myocyte proliferation in transgenic mice. *Proceedings of the National Academy of Sciences USA* 93, 8630–8635.

SAS Institute (1988) *SAS/STAT User's Guide,* Release 6.03. SAS Institute, Cary, North Carolina.

Schöberlein, L. (1989) Die schußbiopsi-eine möglichkeit zur emittlung von qualitätsparametern und muskelstrukturmerkmalen am lebender tier. *Archiv fur Tierzucht, Berlin* 32, 235–244.

Sejrsen, K. (1986) Endocrine mechanisms underlying genetic variation in growth in ruminants. *Proceedings of the Third World Congress on Genetics and Applied Livestock Production, Lincoln,* Vol. XI, pp. 261–267. Lincoln, Nebraska.

Shamay, A., Solinas, S., Pursel, V.G., McKnight, R.A., Alexander, L., Beattie, C., Hennighausen, L. and Wall, R.J. (1991) Production of the mouse whey acidic protein in transgenic pigs during lactation. *Journal of Animal Science* 69, 4552–4562.

Wall, R.J., Pursel, V.G., Hammer, R.E. and Brinster R.L. (1985) Development of porcine ova that were centrifuged to permit visualization of pronuclei and nuclei. *Biology of Reproduction* 32, 645–651.

Zapf, J., Schmid, C. and Froesch, E.R. (1984) Biological and immunological properties of insulin-like growth factors (IGF) I and II. *Clinical Endocrinology and Metabolism* 13, 3–12.

11

Production and Analysis of Transgenic Pigs Containing a Metallothionein Porcine Growth Hormone Gene Construct

M.B. Nottle[1], H. Nagashima[1], P.J. Verma[1], Z.T. Du[1], C.G. Grupen[1], S.M. McIlfatrick[1], R.J. Ashman[1], M.P. Harding[1], C. Giannakis[1], P.L. Wigley[1], I.G. Lyons[1], D.T. Harrison[2], B.G. Luxford[2], R.G. Campbell[2], R.J. Crawford[1] and A.J. Robins[1]

[1]*BresaGen Ltd, Adelaide, Australia;* [2]*Bunge Meat Industries, Corowa, Australia*

While initial studies with growth hormone (GH) fusion genes demonstrated that transgenesis could be used to enhance growth performance in the pig, they also highlighted the need to be able to control expression in order to avoid pathological problems associated with high-level expression. We have produced transgenic pigs containing a GH construct consisting of a modified human metallothionein IIA (MT) promoter fused to the cDNA sequence for the porcine growth hormone gene. A total of 289 pigs were born live of which 88 (2.8% of embryos injected) were transgenic. Founders were reared on diets containing 100 ppm of zinc. Induction of transgene expression was assessed by feeding 1000 ppm of zinc in the diet (high zinc) for 3 weeks and measuring plasma IGF-I as a marker of GH production, before, during and after the high zinc diet. Evidence to suggest that transgene expression could be induced was obtained in 12/36 founders tested. Founders were mated to non-transgenic animals to produce transgenic progeny. Twenty-two per cent (4/18) of male founders did not transmit the transgene and 39% (7/18) transmitted the transgene at frequencies of less than 30%. The effect of transgene expression on growth performance was evaluated by feeding transgenic and non-transgenic progeny the high zinc diet from 20–100 kg liveweight. Rate of gain, feed intake and estimates of carcass fat and muscle were compared between the two groups of progeny of 60–100 kg liveweight. Analysis of transgenic progeny growth performance was confounded by considerable individual variation between transgenic progeny and

© CAB INTERNATIONAL 1999. *Transgenic Animals in Agriculture* (eds J.D. Murray, G.B. Anderson, A.M. Oberbauer and M.M. McGloughlin)

the relatively few transgenics available for evaluation from mosaic founders. A number of transgenic progeny exhibited enhanced growth performance and have been selected for further breeding and analysis.

Introduction

The ability to manipulate the genome of domestic livestock has the potential to revolutionize animal production in the coming decades (reviewed by Brem and Muller, 1994; Wall, 1996). The mouse experiments of Palmiter *et al.* (1982) were the first to demonstrate that growth hormone (GH) fusion genes could dramatically improve animal growth. Since these initial studies, a number of groups have examined the potential of transgenesis to improve growth performance in the pig (reviewed by Pursel *et al.*, Chapter 10, this volume, 1990a,b; Brem and Muller, 1994; Table 11.1). While many of these studies demonstrated that transgenesis could be used to enhance growth performance, they also highlighted the need to be able to control transgene expression to avoid pathological problems associated with high level expression, including lameness and infertility (Pursel *et al.*, 1987; Ebert *et al.*, 1988; Wieghart *et al.*, 1990). Our own experience in producing GH transgenic pigs is discussed in this chapter.

Production of Transgenic Founders

Transgenic pigs are currently produced by injecting hundreds to thousands of copies of a transgene into the pronucleus of a recently fertilized egg. The injected DNA then becomes incorporated at random, normally in head-to-tail arrays at a single genomic site (Palmiter *et al.*, 1982; Hammer *et al.*, 1985a; Burdon and Wall, 1992). A number of GH constructs and growth-related constructs have been used in pigs (reviewed by Pursel *et al.*, Chapter 10, this volume, 1990a,b; Brem and Muller, 1994; Table 11.1). The majority of these have used elements from the mouse or human metallothionein (MT) promoter fused to genomic or cDNA clones of the pig, bovine or human GH gene. The MT promoter appears to have been used because of earlier evidence obtained in mice suggesting that expression of thymidine kinase and GH transgenes could be induced by the addition of zinc to the diet (Brinster *et al.*, 1981; Palmiter *et al.*, 1982; reviewed by Seamark and Wells, 1993).

The initial aim of our study was to produce transgenic pigs in which GH expression could be regulated by manipulating the level of zinc in the diet, with the overall goal of producing commercial lines of transgenic pigs with enhanced growth performance. The transgene used in our studies consisted of a modified human MT II-A promoter fused to the cDNA sequence for the porcine growth hormone gene (Fig. 11.1). Transgenic

Table 11.1. Growth hormone transgenic pig studies.

Fusion gene	Embryos transferred	Piglets born	Number of transgenics	Reference
mMT-hGH	286	15 (5.2)	1 (0.4)	Brem *et al.* (1985)
mMT-hGH	2035	192 (9.4)	20 (1)	Hammer *et al.* (1985a)
mMT-bGH	2330	150 (6.4)	9 (0.4)	Pursel *et al.* (1987)
mMT-hGH	1014	21 (2.1)	4 (0.4)	Brem *et al.* (1988)
MLV-rGH	59	15 (25.4)	1 (1.7)	Ebert *et al.* (1988)
hMT-pGH	423	17 (4.0)	6 (1.4)	Vize *et al.* (1988)
WAP-hGH	1028	51 (5.0)	7 (0.7)	Brem *et al.* (1988)
bPRL-bGH	289	20 (6.9)	5 (1.7)	Polge *et al.* (1989)
MLV-pGH	410	59 (14.4)	6 (1.5)	Ebert *et al.* (1990)
CMV-pGH	372	32 (8.6)	15 (4.0)	Ebert *et al.* (1990)
PEPCK-bGH	1057	124 (11.7)	7 (0.7)	Wieghart *et al.* (1990)
hMT-pGH	1327	148 (11.1)	43 (3.2)	Nottle *et al.* (1994)
hMT-pGH	1835	141 (7.6)	45 (2.5)	Nottle *et al.* (1994)
Total	12465	985 (7.9)	169 (1.3)	

Numbers in brackets are values expressed as a percentage of embryos injected.

founders were produced by pronuclear microinjection using procedures described previously (Nottle *et al.*, 1994). Two groups of transgenic founders were produced (Table 11.1). Previously reported studies have shown that 0.3–4.3% of injected embryos resulted in the birth of a transgenic pig (Pursel and Rexroad, 1993). In our study, the number of live-born piglets that were transgenic was 2.8% of embryos injected. In the mouse, the concentration at which the DNA is injected does not appear to influence the integration rate between 1 and 10 ng μl^{-1} (Brinster *et al.*, 1985). However, our experience over several years with a variety of constructs suggests that the concentration at which DNA is injected in this range may influence integration rates in the pig (Nottle *et al.*, 1997). In particular, we have found

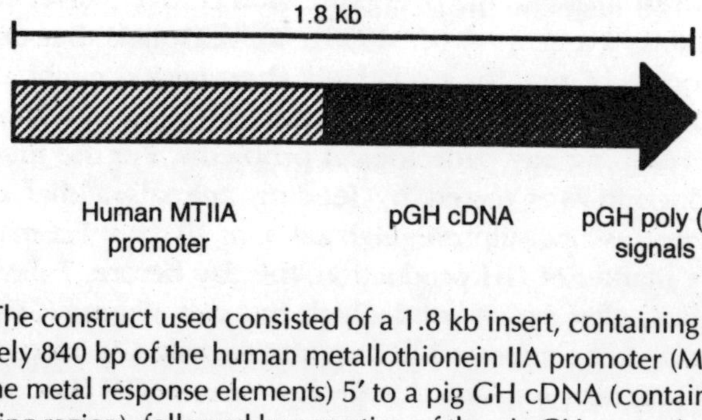

Fig. 11.1. The construct used consisted of a 1.8 kb insert, containing approximately 840 bp of the human metallothionein IIA promoter (MTIIA; including the metal response elements) 5′ to a pig GH cDNA (containing the entire protein coding region), followed by a portion of the pig GH genomic DNA containing polyadenylation signals.

that DNA injected at 10 ng μl^{-1} consistently results in 2–4% of embryos injected or 20–30% of live-born pigs being transgenic.

Transgene expression has been shown to vary depending on where in the genome the transgene becomes incorporated (so called 'position effects'; reviewed by Bishop, 1997). In the majority of reported studies with GH transgenic pigs, GH was constitutively expressed, albeit at variable levels, in sufficient amounts to have a number of deleterious side-effects including lameness and infertility (Pursel *et al.*, 1987; Ebert *et al.*, 1988). Attempts to obtain better control over GH transgene expression using different promoters such as phosphoenolpyruvate carboxykinase (Wieghart *et al.*, 1990) have also proven to be unsatisfactory. To avoid any deleterious effects associated with high-level constitutive expression, we measured plasma GH in our transgenic founders prior to weaning and culled animals with GH levels outside the range of those measured in the non-transgenics. Eight of the 88 founders produced were identified as having high-level constitutive expression and were euthanased.

Induction of Transgene Expression in Founder Populations

The metallothionein promoter contains a complex array of metal responsive elements (Lee *et al.*, 1987) and can be induced by metals such as zinc. In MT-bGH transgenic mice, concentrations of bGH were elevated more than tenfold after zinc was added to their drinking water (Hammer *et al.*, 1985b). In pigs containing the same constructs the addition of 1000–3000 ppm of zinc to the feed approximately doubled bGH expression (Pursel *et al.*, 1990a).

Because we produced a relatively large number of founder transgenics (approximately half of all GH transgenic founders reported; Table 11.1) we decided to screen our founder populations for animals in which expression could be induced. However, it was apparent from earlier work that animals maintained on high zinc diets for long periods may develop pathological problems as a result of chronic overexpression of GH (Pursel *et al.*, 1987; Ebert *et al.*, 1988; Wieghart *et al.*, 1990). We reasoned that exposure to increased amounts of zinc for a relatively short period might allow us to identify founders in which the transgene could be induced without the risk of animals developing any pathological problems. For the first group of founders, induction was tested by feeding animals a diet containing 1000 ppm of zinc (as zinc sulphate; high zinc) for 10 days. Plasma IGF-I was measured as a marker of GH production, the day before, 7 days after the start of and 7 days after the end of the high zinc diet. Plasma IGF-I has been shown previously to be increased in response to daily GH injection (Owens *et al.*, 1990), in animals implanted with slow release GH (Buonomo *et al.*, 1995) and in transgenic pigs expressing GH (Miller *et al.*, 1989). No increase in IGF concentration was detected in this experiment. As a consequence of this finding, the period over which the high zinc diet was fed was increased to 3 weeks for the second group of founders. Increases in plasma IGF-I, of 25% or more above the concentration of plasma IGF-I measured prior to induction, were demonstrated in five of the 24 founders.

On the basis of this finding we retested the group 1 founder males for evidence of transgene induction. Seven of the 12 founders exhibited evidence of being inducible when tested at around 70 weeks of age (Fig. 11.2; Table 11.2). In the majority of these founders, IGF-I had returned to pre-induction levels when measured 6 weeks after the end of the high zinc diet. These findings suggest that transgene expression could be regulated by manipulating the level of zinc in the diet.

Transgene Transmission by Founders

Each transgenic founder produced by pronuclear microinjection is unique in terms of its expression. Assessment of growth performance of GH transgenic pigs has been limited mostly to comparisons between transgenic founders

Table 11.2. Induction status and transgene transmission frequency for group 1 and 2 male founders.

Founder	Inducible	Transgenic/ total progeny	% Transgenic
Group 1			
50402	No	22/53	42
50403	Yes	23/45	51
50404	No	0/36	0
50405	No	24/63	38
50406	Yes	19/103	18
50408	Yes	15/62	24
50409	Yes	0/72	0
50410	Yes	6/122	5
50411	No	13/26	50
50413	Yes	0/52	0
50414	No	0/55	0
50415	Yes	21/56	38
Group 2			
51201	Yes	Not mated	
51202	No	6/67	9
51203	No	30/94	32
51204	No	6/91	7
51205	No	11/62	18
51206	No	26/98	27
51207	Yes	Not mated	
51208	Yes	30/71	42

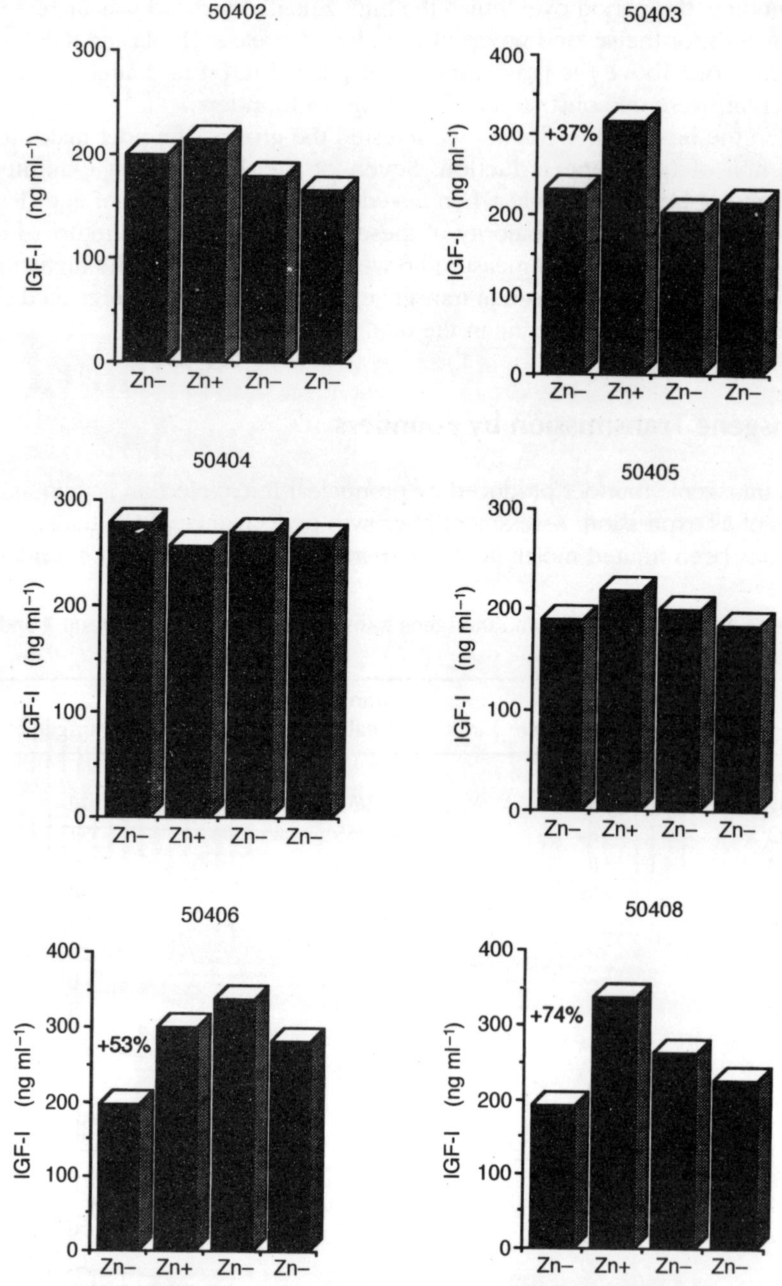

Fig. 11.2. Plasma IGF-I concentration in 12 group 1 founder males (continued opposite). Animals were tested for induction at approximately 70 weeks of age. A diet containing 1000 ppm of zinc (as zinc sulphate) was fed for 3 weeks. Animals were bled twice daily (a.m. and p.m.) on the day before, last day of, and 3 and 6 weeks after the high-zinc diet. Plasma IGF-I was measured according to methods described by Owens *et al.* (1990).

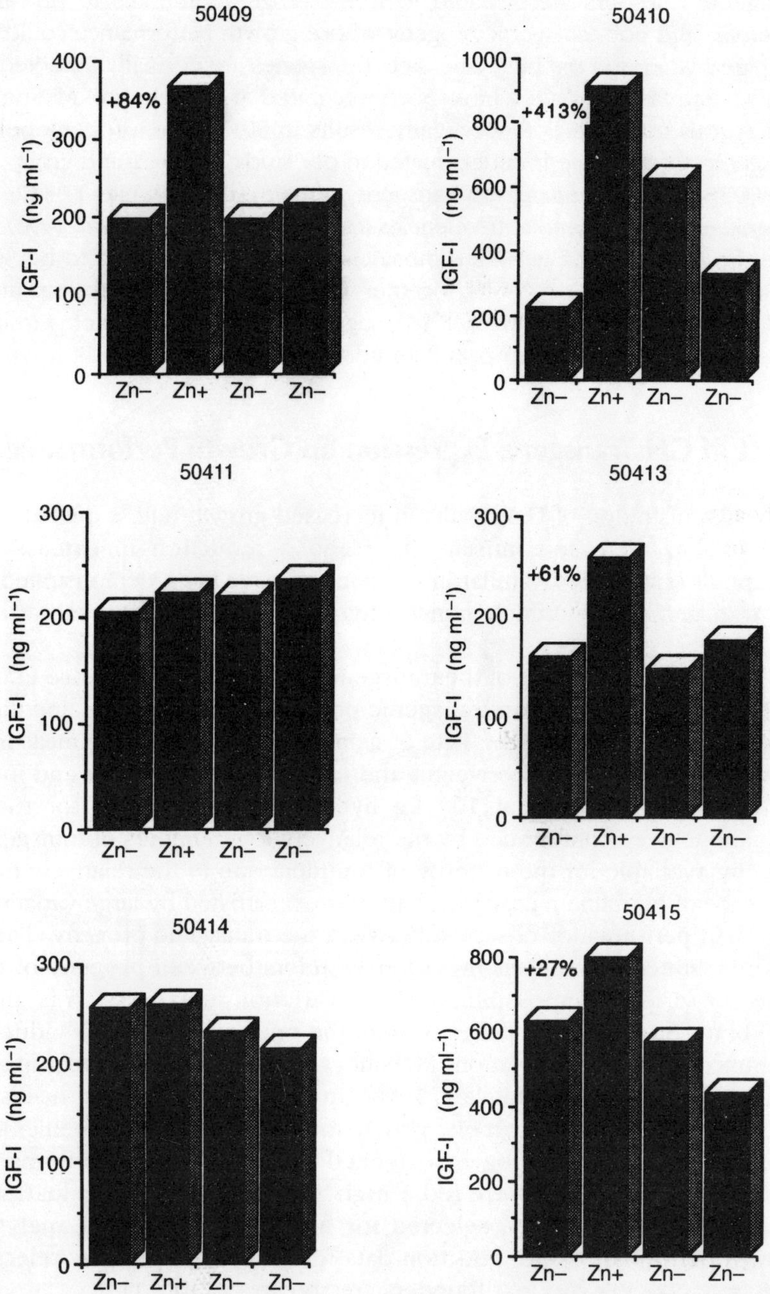
50409
IGF-I (ng ml⁻¹)
0
100
200
300
400
+84%
Zn– Zn+ Zn– Zn–
50410
IGF-I (ng ml⁻¹)
0
200
400
600
800
1000
+413%
Zn– Zn+ Zn– Zn–
50411
IGF-I (ng ml⁻¹)
0
100
200
300
Zn– Zn+ Zn– Zn–
50413
IGF-I (ng ml⁻¹)
0
100
200
300
+61%
Zn– Zn+ Zn– Zn–
50414
IGF-I (ng ml⁻¹)
0
100
200
300
Zn– Zn+ Zn– Zn–
50415
IGF-I (ng ml⁻¹)
0
200
400
600
800
+27%
Zn– Zn+ Zn– Zn–

and non-transgenic littermates (reviewed by Pursel *et al.*, 1990a). To evaluate the effect of the transgene used in our study on growth performance, transgenic founders were mated to non-transgenic animals to produce transgenic and non-transgenic progeny whose growth performance could be compared when fed the high zinc diet. Transgenes are normally inherited in a Mendelian fashion if they have been integrated at a single site. Mating of hemizygous transgenics theoretically results in 50% of the progeny being transgenic. Of the male founders mated in our study (group 1 and group 2), 22% (4/18) did not transmit the transgene to their progeny while 39% (7/18) transmitted the transgene at frequencies less than 30% (Nottle *et al.*, 1996). In mice the incidence of germline mosaicism has been reported to be 30% (Wilkie *et al.*, 1986). Our results, together with those of other workers (Pursel *et al.*, 1990a,b; Brem and Muller, 1994), suggest that the incidence of germline mosaicism may be higher in pigs than in mice.

Effect of GH Transgene Expression on Growth Performance

Daily administration of GH results in increased growth rate, a decrease in feed intake, increased muscle mass and a reduction in carcass fat (Campbell *et al.*, 1989). Similar improvements have been demonstrated in GH transgenic pigs which constitutively express GH (Pursel *et al.*, 1990a,b).

To evaluate the effect of the transgene on growth performance in our studies transgenic and non-transgenic progeny were fed high zinc diets from 20 to 100 kg liveweight. Rate of gain and feed intake were measured between 60 and 100 kg liveweight and estimates of carcass fat and meat content were obtained at 100 kg liveweight. The results for these evaluations were confounded by the relatively low numbers of transgenic progeny available for the majority of founders due to the relatively high incidence of germline mosaicism. This was exacerbated by large variations in growth performance observed between the transgenic progeny. These factors made any comparison within founders between progeny of the same sex virtually impossible. In mice, variation in expression is often seen between transgenic progeny from the one founder possibly due to differences in DNA methylation (Mehtali *et al.*, 1990) and heterochromatin formation (Martin and Whitelaw, 1996). In such cases, selection needs to be carried out for more than one generation to generate a transgenic line. A number of transgenic progeny exhibited enhanced growth performance during the period they were fed a high zinc diet in these evaluations. These animals have been selected for further breeding and analysis. Growth performance and induction data for three male progeny selected from two inducible group 1 founders are shown in Table 11.3.

Table 11.3. Growth performance data for six male F1 progeny selected from two inducible group 1 founder males.

Founder	Founder IGF-I (% increase)	Progeny	Liveweight gain (g day^{-1})[a]	Feed conversion[a]	P2[b] (mm)	Muscle depth[c] (mm)	Progeny IGF-I[d] (% increase)
50406	52	2774	820	2.14	11	39	42
		2968	890	2.26	8	35	3
		2951	790	2.41	8	38	44
Non-transgenic male littermates (mean ± SEM)		745 ± 36	2.68 ± 0.14	10 ± 1	34 ± 1		
50408	74	2296	880	1.86	8	35	15
		2990	900	1.87	6	29	7
		2844	950	2.04	10	35	26
Non-transgenic male littermates (mean ± SEM)		836 ± 32	2.21 ± 0.07	13 ± 7	36 ± 1		

[a] Progeny were fed high-zinc diets from approximately 25 to 100 kg liveweight. Liveweight gain and feed conversion (feed consumed/liveweight gain) was measured from 60 to 100 kg liveweight.
[b] P2 is fat depth measured over the last rib 6.5 cm off the midline at 100 kg liveweight.
[c] Muscle depth is the depth of the longissimus dorsi measured over the last rib, 6.5 cm off the midline at 100 kg liveweight.
[d] Plasma IGF-I was measured using the induction protocol described in the text.

Conclusions

The major finding from our study was that MT–pGH transgenic pigs can be produced using pronuclear microinjection in which expression can be regulated by manipulating the level of zinc in the diet. While we have been able to generate potentially useful genotypes using this technique, our experience (and that of other groups) demonstrates that this is a major undertaking. The low efficiency with which founder transgenics are produced, the variation in the level of expression between founders and the relatively high degree of mosaicism, are all major drawbacks. Furthermore, the large variation between transgenic progeny in their growth performance suggests that the production of commercial lines of transgenic animals may require a number of generations of selection. In order for the full potential of transgenesis to be realized in pigs as well as in other livestock species, methods will be required which allow a single copy of the transgene to be inserted at high efficiency at a predetermined site in the genome which does not interfere with expression.

Acknowledgements

We are grateful to the numerous BresaGen and Bunge staff who made invaluable contributions during the course of this study.

References

Bishop, J.O. (1997) Chromosomal insertion of foreign DNA. *Reproduction, Nutrition and Development* 36, 607–618.

Brem, G. and Muller, M. (1994) Large transgenic mammals. In: Maclean, N. (ed.) *Animals with Novel Genes.* Cambridge University Press, Cambridge, pp. 179–224.

Brem, G., Brenig, B., Goodman, H.M., Selden, R.C., Graf, F., Kruff, B., Springmann, K., Hondele, J., Meyer, J., Winnacker, E.L. and Krausslich, H. (1985) Production of transgenic mice, rabbits and pigs by microinjection into pronuclei. *Zuchthygiene* 20, 251–252.

Brem, G., Brenig, B., Muller, M., Krausslich, H., Springmann, K. and Winnacker, E.L. (1988) Gene transfer by DNA microinjection of growth hormone genes in pigs. *Proceedings of the 11th International Congress on Animal Reproduction and Artificial Insemination* 4, 46.

Brinster, R.L., Chen, H.Y., Trumbauer, M.E., Senear, A.W., Warren, R. and Palmiter, R.D. (1981) Somatic expression of herpes thymidine kinase in mice following injection of a fusion gene into eggs. *Cell* 27, 223–231.

Brinster, R.L., Chen, H.Y., Trumbauer, M.E., Yagle, M.K. and Palmiter, R.D. (1985) Factors affecting the efficiency of introducing foreign DNA into mice by microinjecting eggs. *Proceedings of the National Academy of Sciences USA* 82, 4438–4442.

Buonomo, F.C., Klindt, J. and Yen, J.T. (1995) Administration of porcine somatotropin by sustained-release implant: growth factor and metabolic responses in crossbred white and genetically lean and obese boars and gilts. *Journal of Animal Science* 73, 1318–1326.

Burdon, T.G. and Wall, R.J. (1992) Fate of microinjected genes in preimplantation mouse embryos. *Molecular Reproduction and Development* 33, 436–442.

Campbell, R.G., Steele, N.C., Caperna, T.J., McMurty, J.P., Solomon, M.B. and Mitchell, A.D. (1989) Interrelationships between sex and exogenous growth hormone administration on performance, body composition and protein and fat accretion of growing pigs. *Journal of Animal Science* 67, 177–186.

Ebert, K.M., Low, M.J., Overstrom, E.W., Buonomo, F.C., Baile, C.A., Roberts, T.M., Lee, A., Mandel, G. and Goodman, R.H. (1988) A Moloney MLV-rat somatotropin fusion gene produces biologically active somatotropin in a transgenic pig. *Molecular Endocrinology* 2, 277–283.

Ebert, K.M., Smith, T.E., Buonomo, F.C., Overstrom, E.W. and Low, J. (1990) Porcine growth hormone expression from viral promoters in transgenic swine. *Animal Biotechnology* 1, 145–159.

Hammer, R.E., Pursel, V.G., Rexroad, C.E. Jr, Wall, R.J., Bolt, D.J., Ebert, K.M., Palmiter, R.D. and Brinster, R.L. (1985a) Production of transgenic rabbits, sheep and pigs by microinjection. *Nature* 315, 680–683.

Hammer, R.E., Brinster, R.L. and Palmiter, R.D. (1985b) Use of gene transfer to increase animal growth. *Cold Spring Harbour Symposium on Quantitative Biology* 50, 379–387.

Lee, W., Haslinger, A., Karin, M. and Tijan, R. (1987) Activation of transcription by two factors that bind promoter and enhancer sequences of the human metallothionein gene and SV40. *Nature* 325, 368–372.

Martin, D.I.K. and Whitelaw, E. (1996) The vagaries of variegating transgenes. *BioEssays* 18, 919–923.

Mehtali, M., LeMeur, M. and Lathe, R. (1990) The methylation-free status of a housekeeping transgene is lost at high copy number. *Gene* 91, 179–184.

Miller, K.F., Bolt, D.J., Pursel, V.G., Hammer, R.E., Pinkert, C.A., Palmiter, R.D. and Brinster, R.L. (1989) Expression of human or bovine growth hormone gene with a mouse metallothionein-1 promoter in transgenic swine alters the secretion of porcine growth hormone and insulin-like growth factor-I. *Journal of Endocrinology* 120, 481–488.

Nottle, M.B., Nagashima, H., Verma, P.J., Ashman, R.J., Du, Z.T, Grupen, C.G, McIlfatrick, S.M., Harding M.P., Cheah, C., Crawford, R.J. and Robins, A.J. (1994) Production of pigs containing a metallothionein porcine growth hormone gene construct. In: *Proceedings of the 26th Annual Conference of The Australian Society for Reproductive Biology, Brisbane*, p. 33 (abstract).

Nottle, M.B., Nagashima, H., Verma, P.J., Ashman, R.J., Du, Z., Grupen, C.G., McIlfatrick, S.M., Harding, M.P., Cheah, C., Harrison, D.T., Luxford, B.G., Campbell, R.G., Crawford, R.J. and Robins, A.J. (1996) Inheritance of a metallothionein porcine growth hormone transgene in pigs. *Proceedings of the 13th International Congress on Animal Reproduction, Sydney*, 3 P26–2 (abstract).

Nottle, M.B., Nagashima, H., Verma, P.J., Ashman, R.J., Du, Z.T., Grupen C.G. and McIlfatrick, S.M. (1997) Developments in transgenic techniques in pigs. *Journal of Reproduction and Fertility* 52 (Suppl.), in press.

Owens, P.C., Johnson, P.C., Campbell, R.G. and Ballard, F.J. (1990) Growth hormone increases insulin-like growth factor-I (IGF-I) and decreases IGF-II in plasma of growing pigs. *Journal of Endocrinology* 124, 269–275.

Palmiter, R.D., Brinster, R.L., Hammer, R.E., Trumbauer, M.E., Rosenfeld, M.G., Birnberg, N.C. and Evans, R.M. (1982) Dramatic growth of mice that develop from eggs microinjected with metallothionein-growth fusion genes. *Nature* 300, 611–615.

Polge, E.J.C., Barton, S.C., Surani, M.H.A., Miller, R., Wagner, T., Elsome, K., Davis, A.J., Goode, J.A., Foxcroft, G.R. and Heap, R.B. (1989) Induced expression of a bovine growth hormone construct in transgenic pigs. In: Heap, R.B., Prosser, C.G. and Lamming, G.E. (eds) *Biotechnology in Growth Regulation*. Butterworths, London, pp. 189–199.

Pursel, V.G. and Rexroad, C.E. Jr (1993) Recent progress in the transgenic modification of swine and sheep. *Molecular Reproduction and Development* 36, 251–254.

Pursel, V.G., Rexroad, C.E. Jr, Bolt, D.J., Miller, K.F., Wall, R.J., Hammer, R.E., Pinkert, C.A., Palmiter R.D. and Brinster R.L. (1987) Progress on gene transfer in farm animals. *Veterinary Immunology and Immunopathology* 17, 303–312.

Pursel, V.G., Bolt, D.J., Miller, K.F., Pinkert, C.A., Hammer, R.E., Palmiter, R.D. and Brinster, R.L. (1990a) Expression and performance in transgenic pigs. *Journal of Reproduction and Fertility* 40 (Suppl.), 235–245.

Pursel, V.G., Hammer, R.L., Bolt, D.J., Palmiter, R.D. and Brinster, R.L. (1990b) Integration, expression and germline transmission of growth-related genes in pigs. *Journal of Reproduction and Fertility* 41 (Suppl.), 77–87.

Seamark, R.F. and Wells, J.R.E. (1993) Biotechnology and Reproduction. In: King, G.J. (ed.) *Reproduction in Domesticated Animals*, Vol. 14. World Animal Science, Series B9, Elsevier, Amsterdam, pp. 345–363.

Vize, P.D., Michalska, A.E., Ashman, R.J., Lloyd, B., Stone, B.A., Quinn, P., Wells, J.R.E. and Seamark, R.F. (1988) Introduction of a porcine growth hormone fusion gene into transgenic pigs promotes growth. *Journal of Cell Science* 90, 295–300.

Wall, R.J. (1996) Transgenic livestock: progress and prospects for the future. *Theriogenology* 45, 57–68.

Wieghart, M., Hoover, J.L., McGrane, M.M., Hanson, R.W., Rottman, F.M., Holtzman, S.H., Wagner, T.E. and Pinkert, C.A. (1990) Production of transgenic pigs harbouring a rat phosphoenolpyruvate carboxykinase-bovine growth hormone fusion gene. *Journal of Reproduction and Fertility* 41 (Suppl.), 89–96.

Wilkie, T.M., Brinster, R.L. and Palmiter, R.D. (1986) Germline and somatic mosaicism in transgenic mice. *Developmental Biology* 118, 9–18.

12

The Utilization of Bacterial Genes to Modify Domestic Animal Biochemistry

K.A. Ward, Z. Leish, A.G. Brownlee, J. Bonsing, C.D. Nancarrow and B.W. Brown

CSIRO Division of Animal Production, Prospect, Australia

The ability to transfer genes between organisms in a way that maintains their inheritance and their function makes it possible to modify the biochemistry of domestic animals to improve productivity. One example of this is the modification of sheep biochemistry to provide an increased supply of the amino acid cysteine which is the first limiting amino acid in wool growth. The *cysE* and *cysK* genes were isolated from *Escherichia coli* and modified for expression in animals. In transgenic mice, the genes were transcribed and translated to produce two functional enzymes, serine transacetylase and *O*-acetylserine sulphydrylase which together were able to catalyse the biosynthesis of cysteine in transgenic mice provided that the mice were fed a diet containing a small amount of sodium sulphide. When placed on a diet that was deficient in sulphur-containing amino acids, non-transgenic control mice lost substantial amounts of hair and also lost body weight while the transgenic mice showed no hair loss and continued to grow at normal rates.

The modified genes have been transferred to sheep where they have been shown to integrate with the host genome and can be passed to progeny in normal ratios of Mendelian inheritance. However, the genes were only expressed in one animal and this expression was very low. The results are similar to those obtained in parallel experiments carried out in another laboratory and suggest that a functional cysteine biosynthesis pathway might interfere with the intermediary metabolism of sheep whereas apparently it can operate effectively in transgenic mice.

Introduction

While scientists and farmers generally agree that genetic engineering has the potential to improve domestic animal productivity, few projects have yet achieved commercial viability. Many laboratories have for some time been

© CAB INTERNATIONAL 1999. *Transgenic Animals in Agriculture* (eds J.D. Murray, G.B. Anderson, A.M. Oberbauer and M.M. McGloughlin)

engaged in commercially oriented transgenic animal projects and this poor strike rate suggests that there is some underlying factor, so far largely ignored, that is influencing the process in domestic animals. Technical difficulties provide part of the explanation because the procedure of gene transfer by pronuclear microinjection in domestic animals is less efficient than in laboratory mice, thus reducing the number of transgenic animals available. However, with sufficient effort it is possible to produce enough transgenic domestic animals to adequately test for appropriate phenotypes, so the difficulty of achieving commercial goals is due to more than problems of technique.

A more insidious problem may be associated with the actual goals of the research to date. Most domestic animal-oriented genetic engineering research projects are directed towards improved animal productivity, either by increasing growth rates, altering body composition or the composition of an animal product, or by improving the health or husbandry requirements of the animals. To achieve this it is usually necessary to modify some component of the animal's physiology, thus potentially altering the existing delicate balance of nutrition, endocrinology and metabolism. Since this balance has been established through many generations of selection for superior performance and environmental compatibility, it represents a wide range of optimized gene combinations that are difficult to perturb without causing unexpected deleterious effects on animal phenotype. This suggests that the foreign genes that are introduced in domestic animal genetic engineering projects should be designed to make small changes to animal homeostasis so that the existing equilibrium is only slightly altered. The possibility exists that most projects so far attempted in this new and rapidly developing field of endeavour have attempted changes of a magnitude too great for assimilation by the optimized target animal genotype, thus reducing productivity and, in consequence, commercial viability of the resulting animals. Genetic manipulations that have as their goal the modification of target animal biochemistry might be expected to be particularly sensitive to this problem because they have the potential to alter the complex balance of substrates and co-enzymes that are integral to the smooth operation of intermediary metabolism. However, there are cogent reasons for attempting such biochemical alterations in some of our domestic animals and hence it is worth the effort to examine whether the problems of homeostatic imbalance can be overcome.

During the long evolutionary history of each of our major domestic species there has been a significant loss of biochemical capacity when compared with simple auxotrophs, resulting in the inability of animals to synthesize a wide range of enzyme substrates and cofactors. Whenever these are crucial to animal survival, they must be supplied in the diet as essential nutrients and vitamins. It can be argued that the loss of biosynthetic capacity provides an overall advantage to the animal as long as the nutrients in question are freely available in the diet, because the energy associated with *de novo* synthesis is available for other purposes. Thus during any prolonged selection regime, be it natural or artificial, animals would, in general, perform better and hence be preferred for selection when their available energy was directed towards those compounds that could not be adequately provided by diet, with any excess being utilized for growth and survival in the wild or for production when domesticated. If this is correct, it may be necessary to trade some growth or fitness qualities for a gain in a particular production characteristic if specific biochemical pathways are replaced, provided that this trade-off is not to the detriment of the animal's health and welfare. Until recently, this concept has only been of theoretical interest, but with the advent of genetic engineering techniques, the restoration of lost biochemistry in animals is a genuine possibility. The inability to synthesize nutrients *de novo* normally results from the loss of the genes encoding critical enzymes of the relevant biosynthetic pathways, and functional counterparts to these genes can be readily identified in most auxotrophs. By modifying the functional genes for expression in a target domestic animal species, the missing enzymes can in theory be made available to catalyse the appropriate intermediary metabolism.

The Application of Biochemical Manipulation to Wool Production

The Australian wool industry might be helped significantly by the addition of a new biochemical pathway to the Merino sheep. The primitive breeds of sheep from which the Merino has been derived were poor wool producers. Their coats were characterized by short fibres which underwent cyclic annual growth and were generally shed each year. From this stock has been bred the modern Merino sheep which grows wool continuously and shows little evidence of annual cyclic growth. Under average conditions Merino strains in use today can produce about 3.5 kg of wool each year. However, this imposes a substantial demand on the supply of the amino acid cysteine, because wool fibres are composed of a complex mixture of keratin proteins which are characterized by a high cysteine content. For example, to produce 3–3.5 kg of wool each year requires a supply of 0.7–0.9 g day^{-1} of cysteine for wool growth alone and this represents about 50% of the total cysteine utilized by the sheep (Lee *et al.*, 1993). This cysteine requirement must be satisfied either by transsulphuration from methionine or from the diet of the sheep (Black and Reis, 1979; Lee *et al.*, 1993). Normal pastures are able to satisfy this cysteine demand but under conditions where nutrient supply is poor, cysteine is the first limiting substrate for wool growth (Reis, 1979).

A possible solution to this substrate limitation is to provide sheep with the enzymes necessary for cysteine biosynthesis, provided that the enzymes can be synthesized in tissues which also contain the appropriate substrates for the operation of this pathway. These substrates consist of the amino acid serine, the enzyme cofactor acetyl-coenzyme A (acetyl-CoA) and a source of inorganic sulphur. In the ruminal and intestinal epithelia, all three of these substrates are present, the sulphur source being in the form of sulphide produced by bacterial fermentation in the rumen.

Isolation and Modification of Bacterial Genes Encoding Cysteine Biosynthesis

The genes encoding the enzymes for cysteine biosynthesis are functional in the bacterium *Escherichia coli* and it is therefore theoretically possible to isolate the relevant coding sequences from this organism and modify them for expression in eukaryotes. The cysteine biosynthetic pathway in *E. coli* is a complex pathway in which inorganic sulphur is reduced to sulphide (Umbarger, 1997) and then combined with the amino acid serine in two biosynthetic steps to produce cysteine. To avoid the pathway being constitutively active, it is regulated at the enzyme level by cysteine and *O*-acetylserine concentrations and at the transcriptional level by cysteine repression and *O*-acetylserine induction (Umbarger, 1997).

Since the sheep already has a sulphide source available, the portion of the pathway that is required is the component that catalyses the conversion of serine to *O*-acetylserine and *O*-acetylserine to cysteine. The enzymes required are serine transacetylase (SAT) and *O*-acetylserine sulphydrylase (OAS), encoded by the *cysE* and *cysK* genes in *E. coli*. Their isolation and characterization have been described in detail elsewhere (Denk and Bock, 1987; Byrne *et al.*, 1988) as has the modification of the genes for expression in eukaryotes (Sivaprasad *et al.*, 1989; Leish *et al.*, 1993; Bawden *et al.*, 1995). In summary, the coding sequence for the *cysE* gene and for the *cysK* gene were each fused to sheep metallothionein-Ia (Mt-Ia) promoter sequences and exon 5 of the sheep growth hormone gene then fused 3′ to each coding sequence. The two genes were then joined to provide a single piece of DNA (*MTCEK1*) which encoded both the SAT and OAS enzymes as shown in Fig. 12.1a.

In parallel experiments in another laboratory, similar gene constructions were also prepared using the equivalent genes isolated from the bacterium *Salmonella typhimurium* (Bawden *et al.*, 1995). In some experiments, the *cysM* gene, which encodes a slightly different OAS enzyme in *E. coli* and *S. typhimurium*, was used in place of the *cysK* gene. A range of gene constructions have been made with these *Salmonella* genes (Bawden *et al.*, 1995), the major difference between these and the gene *MTCEK1* being in the use of different eukaryotic promoter sequences.

In order to establish that these genes are able to direct the synthesis of the enzymes needed to catalyse cysteine biosynthesis in eukaryotic cells, they were used to transfect cells in culture. The results obtained with the

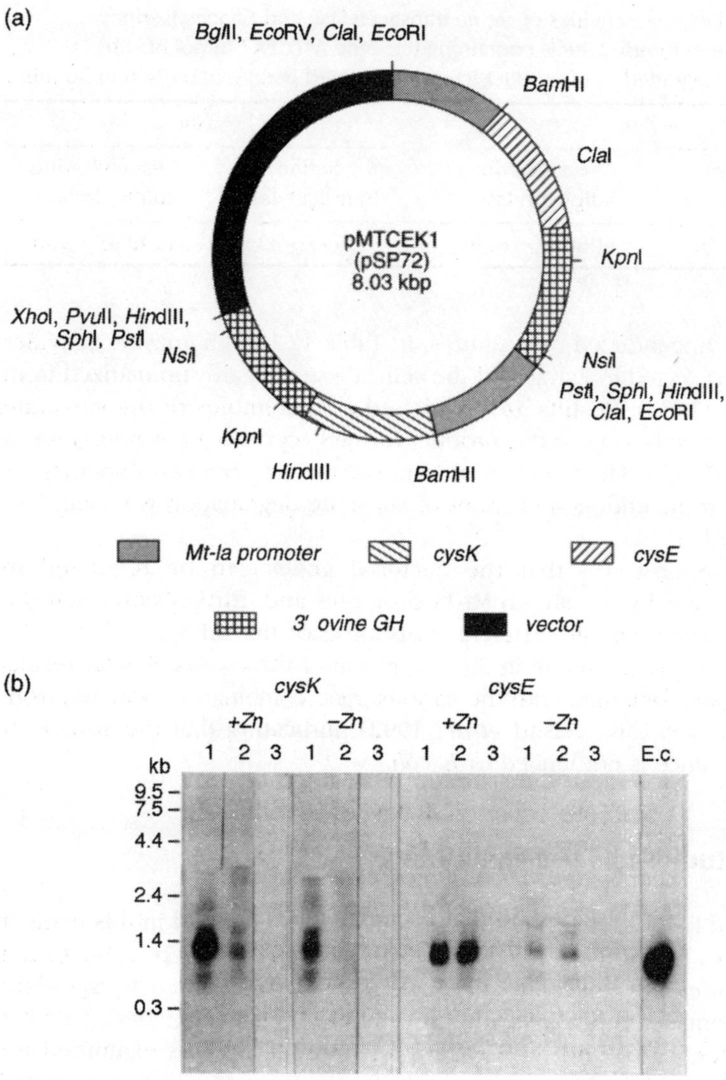

Fig. 12.1. (a) Structure of the plasmid MTCEK1. (b) Northern analysis of RNA from cells transfected with the gene *MTCEK1* (lane 1), with a mixture of two plasmids encoding *cysE* and *cysK* (Leish *et al.*, 1993) (lane 2) or with vector pSP72 (lane 3). All transfections included the plasmid pSV2neo. RNA (40 μg) from transfected L-cells was fractionated on a 1.5% agarose formamide gel, transferred to a nylon membrane and hybridized with a *cysK* or *cysE* ^{32}P-labelled probe. RNA from *E. coli* induced for cysteine biosynthesis (lane E.c.) was used as a control.

gene *MTCEK1* inserted stably into mouse L-cells is shown in Fig. 12.1b and Table 12.1 (Leish *et al.*, 1993). In Fig. 12.1b, the mRNAs specific for the *cysE* and *cysK* genes are readily detected in Northern blots of RNA preparations

Table 12.1. Enzyme activities of serine transacetylase and *O*-acetylserine sulphydrylase in mouse L-cells containing the gene *MTCEK1* (nmol of substrate (acetyl-CoA) degraded or of product (cysteine) formed per mg of protein in 30 min).

+ Zinc		– Zinc	
Serine transacetylase	*O*-acetylserine sulphydrylase	Serine transacetylase	*O*-acetylserine sulphydrylase
268 +/– 92	6960 +/– 829	86 +/– 33	1242 +/– 160

made from zinc-induced cell cultures. In Table 12.1, both enzyme activities are readily detected in extracts of the cells; these data are summarized from Leish *et al.* (1993). Results are expressed as nanomoles of the substrate degraded (acetyl-CoA) or the product formed (cysteine) per milligram of protein in 30 min. The values represent the means from two experiments corrected for the endogenous rates of substrate degradation with standard errors included.

This demonstrates that the bacterial genes can be regulated in eukaryotic cells by the sheep Mt-Ia promoter and, furthermore, that the bacterial codons enable efficient translation of the mRNA and that the bacterial proteins are stable in the cytoplasm of these cells. Similar results have also been obtained with the various gene combinations isolated from *S. typhimurium* (Sivaprasad *et al.*, 1992), indicating that the source of coding sequence is not limited to *E. coli.*

In vivo Studies in Transgenic Mice

Once it had been established that bacterial genes modified in this manner were able to function effectively in eukaryotic cells, it was possible to tèst their activities in transgenic mice. All genes were inserted by standard embryo pronuclear microinjection procedures (Hogan *et al.*, 1986; Palmiter and Brinster, 1986) and the progeny of founder animals examined for expression in various organs. The results for the gene *MTCEK1* are reported here but similar results have been obtained for some of the *S. typhimurium* gene constructions (Bawden *et al.*, 1995). For the gene *MTCEK1*, nine primary transgenic mice were produced and four were bred to stable lines. Figure 12.2 shows the Southern blot analysis of three of these lines, showing the expected *Bam*H1 fragments of 2.69 kb for the *cysE*-containing fragment and 2.85 kb for the *cysK*-containing fragment. The *MTCEK1* gene was stable in these animals from generation to generation, showing no signs of rearrangement despite the significant stretches of identical sequence in each copy of the inserted gene.

Transcription of the gene *MTCEK1* was measured in intestinal epithelium by Northern blot analysis of total RNA (Fig. 12.3a) and in intestinal

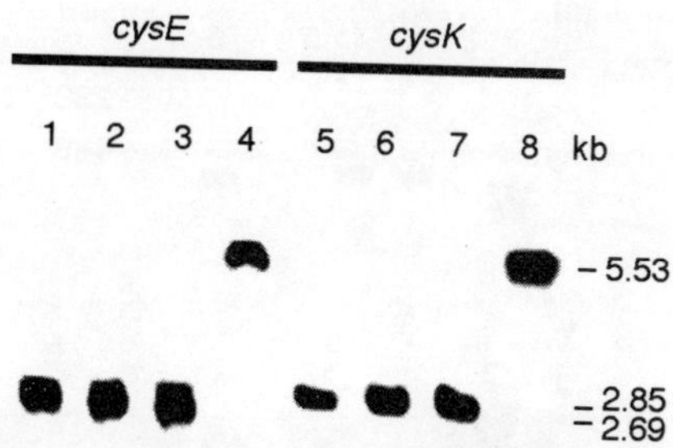

Fig. 12.2. Southern blot identification of transgenic mice carrying the insert from plasmid pMTCEK1. DNA was extracted from tail tissue by conventional techniques and digested with *Bam*H1. Membrane filters were probed with ^{32}P-labelled coding sequence for *cysE* (lanes 1–4) or *cysK* (lanes 5–8) genes. Lanes 1 and 5, mouse line EK8; lanes 2 and 6, mouse line EK28; lanes 3 and 7, mouse line EK46; lanes 4 and 8, pMTCEK1 insert DNA.

epithelium, skin, liver and kidney by reverse transcriptase (RT)-PCR analysis (Fig. 12.3b). The level of expression of the gene was different in the different lines, with EK28 showing the highest levels of mRNA. In Northern blots, the sizes of the mRNA transcripts were measured and shown to be 1.25 kb and 1.3 kb for the *cysE* and *cysK* genes. These sizes are slightly larger than the corresponding transcripts in *E. coli* (0.9 kb and 0.95 kb) and are the sizes predicted on the basis of the sizes of the bacterial coding sequences in combination with the 30 bp of Mt-Ia untranslated sequence, the ~250 bp derived from exon 5 of the sheep growth hormone gene and a poly(A) tract added post-transcriptionally. Thus, the gene appeared to be faithfully transcribed *in vivo* in a fashion analogous to that observed in cells in culture.

The demonstration that the mRNA transcripts from *MTCEK1* could also be detected by RT-PCR (Fig. 12.3b) provides a useful additional method for measurement because of the greatly increased sensitivity of this method compared with that of Northern blot analysis when expression is low. The primers CE1/CE3 and CK1/CK3 were designed for the analysis of the cDNAs synthesised from *cysE* and *cysK* mRNA transcripts. An additional primer, MT6, in combination with CE3 or CK3, provided a check for genomic DNA contamination of the mRNA preparations by amplifying a product from within the Mt-Ia promoter region of the gene *MTCEK1* (results not shown). The results in Fig. 12.3b confirm the Northern blot analyses and demonstrate the increased sensitivity of the RT-PCR method.

The translation of the *cysE* and *cysK* mRNAs in the mouse line EK8 is shown by the presence of the serine transacetylase and the *O*-acetylserine sulphydrylase in tissue extracts from zinc-induced mice (Table 12.2).

The highest enzyme levels were usually found in the intestinal epithelium, followed by kidney, liver and the skin. The enzymes were not detectable in non-transgenic animals, consistent with the proposal that they

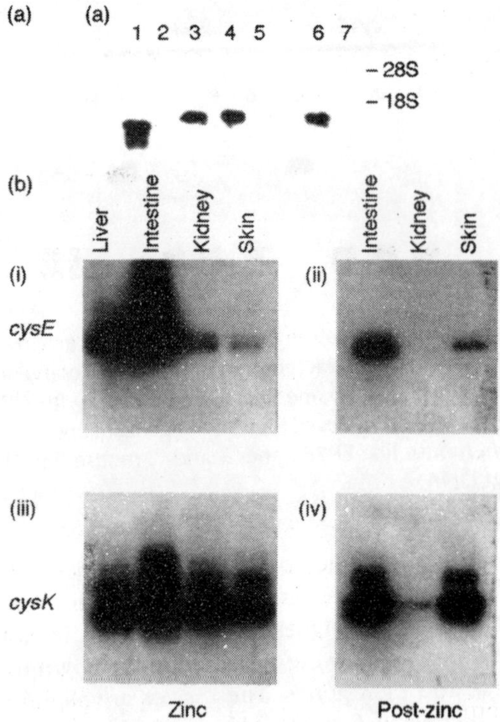

Fig. 12.3. (a) Northern blot analysis of the RNA isolated from intestinal tissue of zinc-fed transgenic mice containing the insert of plasmid pMTCEK1. Total cytoplasmic RNA was isolated from intestinal epithelium by the acid guanidinium thiocyanate–phenol–chloroform method (Chomznski and Sacchi, 1987), separated by electrophoresis on agarose gels containing 14% formamide and transferred to nylon membrane. Probes for *cysE* and *cysK* transcripts were antisense RNA sequences constructed from the coding sequences of the two bacterial genes. Lanes 1–5 were hybridized with a *cysK* probe and lanes 6 and 7 with a *cysE* probe. Lane 1, 4 µg *E. coli* C600 bacterial RNA; lane 2, non-transgenic mouse; lanes 3, 4, EK28 mouse, zinc treated for 3 months (lane 3) or 3 days (lane 4); lane 6, EK28 mouse, zinc treated; lane 7, EK28 mouse, minus zinc. (b) RT-PCR analysis of RNA isolated from EK28 mouse tissues zinc treated (i), (iii), or 3 days after removal of zinc treatment (ii), (iv). RNA was isolated as described above and 10 µg treated with ribonuclease-free deoxyribonuclease and incubated with Superscript reverse transcriptase (Gibco-BRL) to convert it to complementary DNA. The *cysE* and *cysK* transcripts were amplified by PCR in 50 µl volumes using the following primers; *cysE*; (CE1) 5′ATGTCGTGTGAAGAACTGGAA3′ and (CE3) 5′TTAGATCCCATCCCCATACAC3′. *CysK*; (CK1) 5′ATGAGTAAGATTTTTGAAGAT3′ and (CK3) 5′CTGTTGCAATTCTTTCTCAGT3′. The PCR programme consisted of 94°C for 3 min, 30 cycles of 94°C for 1 min, annealing temperature (65°C initially, decreasing by 0.6°C per cycle) for 2 min, 72°C for 1 min, 20 cycles of 94°C for 1 min, 45°C for 2 min, 72°C for 1 min, followed by a soak of 72°C for 7 min. Samples (10 µl) were separated by electrophoresis on 1.5% agarose gels, transferred to nylon membrane and hybridized with *cysE* or *cysK* ^{32}P-labelled probes.

Table 12.2. The specific activities of serine transacetylase and *O*-acetylserine sulphydrylase in tissue homogenates of zinc-induced transgenic mice containing the gene *MTCEK1*. Enzyme assays were performed as described in Leish *et al.* (1993) and results are expressed as in Table 12.1.

Organ	Serine transacetylase	*O*-acetylserine sulphydrylase
Intestine	2152	12,778
Kidney	105	128
Liver	9	3
Skin	3	312

are produced from transcripts of the *MTCEK1* gene and not due to bacterial contamination, an ever-present concern when working with tissues such as intestinal epithelium and skin. However, in order to provide a second line of evidence for the expression of the genes being derived from the transgenic mouse tissues and not from bacterial contamination, the *cysE* and *cysK* mRNAs were identified by *in situ* hybridization to sections of intestinal epithelium (Ward *et al.*, 1994). A strong signal for expression of both mRNA species was obtained in tissues obtained from transgenic animals (Fig. 12.4), with the localization of the signal clearly within a subset of the intestinal epithelial cells themselves. No signal was detected in tissues from non-transgenic animals.

Cysteine biosynthesis in the intestinal epithelium of these animals was demonstrated as follows. The biosynthesis of the enzymes was induced by feeding the mice zinc-supplemented water, after which they were sacrificed and intestinal tissue dissected, rinsed and incubated *in vitro* with a small amount of radioactive $Na_2{}^{35}S$. The incubation medium was then analysed for the presence of radioactive cysteic acid. The results clearly showed the presence of radioactive cysteine in the incubations of intestinal tissue obtained from the transgenic mice but not in those from non-transgenic mice (Fig. 12.5, from Ward *et al.*, 1994).

The Effect of *MTCEK1* Expression on Cysteine-deprived Mice

The level of expression of the SAT and OAS enzymes and the ability of these enzymes to catalyse cysteine from orally administered Na_2S in the transgenic mouse line EK28 suggested that these animals may be able to produce cysteine at levels that might be of dietary significance if they could be provided with a suitable source of sulphide. Accordingly, they were placed on an artificial diet (Leveille *et al.*, 1961; Huovinen and Gustafsson, 1967) that contained only trace amounts of cysteine and methionine but which was supplemented with Na_2S as a sulphur source. The animals were approximately 5 months of age at the start of the experiment. After 2 weeks

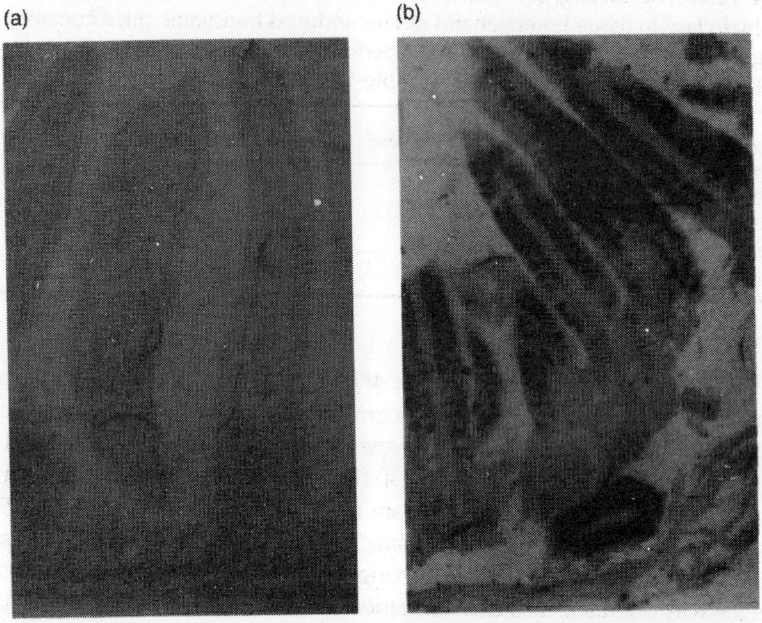

Fig. 12.4. *In situ* analysis of the expression of the gene *MTCEK1* in intestinal epithelia of transgenic mice. Small intestine of zinc-fed normal (a) and transgenic (b) mice was fixed in 4% paraformaldehyde–0.25% formaldehyde and sectioned at a thickness of 8 µm. An antisense *cysK* RNA probe was labelled with DIG (Boehringer Mannheim) and was hybridized and detected according to manufacturer's instructions using nitroblue tetrazolium (NBT).

on the sulphur-deficient diet, the most noticeable phenotypic response was a marked loss of hair in the control animals, while the transgenics showed a normal hair coat (Ward *et al.*, 1994). The hair loss was most noticeable on the regions of the animals which were in the anagen phase of the hair cycle (Fig. 12.6). The experiments were repeated in more detail with a second line of transgenic mice, EK8, with detailed body weight measurements made during the time the animals were on a cysteine-restricted diet. The results of the hair growth measurements were generally consistent with the results obtained with the line EK28 and were also consistent with the observed weight changes (Table 12.3).

Thus, the control animals varied in appearance from overt hair loss to a severely dishevelled appearance and all lost weight during the period of cysteine deprivation. The transgenic animals in both groups I and II generally showed minimal disturbance to hair growth and increased in body weight during the same period, although some hair dishevelment was observed in a few of the transgenic males of group I. When this was observed, it was also correlated with a weight loss on the restricted diet.

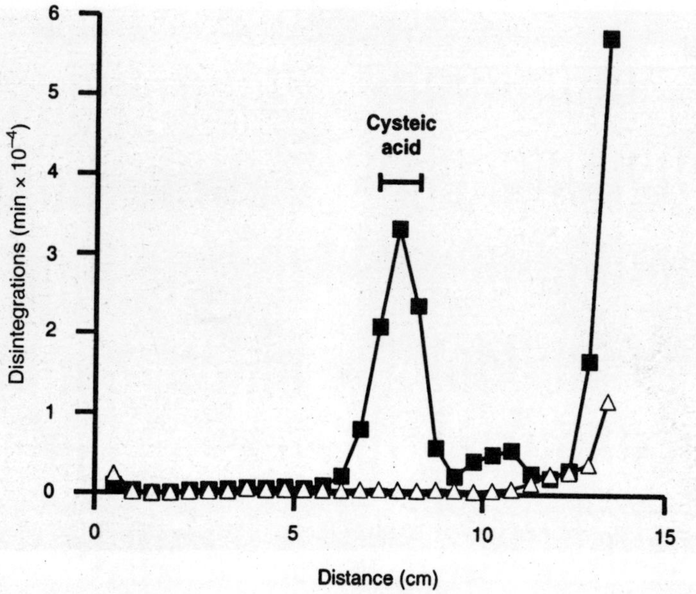

Fig. 12.5. The biosynthesis of cysteine from H_2S *in vitro* in intestinal tissue isolated from transgenic mice containing the gene *MTCEK1*. Small intestine (200 mg) from transgenic and non-transgenic mice was dissected, rinsed with 50 mM Tris buffer, pH 7.4 and incubated in 1 ml of Krebs–Ringer phosphate buffer, pH 7.4 at 37°C for 5 min. $Na_2{}^{35}S$ (5 μCi) was added and incubation continued for a further 15 min. The tissue was then removed and the incubation medium treated with 100 μl of dithiothreitol (100 mM) and 3 ml of performic acid at 4°C overnight. The solution was extracted with chloroform/mehanol (1:1) and the aqueous phase lyophilized and dissolved in 0.5 ml water. Amino acids were separated by paper electrophoresis at 10 V cm^{-1} in pyridine:acetic acid:water (1:10:90) (Huovinen and Gustafsson, 1967). The paper was cut into 1 cm strips and radioactivity determined.

Table 12.3. Weight changes associated with dietary deprivation of sulphur amino acids in transgenic mice containing the gene *MTCEK1*. Dietary details are as described in Fig. 12.6.

	Body weight change (g) (no. of animals)	
Experiment	Transgenic	Non-transgenic
I	+ 1.08 (10)	− 8.7 (14)
II	+ 1.9 (15)	− 7.6 (16)

Insertion of the Cysteine Biosynthetic Pathway into Sheep

The results obtained with the various cysteine gene constructions when inserted into transgenic mice supported the concept that functional

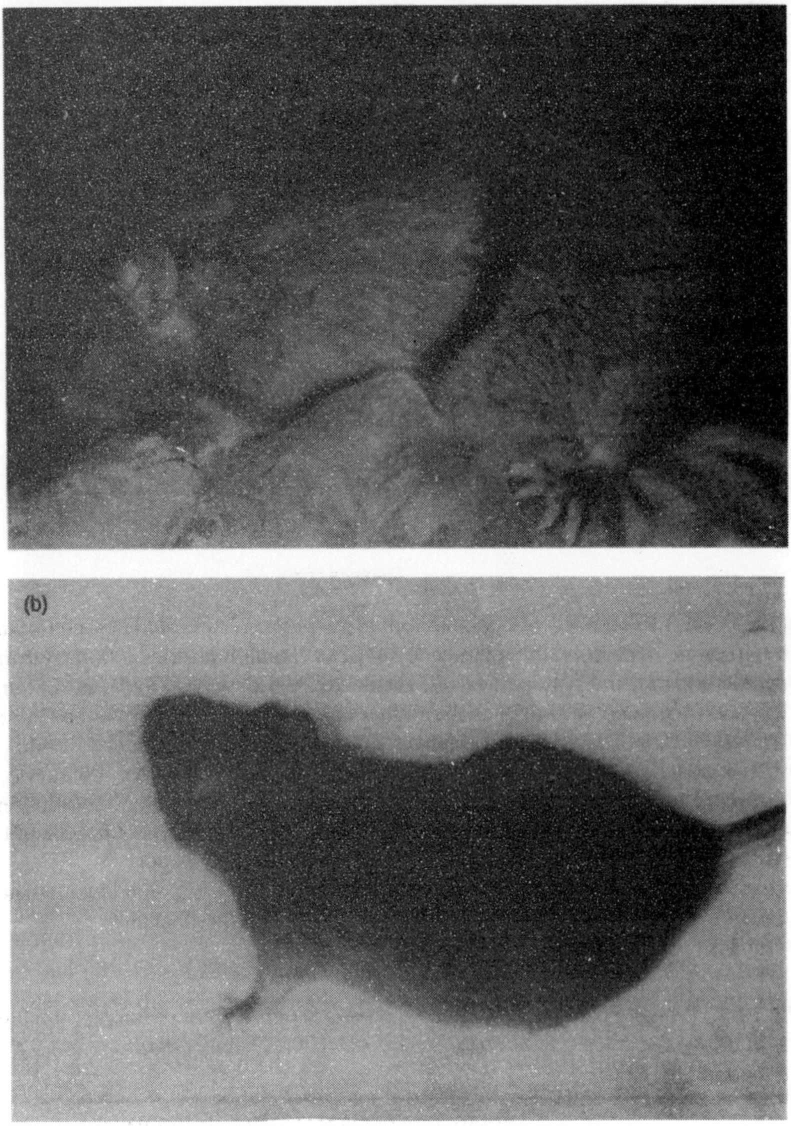

Fig. 12.6. The protective effect of the cysteine biosynthetic pathway on dietary-induced hair loss in mice. Age-matched mice, (a) non-transgenic control mice and (b) transgenic line EK28 were placed on a synthetic diet prepared essentially as described (Huovinen and Gustafsson, 1967). Initially, the mice were accustomed to the synthetic diet supplemented with the sulphur amino acids cysteine and methionine. The animals were then placed on a diet in which the combined sulphur amino acids were reduced to 0.1% (w/w) of the total diet (Leveille *et al.*, 1961), but supplemented with 0.12% (w/w) Na_2S. Throughout the experiment, the drinking water of the mice was supplemented with 25 mM $ZnSO_4$. Photographs were taken after 7 days. Hair loss was observed in the non-transgenic mice but not in the transgenic animals.

metabolic pathways could be transferred from bacteria to animals. The work was then extended to the main aim of the research, namely the transfer of the cysteine biosynthetic pathway to sheep. This has proved to be more difficult than for mice.

In Table 12.4 is provided a list of transgenic sheep that have been produced containing various constructions encoding the SAT and OAS enzymes of *E. coli* or *S. typhimurium.*

The copy number in these various primary animals varies from 2 to >200 and there appears to be no unusual rearrangement of the genes compared with the insertions seen in transgenic mice. A Southern blot of some of the progeny bred from two transgenic sheep containing the gene *MTCEK1* is shown in Fig. 12.7. The results shown are those obtained with a *cysE* probe and are identical to those obtained with a *cysK* probe except that the *Bam*H1 fragment is then 2.8 kb in size. In the progeny of sheep no. 1830 (Fig. 12.7a), the copy number is estimated at about 20 and the predominant 2.65 kb band observed in *Bam*H1 digests indicates that most of the genes in the tandem array are arranged head-to-tail. However, there is also a prominent band at approx. 4 kb indicating that some of the array is arranged tail-to-tail. There is no evidence of rearrangement of the gene array during breeding, which supports the extensive body of evidence in the transgenic mouse lines showing that the gene is stable during prolonged breeding. The other animal in Fig. 12.7 shows evidence of two insertions in the F0 animal. The primary animal, no. 1290, contains a dominant tandem array of approximately 200 gene copies. *Bam*H1 digests of progeny from this animal (Fig. 12.7b) show that some animals retain this copy number but that an equal number of animals have a tandem array of about ten copies which is hard to detect in the presence of the larger insertion. While one explanation for this change in copy number could be that the large gene array in the F0 animal is unstable during breeding, this is unlikely since the ratio of transgenic to non-transgenic progeny is about 2:1. Final proof will be the demonstration that

Table 12.4. A summary of transgenic sheep containing various DNA constructs encoding the cysteine biosynthetic pathway. Much of the information in this table has been produced in the laboratory of Professor G.E. Rogers and these data have been summarized from Bawden *et al.* (1995).

Laboratory	Gene	TG Sheep (live)	Live TGs (% lambs born)	Animals expressing	Genes expressed
Prospect	*MTCEK1*	9	4	No. 1830	*cysE*, *cysK* (low)
Adelaide	RSVLTR–*cysEM*	2	3	No. 208	*cysE*
Adelaide	RSVLTR–cysME	8	17	No. 196	*cysE* (low)
				No. 199	*cysE* (low)
Adelaide	mPgk–*cysKE*	6	1.0	No. 34	*cysE* (low)
Adelaide	mPgk–cysME	3	2.4	None	

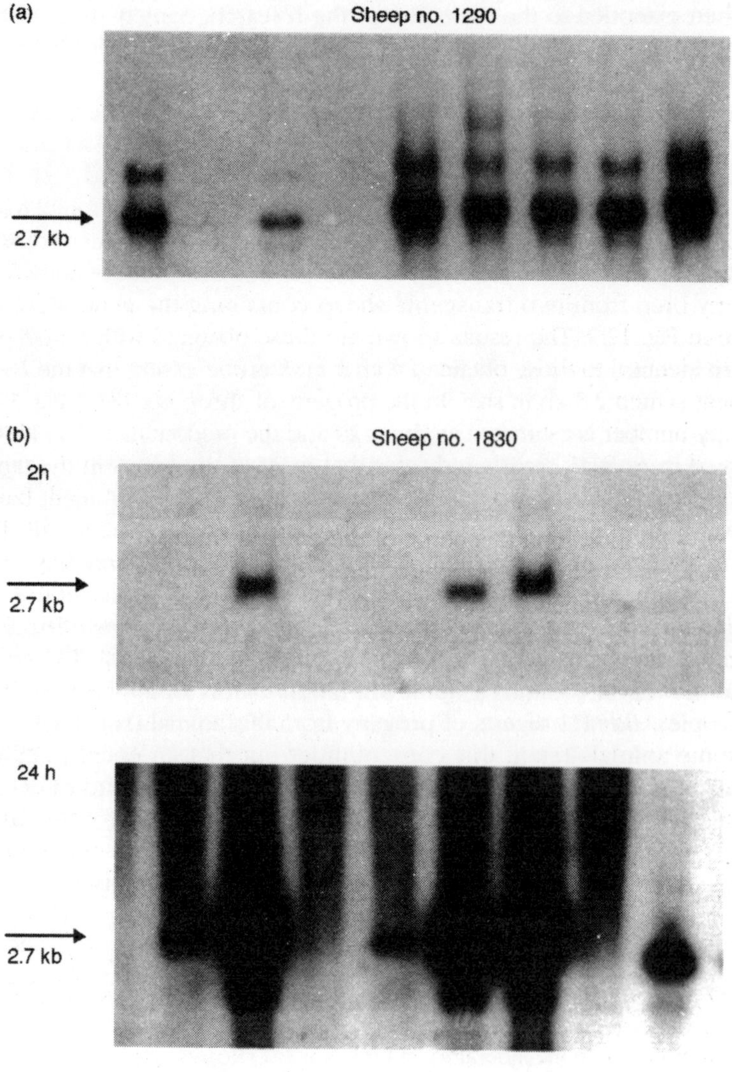

Fig. 12.7. Southern blot analysis of the progeny of (a) sheep no. 1290 and (b) sheep no. 1830 containing the gene *MTCEK1*. DNA was extracted from tail tissue, digested with *Bam*H1, separated on agarose gels and transferred to nylon membrane. Hybridization was with a ^{32}P-labelled *cysE* probe. Identical results were obtained with a *cysK* probe except that the *Bam*H1 band was 2.8 kb instead of 2.65 kb because of the slightly longer *cysK* coding sequence. Longer exposure of the autoradiograph identifies some progeny of no. 1830 with a lower copy number than the primary animal.

low-copy F1 and high-copy F1 generation lambs breed true to copy number in the F2 generation, and these experiments are in progress.

The expression of the genes in the transgenic sheep produced so far is summarized in Table 12.4. This expression is disappointingly low in comparison with that obtained in transgenic mice. Only five sheep have shown any expression of the genes and in each animal, at least one of the genes is expressed at a very low level. One of the sheep containing the gene *MTCEK1* has been shown to express both the *cysE* and *cysK* genes using the highly sensitive technique of RT-PCR to detect specific mRNA transcripts. As shown earlier, RT-PCR can detect the *cysE* and *cysK* gene transcripts in transgenic mice with great sensitivity and can readily detect expression in the skin of these animals. Accordingly, skin biopsies from the transgenic sheep containing the gene *MTCEK1* were analysed for the presence of the two transcripts. Animal no. 1830 contained a readily detectable *cysE* transcript in RNA from skin but the *cysK* transcript was barely detectable (Fig. 12.8). This is probably a real transcript, since no hybridization was detected in the appropriate control samples and genomic contamination is unlikely because the primer pair MT4/CE3, which is designed to amplify a piece of DNA starting in the sheep metallothionein promoter and finishing in the *cysE* coding sequence, failed to amplify any product. Enzyme assays of skin extracts from this animal failed to detect any SAT or OAS activities. In animals containing the *Salmonella* genes, four sheep expressed the *cysE* gene at low levels but not the *cysK* gene, although fibroblast cell cultures from two of the animals were reported to express low levels of *cysK* in addition to the *cysE* gene (Table 12.4).

Discussion of Results in Sheep and Mice

The results described above demonstrate that it is possible to introduce new biochemical pathways into animals by the judicious use of functional bacterial genes. Since genes required for the synthesis of most metabolites are to be found somewhere in the plethora of bacterial species that have been identified, the ability to use these pieces of DNA in transgenic animals provides the opportunity to remove many substrate limitations to specific production characteristics of domestic animals. However, it is equally clear that a new biochemical pathway that operates well in one species may not do so in a second species. In the example given above, the cysteine biosynthetic pathway appears to operate smoothly in transgenic mice but appears to interfere with the embryonic development of sheep. We can only speculate on the possible causes of these observations. One obvious explanation that cannot easily be ignored is that in the sheep the new pathway may be disturbing the concentration of an essential substrate or coenzyme crucial to embryonic development.

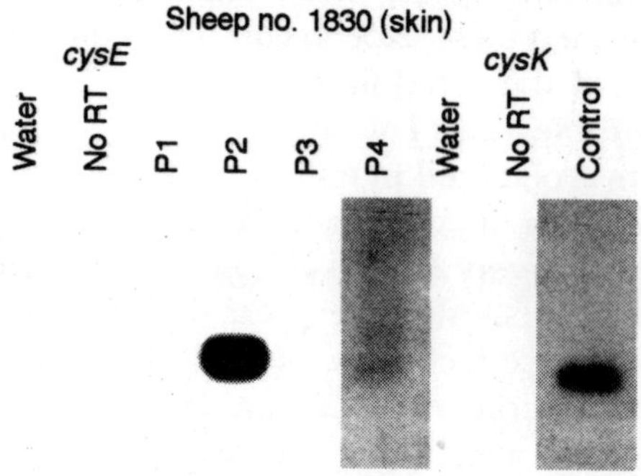

Fig. 12.8. RT-PCR analysis of RNA extracted from skin biopsies of transgenic sheep no. 1830. RNA extraction and RT-PCR analysis were as described for Fig. 3b. Lane 1, water control with primer pair CE1/CE3; lane 2, no reverse transcriptase control with primer pair CE1/CE3; lane 3, cDNA template with primer pair MT4/CE3; lane 4, cDNA template with primer pair CE1/CE3; lane 5, cDNA template with primer pair MT4/CK3; lane 6, cDNA template with primer pair CK1/CK3; lane 7, water control with primer pair CK1/CK3; lane 8, no reverse transcriptase control with primer pair CK1/CK3; lane 9, genomic DNA from sheep no. 1830 with primer pair CK1/CK3.

The substrate acetyl-CoA is one possible candidate. Sheep and mice differ in the way they utilize carbohydrates for the provision of cellular energy. The sheep is a ruminant and hence much of its metabolizable energy is supplied in the form of volatile acids such as acetate, propionate and butyrate, while mice utilize carbohydrates such as glucose (Armstrong *et al.*, 1957, 1960). This results from the biota of the rumen that converts essentially all fermentable carbohydrate into the volatile acids, with acetate accounting for more than 90% in the peripheral circulation (Van Soest, 1982). The enzymes for intermediary metabolism in the sheep are adapted to the utilization of these volatile acids instead of glucose as a source of energy (Van Soest, 1982) with glucose being used only in those reactions which are not replaceable by volatile acids.

Acetyl-CoA holds a crucial position in this volatile fatty acid-oriented metabolism, since it provides the point of entry for acetate into intermediary metabolism where it can be oxidized for energy in the tricarboxylic acid (TCA) cycle. Its levels are also crucial to early embryonic development both in ruminants and non-ruminants, providing 80–95% of the embryo's ATP via oxidative phosphorylation (Thompson *et al.*, 1996). It is apparent that in mice, the insertion of the gene *MTCEK1* does not result in a deleterious alteration to the supply of this vital substrate during development, since transgenic mice are produced in expected numbers. The explanation for this may be as simple as a lack of expression of the cysteine-encoding transgene in the early mouse embryo. However, earlier work has shown that a MT-Ia–growth hormone transgene is expressed in the developing mouse (Shanahan *et al.*, 1989) and it is therefore probable that the *MTCEK1* gene is also expressed at this time. In sheep, it has proved very difficult to produce transgenic animals expressing *MTCEK1* or any other of the cysteine-encoding genes that have been constructed. The development of ruminant embryos differs significantly from that of the mouse at the time of blastulation. At this stage, the energy required for the development of the mouse is supplied almost entirely by glucose catabolism through glycolysis to lactate while that of the ruminant is mainly derived from externally absorbed pyruvate which is converted to acetyl-CoA and oxidized in the TCA cycle (Reiger and Guay, 1988; Waugh and Wales, 1993; Thompson *et al.*, 1996). Clearly, any decrease in the level of acetyl-CoA would have the potential to interfere significantly with the energy supply of the sheep embryo but not with that of the mouse.

The levels of cytosolic acetyl-CoA are unlikely to fall in adult ruminants because it can be readily synthesized from acetate in the presence of the enzyme acetyl-CoA synthase. This may not be the case, however, for the ruminant embryo. Like most embryos, the ruminant at this stage utilizes exogenous acetate very poorly and the substrate has been calculated to make no significant contribution to the embryo's energy supply (Waugh and Wales, 1993). This has been ascribed to a very low level of acetyl-CoA synthase in embryos. It follows that any extensive depletion of cytosolic acetyl-CoA in ruminant embryos would cause a flow of acetyl-CoA from the mitochondrion, reducing the potential energy supply. The transfer of acetyl-CoA from mitochondria to cytosol is possible in ruminant embryos because ATP–citrate lyase, essentially non-functional in adult ruminant tissues, is active during development. Cytosolic depletion of acetyl-CoA would be a consequence of the unregulated action of the enzyme SAT in the embryo cytoplasm. This enzyme is allosterically regulated in bacteria by the concentration of cysteine, with complete inhibition of activity at approximately 100 μM (Kredich and Tomkins, 1966). Since the introduced cysteine biosynthetic pathway would be unable to synthesize cysteine in the embryo because there is no available source of sulphide, the conversion of serine and acetyl-CoA to *O*-acetyl-serine could proceed normally unless the SAT was inhibited by the normal intracellular levels of cysteine. These are not known in the developing sheep embryo, but circulating cysteine levels in the adult sheep are very low, with measurements varying between 12 μM and 37 μM being reported (Lee *et al.*, 1993). In addition, the glutathione levels in sheep are also very low in comparison with monogastric animals (Lee *et al.*, 1993). These data suggest that overall cysteine supply in the sheep is low and that the intracellular levels in the embryo are unlikely to reach those required for the inhibition of the enzyme SAT. This being so, the possibility exists that in any embryo expressing the SAT at high levels, acetyl-CoA concentrations might be reduced below the level needed for embryo survival. This would result in most transgenic sheep being animals with no or very low levels of expression of the SAT and hence by default the OAS, since both are linked in a common gene construct.

One way of determining this is to produce transgenic animals containing either the *cysE* gene or *cysK* genes and to mate such animals to produce progeny with an intact pathway. A second approach is to modify the gene constructs so that expression is confined to the adult animal and to those tissues where a source of sulphide might be expected. These experiments would benefit greatly from an improved method of producing transgenic sheep since they would require many animals for appropriate testing of the various genes. The recent advances in the production of viable embryos from the fusion of enucleated embryos and cultured cells (Wilmut *et al.*, 1997) might be advantageous in pursuit of this goal.

Acknowledgements

The expert technical help of Mrs Gina Hardwicke, Mrs Rose White, Mrs Astrid Dafter, Ms Tania Radziewic, Mr Jim Marshall and Mr Peter Mitchell is gratefully acknowledged. Mr John O'Grady and Mrs Jenny Gordon provided the expert care needed to breed the various mouse lines. We are also very grateful for the services of the entire Prospect Farm staff for the expert care and attention given to the transgenic sheep.

The numerous discussions and helpful advice provided by Professor George Rogers and his group at the University of Adelaide has been essential for the progress of the work and is gratefully acknowledged.

This research has been partially funded by the International Wool Secretariat.

References

Armstrong, D.G., Blaxter, K.L. and Graham, N.M. (1957) Utilization of the end products of ruminant digestion. In: *Proceedings of the British Society of Animal Production*, pp. 3–15.

Armstrong, D.G., Blaxter, K.L. and Graham, N.M. (1960) Fat synthesis from glucose by sheep. In: *Proceedings of the Nutrition Society* 19, 31.

Bawden, C.S., Sivaprasad, A.V., Verma, S.K., Walker, S.K. and Rogers, G.E. (1995) Expression of bacterial cysteine biosynthesis genes in transgenic mice and sheep: toward a new *in vivo* amino acid biosynthesis pathway and improved wool growth. *Transgenic Research* 4, 87–104.

Black, J.L. and Reis, P.J. (1979) Speculation on the control of nutrient partition between wool growth and other body functions. In: Black, J.L. and Reis, P.J. (eds) *Physiological and Environmental Limitations to Wool Growth*. University of New England, Armidale, Australia, pp. 269–293.

Byrne, C.R., Monroe, R.S., Ward, K.A. and Kredich, N.M. (1988) DNA sequences of the *csyK* regions of *S. typhimurium* and *E. coli* and linkage of the *cysK* regions to *ptsH*. *Journal of Bacteriology* 170, 3150–3157.

Chomznski, P. and Sacchi, N. (1987) *Analytical Biochemistry* 162, 156–159.

Denk, D. and Bock, A. (1987) L-cysteine biosynthesis in *Escherichia coli*: nucleotide sequence and expression of the serine acetyltransferase (*cysE*) gene from the wild-type and a cysteine-excreting mutant. *Journal of General Microbiology* 133, 515–525.

Hogan, B., Costantini, F. and Lacy, E. (1986) In: *Manipulating the Mouse Embryo. A Laboratory Manual*. Cold Spring Harbor Laboratory Press, Cold Spring Harbor, New York.

Huovinen, J.A. and Gustafsson, B.E. (1967) Inorganic sulphate, sulphite and sulphide as sulphur donors in the biosynthesis of sulphur amino acids in germ-free and conventional rats. *Biochimica Biophysica Acta* 136, 441–447.

Kredich, N.M. and Tomkins, G.M. (1966) The enzymic synthesis of L-cysteine in *Escherichia coli* and *Salmonella typhimurium*. *Journal of Biological Chemistry* 241, 4955–4965.

Lee, J., Harris, P.M., Sinclair, B.R. and Treloar, B.P. (1993) Whole body metabolism of cysteine and glutathione and their utilization in the skin of Romney sheep: consequences for wool growth. *Journal of Agricultural Science, Cambridge* 121, 111–124.

Leish, Z., Byrne, C.R., Hunt, C.L. and Ward, K.A. (1993) Introduction and expression of the bacterial genes *cysE* and *cysK* in eukaryotic cells. *Applied Environmental Microbiology* 59, 892–898.

Leveille, G.A., Sauberlich, H.E. and Shockley, J.W. (1961) *Journal of Nutrition* 75, 455–458.

Palmiter, R.D. and Brinster, R.L. (1986) Germline transformation of mice. *Annual Review of Genetics* 20, 465–500.

Reiger, D. and Guay, P. (1988) Measurement of the metabolism of energy substrates in individual bovine blastocysts. *Journal of Reproduction and Fertility* 83, 585–591.

Reis, P.J. (1979) Effect of amino acids on the growth and properties of wool. In: Black, J.L. and Reis, P.J. (eds) *Physiological and Environmental Limitations to Wool Growth*. University of New England, Armidale, Australia, pp. 223–242.

Shanahan, C.M., Rigby, N.W., Murray, J.D., Marshall, J.T., Townrow, C.A., Nancarrow, C.D. and Ward, K.A. (1989) Regulation of expression of a sheep metallothioenin 1a-sheep growth hormone fusion gene in transgenic mice. *Molecular and Cellular Biology* 9, 5473–5479.

Sivaprasad, A.V., D'Andrea, R.J., Bawden, C.S., Kuczek, E.S. and Rogers, G.E. (1989) Towards a new sheep genotype with increased wool growth by transgenesis with microbial genes for cysteine synthesis. *Journal of Cellular Biochemistry* 13B (Suppl.), 183.

Sivaprasad, A.V., Kuczek, E.S., Bawden, C.S. and Rogers, G.E. (1992) Coexpression of the *cysE* and cysM genes of *Salmonella typhimurium* in mammalian cells: a step towards establishing cysteine biosynthesis in sheep by transgenesis. *Transgenic Research* 1, 79–92.

Thompson, J.G., Partridge, R.J., Houghton, F.D., Cox, C.I. and Leese, H.J. (1996) Oxygen uptake and carbohydrate metabolism by *in vitro* derived bovine embryos. *Journal of Reproduction and Fertility* 106, 299–306.

Umbarger, H.E. (1997) Amino acid biosynthesis and its regulation. *Annual Review of Biochemistry* 47, 533–606.

Van Soest, P.J. (1982) Intermediary metabolism. In: *Nutritional Ecology of the Ruminant.* Durham and Downey, Portland, Oregon, pp. 249–259.

Ward, K.A., Leish, Z., Bonsing, J., Nishimura, N., Cam, G.R., Brownlee, A.G. and Nancarrow, C.D. (1994) Preventing hairloss in mice. *Nature* 371, 563–564.

Waugh, E.E. and Wales, R.G. (1993) Oxidative utilization of glucose, acetate and lactate by early implantation sheep, mouse and cattle embryos. *Reproduction, Fertility and Development* 5, 123–133.

Wilmut, I., Schnieke, A.E., McWhir, J., Kind, A.J. and Campbell, K.H.S. (1997) Viable offspring derived from fetal and adult mammalian cells. *Nature* 385, 810–813.

13

Production of Transgenic Cattle Expressing a Recombinant Protein in Milk

W.H. Eyestone

PPL Therapeutics, Inc., Blacksburg, Virginia, USA

Transgenic technology has been applied with some success in all of the major agricultural species. However, its application in cattle has been relatively slow due to the greater technical, logistical and financial challenges encountered in working with this species. Sustained efforts to produce transgenic cattle have concentrated on producing valuable biomedical proteins in milk rather than modifying agricultural production traits. The overall aim of the work described here was to enhance the nutritional value of cow's milk for human consumption by expressing human α-lactalbumin (the major whey protein in human milk) in the milk of transgenic cattle. To this end, we sought an efficient system in which to generate healthy, transgenic calves from microinjected zygotes. Embryo production and transfer were performed continuously over a period of 5 months. Zygotes for microinjection were generated by *in vitro* maturation and fertilization of abattoir-derived oocytes. To evaluate the effect of fetal calf serum (FCS) on embryo development, pregnancy rate, calving rate and birthweight, some embryos were cultured with or without FCS from day 5 to day 7 or 8 post-insemination (p.i.). Of 11,507 injected zygotes, 1011 (9%) developed to the compact morula or blastocyst stage. After non-surgical embryo transfer of 478 embryos to recipient heifers (one embryo/recipient), 155 recipients (32%) were pregnant at day 30 of gestation and 90 (19%) produced calves (including one set of identical twins). Most (97%; 88/90) calves were delivered by scheduled Caesarean section at about day 274 of gestation. Mean (range) calf birthweight was 41.83 (29.09–60.45) kg. The male:female sex ratio was 1.44:1 (59% male:41% female) and did not differ from the expected ratio of 1:1 (P>0.05). Perinatal calf survival was high (90/91; 99%). Under the conditions of these experiments, the presence of FCS in embryo culture medium did not affect (P>0.05) embryo development, pregnancy rate, calving rate, birthweight or perinatal survival. Nine (10%) of the calves were transgenic. All transgenic calves were born healthy. A female calf, induced to lactate, expressed human α-lactalbumin in her milk at a level of 2.4 mg ml^{-1}.

© CAB INTERNATIONAL 1999. *Transgenic Animals in Agriculture*
(eds J.D. Murray, G.B. Anderson, A.M. Oberbauer and M.M. McGloughlin)

Introduction

Transgenic technology has been applied in domestic livestock to enhance food and fibre production, promote disease resistance and to produce recombinant proteins from the mammary gland. Reports abound on the integration, expression and transmission of various transgenes in pigs, sheep and goats (reviewed in Wall, 1996). In cattle, however, transgenic technology has developed at a much slower pace. While sporadic efforts to produce transgenic cattle have appeared in the literature since the mid-1980s, transgene expression has been reported in only one case (Bowen *et al.*, 1994) and no account of germline transmission has been published.

Several factors have conspired to limit the application of transgenic technology in cattle. Economically important traits (e.g. lactation, growth) are polygenic, and current transgenic technology is limited to the modification of monogenic traits. Single genes with a major influence over production traits have not yet been identified. Technically speaking, cattle are a notoriously difficult species in which to make transgenics, for the following reasons.

1. Collection of zygotes from single ovulating or even superovulated donors is a tedious, labour-intensive, expensive and frequently unrewarding exercise (reviewed by Eyestone, 1994; Wall, 1996), though recently these '*in vivo*' approaches have largely been supplanted by the more tractable method of producing embryos *in vitro* (Krimpenfort *et al.*, 1991).
2. Cattle are a monotocous species, a condition that severely limits the number of embryos that can be transferred to a single recipient; moreover, pregnancy rates obtained after transfer of only one or two DNA-injected embryos per recipient is about half of that for non-injected embryos (reviewed by Eyestone, 1994).
3. The transgene integration frequency reported for cattle (0–6%; reviewed in Eyestone, 1994; Wall, 1996) is lower than that reported for other agricultural species (5–30%; reviewed in Brem, 1992; Wall, 1996). Thus, a rather large and expensive herd of recipients is required to support any serious effort aimed at making transgenic cattle.
4. The relatively long generation interval in cattle (*c.* 2 years) adds to production expenses and delays development of new transgenic strains.

Despite these impediments, transgenic cattle are being developed for applications where single-gene modifications are sufficient, namely for the production of medically important recombinant proteins in the milk. Indeed, one of the first applications suggested for transgenic technology was that of 'genetic farming' or the production of recombinant proteins of biomedical value in the tissues of transgenic animals (Palmiter *et al.*, 1982), particularly the mammary gland (Lathe *et al.*, 1985). The mammary gland is, in many ways, ideally suited for producing recombinant proteins: its capacity for protein production is extremely high; it can perform many of the complex, post-translational modifications required for biological activity of many proteins of biomedical interest; it secretes proteins in an exocrine fashion, thus minimizing the risk of a bioactive recombinant protein upsetting the host's own physiology; and finally, milk is a well-characterized, familiar milieu from which to isolate the protein of interest.

Several groups have reported that calves resulting from *in vitro*-produced embryos tended to be larger than ordinary calves (Sinclair *et al.*, 1994; Behboodi *et al.*, 1995; Kruip and Den Daas, 1997). Similar observations have been made on lambs born from embryos cultured *in vitro* from early cleavage to the blastocyst stage (Thompson *et al.*, 1995; Walker *et al.*, 1996). Thompson *et al.* (1995) presented evidence suggesting that the presence of fetal calf serum (FCS) in embryo culture medium was related to the production of large offspring from cultured embryos. In the present report, we compared birthweights of calves born after culture in medium supplemented or not with FCS from days 5 to 7 or 8 post-insemination (p.i.).

The aim of this paper is to describe some of our experience in the production of transgenic cattle bearing a transgene for human α-lactalbumin, the major whey protein of human milk. Although the objective of this effort was to enhance the nutritional value of cows' milk for human consumption, a so-called 'nutraceutical' application, the methods employed and the results obtained here should apply to the production of transgenic cattle for more classical agricultural applications as well. A final comment will be made regarding the future of transgenic technology in cattle in view of recent breakthroughs in sheep regarding somatic cell nuclear transfer.

Methods

Zygote production

Zygotes were obtained by *in vitro* fertilization of *in vitro*-matured oocytes recovered from the ovaries of slaughtered cows and heifers. The animals from which the ovaries were obtained were primarily of the Holstein breed. Due to the lack of a suitable local abattoir, we imported oocytes from laboratories in Madison, Wisconsin (Bomed, Inc.) and Elizabethtown, Pennsylvania (Emtran, Inc.). Briefly, ovaries were recovéred within 30 min of slaughter and transported to the respective laboratories in physiological saline at 25–35°C. Cumulus–oocyte complexes (COC) were aspirated from 2–8 mm diameter ovarian follicles between 4 and 10 h after slaughter and placed in maturation medium (TCM-199, buffered with 25 mm Hepes, supplemented with 10% FCS, 5 μg ml^{-1} bovine LH, 5 μg ml^{-1} FSH and 50 μg ml^{-1} gentamicin and equilibrated with 5% CO_2 in air; Sirard *et al.*, 1988). Medium containing COC was placed in tightly capped, polystyrene tubes and shipped in a portable, battery-powered incubator (Minitub USA, Middleton, Wisconsin) by overnight express (Long *et al.*, 1994) to our laboratory in Blacksburg, Virginia. Upon arrival, COC were rinsed in Hepes-buffered TALP medium (TL-Hepes; Parrish *et al.*, 1986) and placed in fertilization medium (TALP, supplemented with 2–20 μg ml^{-1} heparin; Parrish *et al.*, 1986). Oocytes were fertilized with frozen–thawed semen from a Holstein bull of proven fertility. After thawing, the proportion of motile sperm was enhanced and standardized by the 'swim-up' technique (Parrish *et al.*, 1986) and added to fertilization medium to a final concentration of 1 million sperm ml^{-1}. Sperm and COC were co-incubated at 39°C in a humidified atmosphere of 5% CO_2 in air, starting at 22 h post-onset of maturation. Fertilization frequency was assessed histologically on a subset of each batch of oocytes as described by Parrish *et al.* (1986).

Pronuclear DNA injection

Zygotes were removed from fertilization medium 16–18 h p.i., placed in 2 ml TL-Hepes in a 15 ml capped, conical tube and agitated for 2–5 min on a laboratory 'vortexing' mixer apparatus. To visualize pronuclei, zygotes were centrifuged at 13,000 *g* for 4–8 min to displace opaque cytoplasmic lipid inclusions (Wall and Hawk, 1988). Microinjection was performed on an Olympus IMT-2 inverted microscope using Nomarski differential interference contrast optics. Zygotes were stabilized by gentle suction on a holding pipette and the pronuclei brought into focus. Several picolitres of a solution containing the transgene construct were then injected into one of the pronuclei through a finely pulled glass needle. Microinjection was considered successful if the pronucleus swelled upon delivery of the DNA solution and remained swollen after removal of the injection needle. Detailed protocols for microtool manufacture and pronuclear injection may be found in Hogan *et al.* (1986) and Pinkert (1994).

Embryo culture

After microinjection, embryos were cultured in modified synthetic oviduct fluid (SOFM, also supplemented with amino acids; Gardner *et al.*, 1994) for 6 or 7 days. In some cases (see below), FCS (10% v/v) was added to the media on day 5 p.i. About 30 non-injected control embryos were cultured in parallel with each set of microinjected embryos to control for the effect of injection on development.

Effect of culture medium and FCS on embryo development calf birthweight

The effect of FCS supplementation on embryo development, pregnancy rate, calving rates and calf birthweights was examined in a subset of microinjected embryos included in this study. FCS has been reported to have biphasic effects on bovine embryo development *in vitro*. Early cleavage stage development was suppressed by the presence of 10% FCS (Bavister *et al.*, 1992; our unpublished observations) while it enhanced blastocyst formation by day 7 p.i. (van Langendonckt *et al.*, 1996; our unpublished observations). In the present experiment, FCS was added to the culture medium on day 5 to avoid early, embryosuppressive effects while exploiting its embryotrophic effect on the blastocyst stage. Embryos were assigned randomly to SOFM medium supplemented or not with 10% FCS.

Embryo transfer

Oestrus was synchronized in recipient heifers by intramuscular injection of prostaglandin F2-α and diagnosed twice daily by visual observation of mounting behaviour among the treated animals. Embryos were evaluated for development to the compact morula and blastocyst stages on day 7–8 p.i. Good and excellent quality embryos were placed in TL-Hepes, loaded into 0.25 cc semen straws (IMV International, Minneapolis, Minnesota) and transported to our recipient facility (30 min transit time) at 30–35°C. Embryos were transferred non-surgically to heifers in which oestrus had occurred within 24 h of the embryos' insemination *in vitro*. Each heifer received only one embryo.

Pregnancy diagnosis and monitoring

Initial pregnancy diagnosis was performed on day 30 of gestation by transrectal ultrasound (Kastelic *et al.*, 1988). Pregnant recipients were monitored in this fashion weekly until the third month of gestation, after which pregnancies were checked by rectal palpation of the uterus. Recipients were maintained on pasture supplemented with medium-quality grass hay, grain, vitamins and minerals throughout gestation.

Calving and perinatal care

To minimize calf morbidity and mortality associated with dystocia, all recipients were scheduled for elective Caesarean section 5–7 days prior to their calculated due dates. Four weeks prior to their calculated due dates, pregnant recipients were taken off pasture and placed in a smaller paddock where they were observed several times a day for signs of impending parturition. Two weeks before their calculated due date, recipients were placed in a box stall in the Large Animal Clinic at the Virginia–Maryland Regional College of Veterinary Medicine where they were observed for signs of impending parturition every 6 h. At the discretion of the attending veterinarian, recipients entering labour during this period were either allowed to deliver vaginally (with assistance as required) or submitted for emergency Caesarean section. Approximately 1 week before their calculated due dates, recipients were induced into labour and subjected to elective Caesarean section on the following day.

Within 20 min of birth, newborn calves were towelled dry, fed 1 l of colostrum, treated around the navel with an iodine solution and given an intramuscular injection of selenium and vitamin E. Calves were fed a total of 8 l of colostrum during the first 48 h of life, then placed in hutch-style housing and fed milk replacer twice a day for 6 weeks.

Screening of calves for transgene integration

The transgenic status of calves was determined by analysing samples of blood and ear tissue. Within 48 h of birth, 15–20 ml of blood was collected by jugular venipuncture into a heparinized vacuum tube. Ear tissue was sampled with the aid of a porcine ear-notching instrument and placed into tissue lysis buffer at room temperature. DNA from both tissues was purified and subjected to PCR and Southern blotting to detect the presence of the transgene.

To test whether blood and ear were representative tissues on which to base transgene screening, we analysed six additional tissues from 30 calves diagnosed as negative for the transgene. Samples of skin, skeletal muscle, liver, lung, rudimentary mammary tissue and gonads were obtained at necropsy and processed as described above for ear tissue. PCR was used to detect the transgene in purified DNA from each tissue.

Evaluation of transgene expression

Transgene expression was evaluated in one transgenic heifer by inducing her into lactation at 6 months of age. Lactation was induced by the method of Smith and Schanbacher (1973). The concentration of human α-lactalbumin in the lacteal secretion was determined by Western blotting that employed a polyclonal antibody to human α-lactalbumin, pre-absorbed with bovine milk proteins to eliminate cross-reactivity with bovine α-lactalbumin.

Statistical analysis

Proportional data were compared by the chi-squared test; continuous data were compared by one-way analysis of variance (ANOVA). Both analyses were performed using the statistical analysis package supplied with Microsoft Excel.

Results

Embryo production and DNA microinjection

The results of a 5-month microinjection campaign are summarized in Table 13.1. Overall, 20,918 presumptive zygotes were examined for the presence of visible pronuclei; of these, 11,507 (55%) were successfully microinjected. Fertilization frequency, assessed histologically, ranged from 78% to 100%. On any given day, between 5% and 30% of the injected zygotes lysed within several hours of injection. Ultimately, 1011 (9%) embryos developed to the compact morula or blastocyst stage by day 7 or 8 p.i. By comparison, 27% of contemporaneously cultured non-injected control embryos developed to the compact morula and blastocyst stages during this period.

Establishment of pregnancies and calves

Of 478 embryo transfer recipients, 155 (32%) were diagnosed pregnant on day 30 of gestation and 90 (19%) produced calves (Table 13.1). Thus, 65 (32%) pregnancies were lost between day 30 and calving. Embryo developmental stage at the time of transfer had no effect on either pregnancy rate or calving rate ($P>0.05$). Mean (±SD) gestation length was 274±3.16 days. While most calves were delivered by scheduled, Caesarean section, three (3%) were delivered spontaneously at days 255, 259 and 262 of gestation. Although the male:female ratio was 1.44:1 (59% male:41% female), the sex ratio did not differ ($P>0.05$) from the expected ratio of 1:1. Mean (±SD) weight of the calves at birth was 41.83±6.02 kg (range: 29.09–60.45 kg). Birth weight of calves was not affected by sex or transgenic status ($P>0.05$). Mean (±SD) birthweight of the three calves that delivered spontaneously was 30.68±2.25 kg.

One heifer delivered normal, healthy twin bull calves, the apparent result of spontaneous twin formation after transfer of a single embryo to the recipient. Two calves (2%) were born with congenital skeletal abnormalities. One (1%) of the spontaneously calving heifers delivered a dead calf; a post-mortem examination concluded that the calf died from complications due to dystocia. None of these three calves was transgenic.

Transgenic calves

Nine (10%) of the 91 calves were transgenic (Table 13.1). The sex ratio (male:female) among transgenic calves was 1.25:1 (56% male:44% female) and did not differ from the expected ratio of 1:1 ($P>0.05$). All transgenic calves were normal and healthy at birth.

Analysis of various somatic tissues taken from 30 calves initially identified as non-transgenic for the presence of the transgene in ear and

Table 13.1. Summary of data from a transgenic cattle production campaign. Embryo production and microinjections were performed over a 5-month period.

No. injected/total (%)	No. developed (% of injected)	No. transferred (% of injected)	No. pregnant/ no. recipients (%) day 30	No. calves/ no. recipients (%)	No. transgenic/ no. calves (%)
11,507/20,918 (55)	1011 (9)	478 (5)	155/478 (32)	91/478 (19)	9/91 (10)

27% of non-injected control embryos developed over this period.

blood revealed no evidence of the transgene in any other tissue examined, indicating that analysis of these two tissues is sufficient for predicting the absence of the transgene in other tissues.

Effect of FCS during embryo culture on embryo development, pregnancy, calving and birthweights

The results of this experiment are summarized in Table 13.2. The presence of FCS in embryo culture medium between days 5 and 7 or 8 p.i. did not affect the proportion of embryos that developed to the compact morula or blastocyst stage by day 7 or 8 (P>0.05). Similarly, FCS did not affect pregnancy rate at day 30 of gestation, nor did it affect calving rate, birthweight or sex ratio (P>0.05).

Transgene expression in milk

A total of 91 ml of a lacteal secretion was obtained over a 7-day induced lactation period. The secretion contained human α-lactalbumin at a concentration of 2.4 mg ml^{-1}.

Discussion

The work presented here confirms previous reports (Krimpenfort *et al.*, 1991; Hill *et al.*, 1992; Bowen *et al.*, 1994; Hyttenin *et al.*, 1994) on the feasibility of generating transgenic cattle on a large scale from relatively inexpensive and plentiful *in vitro*-produced ova. The proportion of transgenic calves born in this study (9/91; 10%) was higher than previously reported for cattle (0–6%: reviewed by Eyestone, 1994). Finally, the human α-lactalbumin transgene was shown to be expressed in the mammary gland of a heifer, demonstrating production of a recombinant protein in bovine milk.

Pronuclei were visible for microinjection in only 55% of the zygotes examined, despite the fact that histological evaluation revealed that between 78% and 100% of the oocytes had been fertilized. The reasons for this discrepancy are unknown, but we have observed that oocytes matured in transit yield zygotes in which pronuclei are more difficult to visualize compared with those matured under the more controlled conditions of a laboratory incubator. Pronuclei are prominent under Nomarski optics only during a rather brief time 'window' (approximately 16–24 h p.i.), whereas pronuclei are histologically identifiable outside of this window. Asynchronous pronuclear development, perhaps caused by suboptimal oocyte maturation conditions in transit, may have led to reduced pronuclear visibility at any one time.

As reported for other species (Wall, 1996), pronuclear injection of bovine zygotes reduced subsequent development to the blastocyst stage compared with non-injected controls (Table 13.1). Moreover, transfer of apparently normal, healthy, day 7 or 8 p.i. embryos to synchronous recipients led to lower pregnancy rates at day 30 of gestation (32%) compared with those generally achieved with non-injected *in vitro*-produced embryos of similar morphological quality (57% in our laboratory). Furthermore, the incidence of fetal loss between day 30 of gestation and calving was higher (42%; 65/155) than that reported for pregnancies from non-injected, *in vitro*-produced embryos (20%; Reichenbach *et al.*, 1991). Thus, pronuclear DNA injection leads to losses throughout early embryo development and gestation in cattle. The reasons for these losses are unknown, but may stem from lethal insertional mutations or other physical damage to DNA resulting from the force of microinjection.

A number of workers have reported that calves resulting from *in vitro*-produced embryos were associated with higher mean birthweights (Behboodi *et al.*, 1995; Kruip and Den Daas, 1997), longer gestation intervals (Kruip and Den Daas, 1997), more frequent dystocia (Behboodi *et al.*, 1995; Kruip and Den Daas, 1997) and higher perinatal mortality (Behboodi *et al.*, 1995; Kruip and Den Daas, 1997) compared with ordinary calves. In view of these reports, and considering that we chose to transfer Holstein embryos into Holstein × Angus or Hereford heifers, we opted to deliver all calves by elective Caesarean section in an attempt to alleviate some of the perinatal problems experienced by other groups. Indeed, the perinatal losses recorded in our study (1%) were at least as low as those expected for ordinary calves delivered naturally (3–7%; Hawk and Bellows, 1980; Behboodi *et al.*, 1995; Kruip and Den Daas, 1997). Mean birthweight reported in this study (41.8 kg) was similar to those reported elsewhere for Holstein calves born after natural mating or artificial insemination (42.3 kg: Salisbury and Vandemark, 1961; 42.8 kg: Kruip and Den Daas, 1997) or conventional embryo transfer (42.7 kg: Kruip and Den Daas, 1997). However, it must be noted that in our study calves were delivered after a mean gestation of 274 days, which is about 5 days shorter than a natural-length gestation for a Holstein calf (Salisbury and Vandemark, 1961; Jainudeen and Hafez, 1980) and that birthweights were no doubt somewhat lighter than if the calves had been delivered after a full, natural gestation.

Considering the relative ease and economy of generating transgenic cattle from *in vitro*-produced embryos, it is difficult to imagine ever contemplating the formerly used approach based on *in vivo*-generated zygotes. Nevertheless, the *in vitro* approach is far from efficient. For example, only 0.08% (9/11, 507) of the injected zygotes in our study yielded transgenic calves. Opportunities for improvement include treatments to increase zygote developmental potential, embryo development in culture and embryo developmental potential after transfer to recipients. Since the mechanism of transgene integration after pronuclear injection is unknown

(Bishop, 1997), it is unlikely that rational approaches will be developed for improving the frequency of integration (Wall and Seidel, 1992).

A major expense in the generation of transgenic cattle lies in the acquisition and maintenance of recipient animals, most of which will never carry a transgenic fetus to term. Attempts to discriminate between transgenic and non-transgenic bovine embryos prior to transfer have included PCR of embryo biopsies (Bowen *et al.*, 1994; Krisher *et al.*, 1994) and inclusion of fluorescent reporter genes into the injected constructs (Menck *et al.*, 1997). However, all of these methods have given rise to substantial proportions of false positives, rendering them unsuitable for routine screening. Analysis of fetal or placental cells recovered by allanto- or amniocentesis (Garcia and Salaheddine, 1997) could provide an alternative means of identifying transgenic pregnancies, though this approach does not eliminate the initial requirement for large recipient herds. Apart from some breakthrough in transgene integration technology, or in development of PCR or other methods that would eliminate false positives (e.g. fluorescence *in situ* hybridization (FISH): Lewis-Williams *et al.*, 1997; constitutively expressed green fluorescent protein: Takada *et al.*, 1997) improvements in the efficiency of microinjection-based methods for making transgenic cattle are likely to be incremental for the foreseeable future.

The presence of FCS in embryo culture medium from days 5 to 7 or 8 did not affect development to the compact morula or blastocyst stage, nor did it affect initial pregnancy rate, calving rate or calf birthweight (Table 13.2). These observations are in contrast to reports in sheep (Thompson *et al.*, 1995) where the presence of FCS was shown to result in larger offspring. However, in that report, embryos were cultured in the presence of serum from the one-cell to blastocyst stage. Several investigators have reported accelerated fetal growth (Farin and Farin, 1995) and the birth of large calves after *in vitro* embryo culture (Sinclair *et al.*, 1994; Behboodi *et al.*, 1995; Kruip and Den Daas, 1997) compared with embryos generated *in vivo* by artificial insemination or conventional embryo transfer. Though no formal experiment has addressed the question of when during embryo development FCS might influence downstream parameters like birthweight, the data cited above suggest that such effects may occur during exposure of embryos to serum prior to day 5 p.i.

Table 13.2. Effect of FCS from day 5 to 7 or 8 p.i. on embryo development, pregnancy and calving rates and birthweight.

FCS	No. developed/ no. injected[a] (%)	No. pregnant/ no. recips[a] (%) day 30	No. calves/ no. recips[a] (%)	Mean BW[a] (SD), kg	BW range, kg
+	139/952 (15)	23/84 (27)	8/84 (10)	42.08 (4.00)	38.18–49.09
–	132/882 (15)	25/61 (41)	11/61 (18)	38.38 (5.22)	31.82–47.73

[a] No difference between treatments (P>0.05).

Recent advances in nuclear transfer technology have made possible the generation of adult sheep from cultured somatic cells. In an unprecedented report, Wilmut *et al.* (1997) transferred nuclei from sheep fetal fibroblasts and even mammary epithelial cells from a lactating adult ewe into enucleated oocytes, then transferred the 'reconstructed' oocytes to recipient ewes that later gave birth to lambs. Thus, fully differentiated somatic cell nuclei were reprogrammed to a state of totipotency upon transfer to an oocyte cytoplasm. This feat has now been repeated using fetal fibroblasts bearing a transgene (Schneike *et al.*, 1997) indicating that differentiated cells can be genetically modified *in vitro* and give rise to viable, cloned transgenic offspring. If this technology proves feasible in cattle, it will be possible to generate clonal somatic cell lines bearing either randomly inserted genes or more importantly, site-specific gene insertions and/or deletions derived by homologous recombination. A major advantage to this approach is that all calves born after transfer of nuclei from a genetically modified cell line would bear the desired modification(s), thus eliminating the burden of recipients harbouring non-transgenic fetuses. Moreover, the variety of genetic modifications possible with homologous recombination ('knock-outs', 'knock-ins', 'knock-out/replacements') expands the range of novel genotypes possible. For example, it has been proposed that the ultimate milk modification for infant formula applications would be to replace all of the bovine milk protein genes with their human counterparts and thus create cows capable of producing milk containing a complete human milk protein profile (Yom and Bremel, 1993). Theoretically at least, infant formulae based on such milk would be more suitable for human infant consumption than formula manufactured from ordinary cows' milk.

Homologous recombination, combined with somatic cell nuclear transfer, creates a real possibility of generating agriculturally important transgenic modifications. For example, it may be possible to 'knock out' genes that encode receptors for specific viruses and thus knock out their susceptibility to economically important viral diseases, e.g. foot-and-mouth disease, bluetongue and bovine viral diarrhoea. In view of the reproductive similarities between the sheep and cow, the transfer of somatic cell nuclear transfer technology to the cow should be relatively straightforward; we may thus look to this new technology as the next major improvement for the generation of transgenic cattle, and one that may play an important role in creating useful modifications for agricultural purposes.

References

Behboodi, E., Anderson, G.B., Bondurant, R., Cargill, S., Kreuscher, B., Medrano, J. and Murray, J. (1995) Birth of large calves that developed from *in vitro*-derived bovine embryos. *Theriogenology* 44, 227–232.

Bavister, B.D., Rose-Hellekant, T.A. and Pinyopummintr, T. (1992) Development of *in vitro* matured/*in vitro* fertilized bovine embryos into morulae and blastocysts in defined culture media. *Theriogenology* 37, 127–146.

Bishop, J.O. (1997) Chromosomal insertion of DNA. *Reproduction, Nutrition and Development* 36, 607–618.

Bowen, R., Reed, M., Schneike, A., Seidel, G., Brink, Z., Stacy, A., Thomas,/W. and Kajikawa, K. (1994) Transgenic cattle resulting from biopsied embryos: expression of c-ski in a transgenic calf. *Biology of Reproduction* 50, 664–668.

Brem, G. (1992) Gene transfer in farm animals. In: Lauria, A. and Gandolfi, F. (eds) *Embryonic Development and Manipulation in Animal Development.* Portland Press, London, pp. 147–164.

Eyestone, W.H. (1994) Challenges and progress in the production of transgenic cattle. *Reproduction, Fertility and Development* 6, 647–652.

Farin, P.W. and Farin, C.E. (1995) Transfer of bovine embryos produced *in vivo* or *in vitro*: survival and fetal development. *Biology of Reproduction* 52, 676–682.

Garcia, A. and Salaheddine (1997) Bovine ultrasound-guided amniocentesis. *Theriogenology* 47, 1003–1008.

Gardner, D.K., Lane, M., Spitzer, A. and Batt, P.A. (1994) Enhanced rates of development from sheep zygotes to the blastocyst stage *in vitro* in the absence of serum and somatic cells: amino acids, vitamins, and culturing embryos in groups stimulate development. *Biology of Reproduction* 50, 390–400.

Hawk, H.H. and Bellows, R.A. (1980) Beef and dairy cattle. In: Hafez, E.S.E. (ed.) *Reproduction in Farm Animals.* Lea and Feibeger, Philadelphia, pp. 337–345.

Hill, K., Curry, J., DeMayo, K., Jones-Diller, K., Slapak, J. and Bondioli, K. (1992) Production of transgenic cattle by pronuclear injection. *Theriogenology* 37, 222 (abstract).

Hogan, B., Constantini, F. and Lacy, E. (1986) *Manipulating the Mouse Embryo: a Laboratory Manual.* Cold Spring Harbor Laboratory Press, Cold Spring Harbor, New York.

Hyttenin, J.-M., Peura, T., Tolvanen, M., Aalto, J., Alhonen, L., Sinervirta, R., Halmekyto, M., Myohannen, S. and Janne, J. (1994) Generation of transgenic dairy cattle from transgene-analysed and sexed embryos produced *in vitro. BioTechnology* 1, 606–608.

Jainudeen, M.R. and Hafez, E.S.E. (1980) Gestation, prenatal physiology and parturition. In: Hafez, E.S.E. (ed.) *Reproduction in Farm Animals.* Lea and Febiger, Philadelphia, pp. 247–283.

Kastelic, J.P., Curran, S. and Ginther, O.J. (1988) Ultrasonic evaluation of the bovine conceptus. *Theriogenology* 29, 39–54.

Krimpenfort, P.A., Rademakers, A., Eyestone, W., van der Schans, A., van den Brook, P., Kooiman, P., Kootwijk, E., Platenburg, G., Pieper, F., Strijker, R. and deBoer, H.A. (1991) Generation of transgenic dairy cattle by *in vitro* embryo production. *BioTechnology* 9, 844–847.

Krisher, R., Gibbons, J., Gwazdauskas, F. and Eyestone, W. (1994) DNA detection frequency in microinjected bovine embryos following extended culture *in vitro. Animal Biotechnology* 6, 15–26.

Kruip, Th. A.M. and Den Daas, J.H.G. (1997) *In vitro* produced and cloned embryos: effect on pregnancy, parturition and offspring. *Theriogenology* 47, 43–52.

Lathe, R., Clark, J.A., Bishop, J., Simons, P. and Wilmut, I. (1985) Novel products from livestock. In: Smith, C., King, W. and McKay, J. (eds) *Exploiting New Technologies in Animal Breeding.* Oxford Scientific Publication, Oxford.

Lewis-Williams, J., Sun, Y., Han, Y., Ziomek, C., Denniston, R.S., Echelard, Y. and Godke, R.A. (1997) Birth of successfully identified transgenic goats using preimplantation stage embryos biopsied by FISH. *Theriogenology* 47, 226 (abstract).

Long, C.R., Diamani, P., Pinto-Correia, C., MacLean, R.A., Duby, R.T. and Robl, J.M. (1994) Morphology and subsequent development in culture of bovine oocytes matured and fertilized *in vitro* under various conditions of fertilization. *Journal of Reproduction and Fertility* 102, 361–369.

Menck, M.C., Mercier, Y., Lobo, R.B., Heyman, Y., Renard, J.-P. and Thompson, E.M. (1997) *In vivo* luminescent selection of putative transgenic bovine blastocysts. *Theriogenology* 47, 227 (abstract).

Palmiter, R., Brinster, R., Hammer, R., Trebaner, M., Birnberg, N. and Evans, R. (1982) Dramatic growth of mice that developed from eggs microinjected with metallothionine-growth hormone fusion genes. *Nature* 300, 611–615.

Parrish, J.J., Susko-Parrish, J.L., Leibfried-Rutledge, L.L., Critser, E.S., Eyestone, W.H. and First, N.L. (1986) Bovine *in vitro* fertilization with frozen–thawed sperm. *Theriogenology* 25, 591–600.

Pinkert, C. (1994) *Transgenic Animal Technology*, Academic Press, San Diego.

Reichenbach, H.-D., Berg, U. and Brem, G. (1991) *In vivo* and *in vitro* development of microinjected bovine zygotes and two-cell stage embryos obtained at slaughter after superovulation. In: *Proceedings of the 7th Scientific Meeting of the European Embryo Transfer Association*, Cambridge, UK, p. 196 (abstract).

Salisbury, G.W. and Vandemark, N.L. (1961) *Physiology of Reproduction and Artificial Insemination of Cattle.* W.H. Freeman, San Francisco.

Schneike, A., Kind, A.J., Ritchie, W.A., Mycock, K., Scott, A.R., Ritchie, M., Wilmut, I., Colman, A. and Campbell, K.H.S. (1997) Human factor IX transgenic sheep produced by transfer of nuclei from transfected fetal fibroblasts. *Science* 278, 2130–2134.

Sinclair, K.D., Broadbent, P.J., Dolman, D.F., Taylor, A.G. and Reaper, J.F. (1994) *In vitro* produced embryos as a means of achieving pregnancy and improving productivity in beef cows. *Theriogenology* 41, 294 (abstract).

Sirard, M.-A., Parrish, J.J., Ware, C.B., Leibfried-Rutledge, L.L. and First N.L. (1988) The culture of bovine oocytes to obtain developmentally competent embryos. *Biology of Reproduction* 39, 546–552.

Smith, K.L. and Schanbacher, F.L. (1973) Hormone induced lactation in the bovine. I. Lactation performance following injection of 17-beta estradiol and progesterone. *Journal of Dairy Science* 56, 738–742.

Takada, T., Iida, K., Awaji, T.K., Takahashi, R., Shibui, A., Yoshida, K., Sugano, S. and Tsujimoto, G. (1997) Selective production of transgenic mice using green fluorescent protein as a marker. *Nature Biotechnology* 15, 458–461.

Thompson, J.G., Gardner, D.K., Pugh, A.P., McMillan, W.H. and Tervit, H.R. (1995) Lamb birthweight is affected by culture system utilized during *in vitro* pre-elongation development of bovine embryos. *Biology of Reproduction* 53, 1385–1391.

van Langendonckt, A., Acquier, P., Donnay, I., Massip, A. and Dessy, F. (1996) Acceleration of *in vitro* bovine development in the presence of fetal calf serum. *Theriogenology* 45, 194 (abstract).

Walker, S., Hartwich, K.M. and Seamark, R.F. (1996) The production of unusually large offspring following embryo manipulation: concepts and challenges. *Theriogenology* 45, 111–120.

Wall, R. (1996) Modification of milk composition in transgenic animals. In: Miller, R.H., Pursel, V.G. and Norman, H.D. (eds) *Beltsville Symposia in Agricultural Research XX: Biotechnology's Role in the Genetic Improvement of Farm Animals.* American Society of Animal Science, Savoy, Illinois, pp. 165–188.

Wall, R. and Hawk, H.H. (1988) Development of centrifuged cow zygotes in cultured rabbit oviducts. *Journal of Reproduction and Fertility* 82, 673–680.

Wall, R. and Seidel, G. (1992) Transgenic farm animals: a critical analysis. *Theriogenology* 38, 337–357.

Wilmut, I., Schneike, A., McWhir, J., Kind, A.J. and Campbell, K.H.S. (1997) Viable offspring derived from fetal and adult mammalian cells. *Nature* 385, 810–813.

Yom, H.-C. and Bremmel, R.D. (1993) Genetic engineering of milk composition: modification of milk components in lactating transgenic animals. *American Journal of Clinical Nutrition* 58 (Suppl.), 3060S.

14

Changing the Composition and Properties of Milk

J.D. Murray and E.A. Maga

Departments of Animal Science and Population Health and Reproduction, University of California, Davis, California, USA

Since the advent of the molecular biology era in the early 1970s, biotechnology has held great promise for improving animal agriculture. Since 1982, genetic engineering has held the promise of being able to significantly improve animal agriculture, with the dairy industry being one of the first industries to see this promise. Since then, the dairy industry has watched as transgenic technology has been applied to express foreign proteins in the mammary gland for the pharmaceutical industry. However, transgenic technology can also be used to alter the functional and physical properties of milk resulting in a milk with novel manufacturing properties. Work over the last decade on expression systems and in model species, such as the mouse, has now set the stage for the application of transgenic technology directly to the dairy animal to change the nutritional, antimicrobial and functional properties of milk.

Introduction

Approximately 30% of dietary protein consumed in the Western world is obtained from milk, with the majority of this milk being derived from cows (Hambraeus, 1982). However, milk derived from dairy cows, goats or sheep differs significantly from human milk, as shown in Table 14.1. The most notable differences in general are the lower overall protein content of human milk and the much higher proportion of whey protein to casein protein. Milk consists of six major proteins, four being α_{s1}-, α_{s2}-, β- and κ-casein which are present in the milk of most mammals. The remaining proteins are found in the whey fraction and include α-lactalbumin, which is common to all milks, and β-lactoglobulin (ruminants), lactoferrin (primates) or whey acidic protein (WAP; rodents, rabbits, camelids). The relative abundance of the various protein constituents of cow's milk varies between

© CAB INTERNATIONAL 1999. *Transgenic Animals in Agriculture* (eds J.D. Murray, G.B. Anderson, A.M. Oberbauer and M.M. McGloughlin)

Table 14.1. Comparison of fat, casein and whey content of milk from ruminants and man (data from Davies *et al.*, 1983).

Species	Casein (g l^{-1})	Whey (g l^{-1})	Fat (g l^{-1})
Cow	28	6	37
Goat	25	4	45
Sheep	46	9	74
Human	4	6	38

breeds and genotypes, with the general proportions being approximately 31.5% α_{s1}-casein, 29.5% β-casein, 8.5% α_{s2}-casein, 11% κ-casein, 10% β-lactoglobulin, 4% α-lactalbumin, and 5.5% serum proteins and immunoglobulins (Davies *et al.*, 1983).

The high reliance on dairy protein has resulted in considerable research being directed towards understanding lactation, the functional properties of milk and the molecular biology of the milk-specific genes. With the identification of various alleles of the major ruminant milk protein genes, genotypic effects on milk yield, composition and functionality were characterized. For example, milk from cows homozygous for the B allele of β-lactoglobulin had a higher content of fat, total solids, casein and whey protein than milk from homozygous A cows, while the amount of β-lactoglobulin was decreased (McLean *et al.*, 1984). Milk from κ-casein B cows has a shorter rennet clotting time and increased curd firmness (Scharr, 1984), and increased natural heat stability (McLean *et al.*, 1987). The variation in composition and functionality associated with natural genetic variants of milk protein genes suggests that making directed changes in these properties should also be possible.

Thus, with the demonstration that a gene construct could be transferred and expressed in mammals (Palmiter *et al.*, 1982) it was natural that researchers should begin to speculate on the types of genetic changes that might usefully be made to the milk protein system. In 1984, Tom Richardson suggested that genetic engineering techniques should be applied to milk protein genes to 'change systematically the primary structure of a protein and to correlate these changes with alterations in protein functionality', with the aim of studying the function of these altered proteins in a bacterial system. Over the next decade Richardson and colleagues followed this initial general suggestion with a number of specific proposals for altering the milk protein system *in vivo* with the overall aim of changing the functional properties of the milk protein system. For example they suggested that the addition of extra copies of the κ-casein gene, in order to overexpress κ-casein, could result in an increase in the thermal stability of casein aggregates in milk (Kang and Richardson, 1985). A more complex proposal was to use site-directed mutagenesis to change isoleucine$_{71}$ to phenalanine in a clone of α_{s1}-casein prior to gene transfer. The presence in milk of 10–20% of the α_{s1}-casein as the mutant form might increase proteolysis and thereby promote the faster ripening of cheese (Jimenez-Flores and Richardson, 1988; Yom and Richardson, 1993). All in all, Richardson and his colleagues put forward specific possible alterations in the properties of milk that might be gained by overexpressing, deleting or adding back a mutated form of most of the major milk protein genes (Kang and Richardson, 1985; Jimenez-Flores and Richardson, 1988; Yom and Richardson, 1993). Following these initial suggestions by Richardson's group, a number of articles have been written discussing the possible alteration of milk by transgenesis (Bremel *et al.*, 1989; Muysson and Verrinder-Gibbins, 1989; Clark, 1992).

There have been a number of recent reviews written describing the major milk protein genes from a variety of mammals and their possible uses in transgenic studies (e.g. Bawden *et al.*, 1994; Maga and Murray, 1995; Clark, 1996). Promoter elements from the α_{s1}-casein, β-casein, α-lactalbumin, β-lactoglobulin and WAP genes, from one or more species, have now been used to drive expression of a transgene, usually in the mammary gland of mice (Table 14.2). Mammary gland-directed transgenes have been expressed, in addition to mice, in sheep, pigs, goats, rabbits and cows.

To date, most of the work on targeting transgene expression to the mammary gland of an animal has focused either on studying promoter function and identifying the important and necessary regions of DNA required for expression, or has been directed at producing biologically important and active proteins (such as pharmaceuticals) in the milk of a transgenic animal with the intent of recovering the protein of interest from the milk (Wilmut *et al.*, 1991; Mercier and Vilotte, 1993; Bawden *et al.*, 1994). It is now also possible to use transgenic animals to alter directly the properties and composition of the milk itself by genetically adding a new, or altered, protein to the milk protein system to cause effects and not for recovery of the protein for other uses. More specifically, we have been studying the addition of human lysozyme or a modified bovine κ-casein to mouse milk in order to affect the functional and physical properties of the milk protein system and thereby alter the manufacturing applications of milk (Maga and Murray, 1995). Work in the mouse also has demonstrated that antisense or ribozyme transgene constructs can be used to decrease the production of a targeted protein in milk (L'Huillier *et al.*, 1996; Sokol *et al.*, 1998).

Milk Gene Products and Functions

The average composition of bovine milk is 86% water, 5% lactose, 4.1% fat, 3.6% protein and 0.7% minerals with a pH of 6.6–6.7. Milk composition remains relatively constant with the exception of fat content, which varies depending upon the breed of cow, feed type and stage of lactation (Johnson, 1974). All the components of milk are secreted by the mammary gland during lactation, however approximately half of the fat secreted into milk is derived from the serum while the remaining half is synthesized *de*

Table 14.2. Mammary gland-specific transgenic animals.

Promoter	To express	Animal	Reference
Mouse WAP	Human t-PA	Mice, goats	Gorden *et al.*, 1987; Ebert *et al.*, 1991
	Human SOD	Mice, rabbits	Hansson *et al.*, 1994; Stromqvist *et al.*, 1997
	Human protein C	Mice, pigs	Velander *et al.*, 1992a,b; Drohan *et al.*, 1994
	Mouse WAP	Pigs, sheep	Wall *et al.*, 1991, 1996
	Human factor VIII	Pigs	Paleyanda *et al.*, 1997
	Bovine TAP	Mice	Yarus *et al.*, 1996
Rabbit WAP	Human α_1-antitrypsin	Mice	Bischoff *et al.*, 1992
	Human growth hormone	Mice	Devinoy *et al.*, 1994
Bovine α-lactalbumin	Bovine α-lactalbumin	Mice	Vilotte *et al.*, 1989; Soulier *et al.*, 1992; Bleck and Bremel 1993
	Ribozyme α-lactalbumin	Mice	L'Huillier *et al.*, 1996
Goat α-lactalbumin	Goat α-lactalbumin	Mice	Soulier *et al.*, 1992
Sheep β-lactoglobulin	Sheep β-lactoglobulin	Mice	Simons *et al.*, 1987; Shani *et al.*, 1992
	Human α_1-antitrypsin	Mice, sheep	Archibald *et al.*, 1990; Wright *et al.*, 1991
	Human serum albumin	Mice	Shani *et al.*, 1992
	Human SOD	Mice	Hansson *et al.*, 1994
Bovine β-lactoglobulin	Bovine β-lactoglobulin	Mice	Hyttinen *et al.*, 1998; Gutierrez-Adan *et al.*, 1999
	Human erythroprotein	Mice, rabbits	Korhonen *et al.*, 1997
Goat β-lactoglobulin	Goat β-lactoglobulin	Mice	Ibanez *et al.*, 1997
Rat β-casein	Rat β-casein	Mice	Lee *et al.*, 1988
	Bacterial CAT	Mice	Lee *et al.*, 1989
Rabbit β-casein	Human interleukin-2	Rabbits	Buhler *et al.*, 1990
Goat β-casein	Goat β-casein	Mice	Persuy *et al.*, 1992; Roberts *et al.*, 1992
	CFTR	Mice	DiTullio *et al.*, 1992
	Bovine κ-casein	Mice	Gutierrez *et al.*, 1995
	Human t-PA	Goats	Ebert *et al.*, 1994
Bovine α_{s1}-casein	Bovine α_{s1}-casein	Mice	Clarke *et al.*, 1994
	Bovine α_{s1}-casein	Mice	Rijnkels *et al.*, 1998
	Bacterial CAT	Mice	Clarke *et al.*, 1994
	Antisense CAT	Mice	Sokol *et al.*, 1998
	Human GM-CSF	Mice	Uusi-Oukari *et al.*, 1997
	Human IGF-1	Rabbits	Brem *et al.*, 1994
	Human urokinase	Mice	Meade *et al.*, 1990
	Human lysozyme	Mice	Maga *et al.*, 1994
	Human lactoferrin	Mice, cow	Krimpenfort *et al.*, 1991; Platenburg *et al.*, 1994

novo in the mammary gland. The six major mammary gland gene products, α_{s1}-, α_{s2}-, β- and κ-casein, β-lactoglobulin and α-lactalbumin, account for the 3.6% or 30–35 g l^{-1} of protein present in cows' milk. The caseins are present at a ratio of $\alpha_{s1}:\alpha_{s2}:\beta:\kappa$ at 3:1:3:1. The caseins contribute the majority of the protein, 80% or 24–28 g l^{-1}, while the whey proteins account for 20% of the total protein or 5–7 g l^{-1}. A typical cow yields 10,000 pounds of milk over a 305-day lactation period.

Caseins

The caseins are single polypeptide chains that are random in structure with little secondary structure, high average hydrophobicity and a net negative charge. Caseins contain many essential amino acids but are low in cystine content (Fox, 1982). Caseins are post-translationally phosphorylated, essential for calcium binding, at accessible serine residues in the amino acid sequence Ser/Thr-X-Glu/Ser-PO_4, where X is any amino acid (Mercier, 1981).

The caseins are present in milk in the form of micelles, or a colloidal suspension of proteins of the order of 20–600 nm in size (Fox, 1982). α_{s1}-, α_{s2}- and β-casein form the core of the micelle while the amphiphilic κ-casein lies on the surface. Due to their unordered and highly hydrophobic nature, the caseins can aggregate with themselves and the other caseins by hydrophobic interactions, hydrogen bonding, electrostatic attractions (Ca to Ser-PO_4) or repulsions (Ser-PO_4 to Ser-PO_4). α_{s1}- and β-casein tend to self associate while α_{s1}- and κ-casein form aggregates with each other.

Micelles function to sequester and transport calcium in a usable form to the newborn for bone development. The structure of the micelle can be disrupted by acid or the enzyme rennin. Acid acts to neutralize the negative charges of the phosphate and promote association leading to isoelectric coagulation of the caseins when the pH reaches 4.6. This happens in the stomach as the acid pH of the stomach causes casein precipitation, which makes the milk protein more digestible. Rennin is a combination of chymosin and pepsin isolated from calf stomach, which specifically cleaves κ-casein at the Phe105–Met106 peptide bond. Once cleaved, para κ-casein, or the hydrophobic N-terminal of the protein, stays associated with the micelle and the hydrophilic C-terminal is released from the micelle. Cleavage of κ-casein causes destabilization of the micelle structure resulting in precipitation of the caseins, thus separating the milk into two distinct fractions, the non-soluble caseins and the soluble whey proteins.

Whey proteins

Whey proteins are usually compact and globular with a relatively uniform distribution of polar, non-polar and charged residues in contrast with the clustering of similar residues that characterize the caseins (Fox, 1982). The whey proteins tend to be high in cysteine content, whose disulphide bonds can contribute to flavour. Whey proteins are more hydrophilic than the caseins and bind more water. The whey proteins are so named due to the fact that they stay in solution after the caseins are precipitated by acid or rennet.

Transgene Expression in the Mammary Gland

Representative alleles of the major milk genes (α_{s1}-, α_{s2}-, β- and κ-casein, α-lactalbumin, β-lactoglobulin, and whey acidic protein) have been cloned from various species and the promoter regions have been isolated and characterized (Table 14.2; for review see Bawden *et al.*, 1994; Maga and Murray, 1995). The promoters from most of these genes have been used to generate transgenic animals, usually mice, but also cattle, sheep, pigs and goats, which produce foreign proteins in the mammary gland (Table 14.2; Maga and Murray, 1995). The sheep β-lactoglobulin, goat β-casein and bovine α_{s1}-casein promoters have been the most efficient at supporting good levels of heterologous protein expression in the mammary gland of transgenic mice (Simons *et al.*, 1987; Meade *et al.*, 1990; Persuy *et al.*, 1992). Better expression is obtained in most cases if genomic DNA sequences, rather than cDNA, are used and the incorporation of untranslated exons and introns may contribute to increased expression of the transgene (Whitelaw *et al.*, 1991).

There are a number of conclusions that can be drawn from the studies to date on expressing transgenes in the mammary gland. First, all of the mammary gland-specific promoters can be used to direct tissue and developmentally correct expression of a transgene to the mammary gland. However, expression is obtained with varying degrees of efficiency that suggests there may be one or more as yet unidentified *cis*-acting control sequences associated with these genes. There are some species differences with respect to how specific promoters will function, e.g. the WAP promoter is tissue specific in pigs but is not in sheep (Wall *et al.*, 1996). Second, the expression of a foreign protein in the mammary gland, with secretion into the milk, in most cases does not appear to affect the functioning of the mammary gland. Exceptions include impaired lactation of sows from three different transgenic lines expressing mouse WAP in the mammary gland (Shamay *et al.*, 1992) and the work of Bleck *et al.* (1995) showing abnormal lactation in mice expressing bovine β-casein. Third, mammary gland-directed transgenes can be successfully inserted and expressed in all species of mammals thus far studied, including mice, rabbits, cattle, sheep, goats and pigs (Table 14.2).

Transgenic Ruminants

A number of laboratories worldwide have successfully produced transgenic sheep (Hammer *et al.*, 1985; Simons *et al.*, 1988; Murray *et al.*, 1989;

Bawden *et al.*, 1995; Schnieke *et al.*, 1997), goats (Ebert *et al.*, 1991) and cattle (Krimpenfort *et al.*, 1991; Hyttinen *et al.*, 1994; Cibelli *et al.*, 1998), although the efficiency of producing transgenic ruminants remains low. All of the research to date with transgene expression directed towards the mammary gland in ruminants has focused on the production of pharmaceutical proteins, such as α_1-antitrypsin in sheep (Wright *et al.*, 1991), tissue plasminogen activator in goats (Ebert *et al.*, 1991), and human lactoferrin in dairy cattle (Krimpenfort *et al.*, 1991) or testing the efficacy of specific heterologous promoters to reliably direct transgene expression in the mammary gland (Wall *et al.*, 1996). Recent advances in cloning sheep (Schnieke *et al.*, 1997; Wells *et al.*, 1997; Wilmut *et al.*, 1997) may result in increased efficiencies in the production of transgenic ruminants as well as allowing for targeted insertion of the transgene sequences. Nevertheless, with sufficient resources it is now possible to reliably produce transgenics in all of these species and obtain tissue and developmentally correct expression of the transgene in the mammary gland.

What Might We Do to Alter the Properties of Milk?

As mentioned in the introduction, a number of papers have been written over the last 14 years discussing the possible alterations that could usefully be made by genetic engineering to improve the value of milk as an agricultural commodity (e.g. Richardson, 1984; Kang and Richardson, 1985; Jimenez-Flores and Richardson, 1988; Bremel *et al.*, 1989; Muysson and Verrinder-Gibbins, 1989; Oh and Richardson, 1991; Clark, 1992, 1996; Yom and Richardson, 1993; Bawden *et al.*, 1994). Box 14.1 lists five basic types of changes in milk that might be usefully considered. Within these broad areas, a wide variety of modifications to milk have been suggested, including: adding extra copies of an existing gene (α_{s1}-, κ- and β-casein), down-regulating the expression of a gene (α-lactalbumin), adding new genes (human lysozyme or lactoferrin), removal of a gene (β-casein, β-lactoglobulin or acetyl-CoA carboxylase), and adding a mutated gene (α_{s1}-, κ- and β-casein). Preliminary research has been carried out using transgenic mice as model systems in the first four of these categories. Given the present state of transgenic technology in ruminants, all of these modifications, and others not listed, could now potentially be made.

Box 14.1. Areas where milk might be usefully manipulated.

- Altering the proteins to change the manufacturing properties of milk.
- Increasing the antimicrobial activity of milk.
- Altering the type and amount of fatty acids in milk.
- Changing the amino acid composition of milk to improve human nutrition.
- Increasing the overall protein content of milk.

What We Have Learned from Transgenic Mouse Models

Adding extra copies of an existing gene

The addition of more κ-casein to the milk protein system could affect the physical properties of the milk since κ-casein is directly involved with micelle formation, structure and size (Waugh, 1971; Fox, 1982; Schmidt, 1982). An increase in κ-casein could increase the thermal stability of casein aggregates (Jimenez-Flores and Richardson, 1988) and act to decrease micelle size (Fox, 1982). A smaller micelle diameter would lead to a larger available surface area, which would result in a more consistent and firmer curd as well as an increase in cheese yield. These modified properties of milk could be of great benefit and interest to the dairy industry. Results on milk from mice expressing the bovine κ-casein gene showed significantly greater rennet gel strength and decreased micelle particle size when compared with milk from non-transgenic, full-sib control mice (Gutiérrez-Adán *et al.*, 1996).

Down-regulating the expression of a gene

The expression of a gene can be down regulated *in vivo* in a developmentally and tissue-specific manner using transgenes expressing antisense or ribozyme messages (for review see Sokol and Murray, 1996). Two experiments have been completed assessing the efficacy of using antisense or ribozyme constructs to inhibit mRNA translation in the mammary gland. In both cases the target was message-derived from the expression of a non-endogenous transgene. In the first case, L'Huillier *et al.* (1996) used a ribozyme construct targeting the mRNA of bovine α-lactalbumin. Using three different lines of transgenic ribozyme mice crossed to bovine α-lactalbumin transgenic mice, these authors were able to show a 50–78% reduction in the level of bovine α-lactalbumin protein in the milk of double hemizygous females. There was no effect of the bovine-based α-lactalbumin ribozyme on the translation of the endogenous mouse α-lactalbumin mRNA. Sokol *et al.* (1998), using the same double hemizygous strategy, evaluated the ability of antisense and antisense/ribozyme constructs to down-regulate the expression of the bacterial chloramphenicol acetyltransferase (CAT) gene in the mammary gland of transgenic mice. In this case, antisense and antisense/ribozyme constructs were equally efficient, leading to approximately an 80% reduction in the amount of CAT protein secreted into the milk.

Adding new genes

Human lysozyme transgenic mice were studied to determine the consequences on some basic rheological and antimicrobial properties of milk as a result of having human lysozyme expressed in the milk. Lysozymes are ubiquitous enzymes found in avian egg whites and mammalian secretions such as tears, saliva and milk that are positively charged at physiological pH (Jolles and Jolles, 1984) and have an inherent antimicrobial activity (Phillips, 1966). If human lysozyme was present in bovine milk at a significant level, two main effects could be expected. First, because of its antimicrobial activity, lysozyme may reduce the overall level of bacteria in milk thus decreasing disease in the udder and overall bacterial levels in the milk. As lysozyme is considered to be part of passive immunity and the natural defence against bacteria, viruses, parasites and fungi in human milk (Chang, 1990), there may be human health advantages as well. Second, due to the net positive charge, lysozyme may be able to interact with the negatively charged caseins to produce a milk with altered functional and physical properties. In studies using transgenic mice expressing human lysozyme in their milk at an average concentration of 0.38 mg ml^{-1} the rennet clotting time of the milk was decreased by 35%, gel strength of rennet induced gels was significantly higher in milk from the transgenic mice than in milk from control mice, while the average size of the micelles tended to be smaller (Maga *et al.*, 1995). Milk from these same transgenic lines was found to be bacteriostatic against two cold spoilage organisms, *Pseudomonas fragi* and *Lactobacillus viscous*, and a mastitis-causing isolate of *Staphylococcus aureus* (Maga *et al.*, 1998).

Removal of a gene

In order to determine the consequences on the milk protein system of deleting a major milk protein, Kumar *et al.* (1994) produced a β-casein knockout line of mice. Mice heterozygous for the disruption of this gene had reduced levels of β-casein in the milk, while animals homozygous for the knocked-out allele lacked β-casein in the milk. In the homozygous knockout animals the overall protein concentration was reduced, although there was an increase in the amount of the other milk proteins. The homozygous knockout females lactated normally, with the milk having correctly assembled micelles, although the diameter of the micelles was reduced. Kumar *et al.* (1994) concluded that β-casein is a non-essential component of the milk protein system, thus illustrating that profound changes can be made in the composition of milk without disrupting the general organization of the micellular system.

Conclusions

Throughout this review we have tried to use examples to illustrate that the science has progressed sufficiently over the last 14 years for us to be optimistic that transgenic technology can, and will, be successfully used to improve the dairy animal. We have not tried to cite every possible paper and apologize if a particularly important study or suggestion has been missed. However, the papers cited do illustrate that we can now construct a transgene that will have a good chance of functioning in the mammary gland of a transgenic female at a high level in a tissue- and developmentally appropriate manner. Transgenic sheep, goats and cattle can now be routinely made, although the efficiency is still quite low and the costs high, particularly for cattle.

The advent of cloning may increase our ability to produce transgenic ruminants on two fronts. First, as the transgene is inserted during the cell culture phase, each offspring born will be transgenic and each one will have the same insertion point. The fact that they are clones is only relevant to the extent that care must be taken to avoid inbreeding. Secondly, the ability to rederive animals from cells in culture finally opens the possibility of doing either gene knockout or gene replacement experiments, thus increasing the options available for altering the composition of milk and the control of lactation. The ability to carry out the full range of genetic modifications routinely produced in mice is now available for use in dairy animals, although much research remains to be done with respect to identifying appropriate target phenotypes for alteration, the construction of the required transgenes, and on improving the efficiency of production of transgenic livestock.

Experiments performed in mice have shown that the mammary gland and milk systems are robust and can be altered and added to using a wide variety of different proteins and still function. Furthermore, the system can be altered such that predictable changes in the functional and antimicrobial nature of milk can be produced. The results in mice suggest that human lysozyme and κ-casein are good candidates for altering properties of milk in a beneficial fashion and demonstrate that transgenic technology can be used for agricultural purposes as well as for studying gene function and the production and recovery of novel proteins in the milk of transgenic livestock. However, here at the beginning there is no lack of good ideas for possible changes to the mammary gland and the milk protein system: one only has to start with the papers of Tom Richardson.

Acknowledgements

Support for our work on modifying the composition of milk was provided by the California Dairy Research Foundation. We would also like to thank all of our colleagues for the many discussions we have had over the years concerning the application of transgenic technology to the dairy industry and their contributions to our collaborative work.

References

Archibald, A.L., McClenaghan, M., Hornsey, V., Simons, J.P. and Clark, A.J. (1990) High level expression of biologically active human α_1-antitrypsin in the milk of transgenic mice. *Proceedings of the National Academy of Sciences USA* 87, 5178–5182.

Bawden, C.S., Sivaprasad, A.V., Verma, P.J., Walker, S.K. and Rogers, G.E. (1995) Expression of bacterial cysteine biosynthesis genes in transgenic mice and sheep: toward a new *in vivo* amino acid biosynthesis pathway and improved wool growth. *Transgenic Research* 4, 87–104.

Bawden, W.S., Passey, R.L. and Mackinlay, A.G. (1994) The genes encoding the major milk-specific proteins and their use in transgenic studies and protein engineering. *Biotechnology, Genetics and Engineering Review* 12, 89–137.

Bischoff, R., Degryse, E., Perraud, F., Dalemans, W., Ali-Hadji, D., Thepot, D., Devinoy, E., Houdebine, L.M. and Pavirani, A. (1992) A 17.6 kbp region located upstream of the rabbit WAP gene directs high level expression of a functional human protein variant in transgenic mouse milk. *FEBS* 305, 265–268.

Bleck, G.T. and Bremel, R.D. (1993) Variation in expression of a bovine α-lactalbumin transgene in milk of transgenic mice. *Journal of Dairy Science* 77, 1897–1904.

Bleck, G.T., Jiménez-Flores, R. and Bremel, R.D. (1995) Abnormal properties of milk from transgenic mice expressing bovine β-casein under control of the bovine α-lactalbumin 5′ flanking regions. *International Dairy Journal* 5, 619–632.

Brem, G., Hartl, P., Besenfelder, U., Wolf, E., Zinovieva, N. and Pfaller, R. (1994) Expression of synthetic cDNA sequences encoding human insulin-like growth factor-1 (IGF-1) in the mammary gland of transgenic rabbits. *Gene* 149, 351–355.

Bremel, R.D., Yom, H.-C. and Bleck, G.T. (1989) Alteration of milk composition using molecular genetics. *Journal of Dairy Science* 72, 2826–2833.

Buhler, T.A., Bruyere, T., Went, D.F., Stranzinger, G. and Burki, K. (1990) Rabbit β-casein promoter directs secretion of human interleukin-2 into the milk of transgenic rabbits. *Bio/Technology* 8, 140–143.

Chang, S.S. (1990) Antimicrobial proteins of maternal and cord sera and human milk in relation to maternal nutritional studies. *American Journal of Clinical Nutrition* 51, 183–187.

Cibelli, J.B., Stice, S.L., Golueke, P.J., Kane, J.J., Jerry, J., Blackwell, C., Ponce de León, F.A. and Robl, J.M. (1998) Cloned transgenic calves produced from nonquiescent fetal fibroblasts. *Science* 280 (5367), 1256.

Clark, A.J. (1992) Prospects for the genetic engineering of milk. *Journal of Cell Biochemistry* 49, 121–127.

Clark, A.J. (1996) Genetic modification of milk proteins. *American Journal of Clinical Nutrition* 63, 633S-638S.

Clarke, R.A., Sokol, D., Rigby, N., Ward, K., Murray, J.D. and Mackinlay, A.G. (1994) Mammary gland specific expression of bovine α_{s1}-casein derived transgenes in mice. *Transgenics* 1, 313–319.

Davies, D.T., Holt, C. and Christie, W.W. (1983) The composition of milk. In: Mepham, T.B. (ed.) *Biochemistry of Lactation*. Elsevier, Amsterdam, pp. 1–11.

Devinoy, E., Thepot, D., Stinnakre, M.-G., Fontaine, M.-L., Grabowski, H., Puissant, C., Pavirani, A. and Houdebine, L.M. (1994) High level production of human growth hormone in the milk of transgenic mice: the upstream region of the rabbit whey acidic protein (WAP) gene targets transgene expression to the mammary gland. *Transgenic Research* 3, 79–89.

DiTullio, P., Cheng, S.H., Marshall, J., Gregory, R.J., Ebert, K.M., Meade, H.M. and Smith, A.E. (1992) Production of cystic fibrosis transmembrane conductance regulator in the milk of transgenic mice. *Bio/Technology* 10, 74–77.

Drohan, W.N., Zhang, D., Paleyanda, R.K., Chang, R., Wroble, M., Velander, W. and Lubon, H. (1994) Inefficient processing of human protein C in the mouse mammary gland. *Transgenic Research* 3, 355–364.

Ebert, K.M., Selgrath, J.P., DiTullio, P., Denman, J., Smith, T.E., Memon, M.A., Schindler, J.E., Monastersky, G.M., Vitale, J.A. and Gorden, K. (1991) Transgenic production of a variant of human tissue type plasminogen activator in goat milk: generation of transgenic goats and analysis of expression. *Bio/Technology* 9, 835–838.

Ebert, K.M., DiTullio, P., Barry, C.A., Schindler, J.E., Ayres, S.L., Smith, T.E., Pellerin, L.J., Meade, H.M., Denman, J. and Roberts, B. (1994) Induction of human tissue plasminogen activator in the mammary gland of transgenic goats. *Bio/Technology* 12, 699–702.

Fox, P.F. (1982) Chemistry of milk protein. In: Fox, P.F. (ed.) *Developments in Dairy Chemistry*, Vols 1 and 2. Applied Science Publishers, New York.

Gorden, K., Lee, E., Vitale, J.A., Smith, A.E., Westphal, H. and Hennighausen, L. (1987) Production of human type plasminogen activator in transgenic mouse milk. *Bio/Technology* 5, 1183–1187.

Gutierrez, A., Meade, H.M., Jimenez-Flores, R., Anderson, G.B., Murray, J.D. and Medrano, J.F. (1995) Expression analysis of bovine κ-casein in the mammary gland of transgenic mice. *Transgenic Research* 5, 271–279.

Gutiérrez-Adán, A., Maga, E.A., Meade, H.M., Shoemaker, C.F., Medrano, J.F., Anderson, G.B. and Murray, J.D. (1996) Alteration of physical characteristics of milk from bovine kappa-casein transgenic mice. *Journal of Dairy Science* 79, 791–799

Gutiérrez-Adán, A., Maga, E.A., Behboodi, E., Conrad-Brink, J.S., MacKinlay, A.G., Anderson, G.B. and Murray, J.D. (1999) Expression of bovine β-lactoglobulin in the milk of transgenic mice. *Journal of Dairy Research*, in press.

Hambraeus, L. (1982) Nutritional aspects of milk protein. In: Fox, P.F. (ed.) *Developments in Dairy Chemistry*. Applied Science Publishers, London, pp. 289–313.

Hammer, R.E., Pursel, V.G., Rexroad, C.E., Jr, Wall, R.J., Bolt, D.J., Ebert, K.M., Palmiter, R.D. and Brinster, R.L. (1985) Production of transgenic rabbits, sheep and pigs by microinjection. *Nature* 315, 680–683.

Hansson, L., Edlund, M., Edlund, A., Johansson, T., Marklund, S.L., Fromm, S., Stromqvist, M. and Tornell, J. (1994) Expression and characterization of biologically active human extracellular superoxide dismutase in milk of transgenic mice. *Journal of Biological Chemistry* 69, 5358–5363.

Hyttinen, J.-M., Peura, T., Tolvanen, M., Aalto, J., Alhonen, L., Sinervirta, R., Halmekytö, M., Myöhänen, S. and Jänne, J. (1994) Generation of transgenic dairy cattle from transgene-analyzed and sexed embryos produced *in vitro*. *Bio/Technology* 12, 606–608.

Hyttinen, J.-M., Korhonen, V.-P., Hiltunen, M.O., Myöhänen, S. and Jänne, J. (1998) High-level expression of bovine β-lactoglobulin gene in transgenic mice. *Journal of Biotechnology* 61, 191–198.

Ibanez, E., Folch, J.M., Vidal, F., Coll, A., Santalo, J., Egozcue, J. and Sanchez, A. (1997) Expression of caprine α-lactoglobulin in the milk of transgenic mice. *Transgenic Research* 6, 69–74.

Jimenez-Flores, R. and Richardson, T. (1988) Genetic engineering of the caseins to modify the behaviour of milk during processing: a review. *Journal of Dairy Science* 71, 2640–2654.

Johnson, A.H. (1974) The composition of milk. In: Webb, B.H., Johnson, A.H. and Alford, J.A. (eds) *Fundamentals of Dairy Chemistry*, 2nd edn. AVI Publishing, Westport, Connecticut, pp. 1–57.

Jolles, P. and Jolles, J. (1984) What's new in lysozyme research. *Molecular and Cellular Biochemistry* 63, 165–189.

Kang, Y. and Richardson, T. (1985) Genetic engineering of caseins. *Food Technology* (October), 89–94.

Korhonen, V.P., Tolvanen, M., Hyttinen, J.M., Uusi-Oukari, M., Sinervirta, R., Alhonen, L., Jauhiainen, M., Janne, O.A. and Janne, J. (1997) Expression of bovine beta-lactoglobulin/human erythroprotein fusion protein in the milk of transgenic mice and rabbits. *European Journal of Biochemistry* 245, 482–489.

Krimpenfort, P., Rademakers, A., Eyestone, W., van der Schans, A., van den Broek, S., Kooiman, P., Kootwijk, E., Platenburg, G., Pieper, F., Strijker, R. and de Boer, H. (1991) Generation of transgenic dairy cattle using *in vitro* embryo production. *Bio/Technology* 9, 844–847.

Kumar, S., Clarke, A.R., Hooper, M.L., Horne, D.S., Law, A.J., Leaver, J., Springbett, A., Stevenson, E. and Simons, J.P. (1994) Milk composition and lactation of beta-casein-deficient mice. *Proceedings of the National Academy of Sciences USA* 91, 6138–6142.

Lee, K.-F., DeMayo, F.J., Atiee, S.H. and Rosen, J.M. (1988) Tissue specific expression of the rat β-casein gene in transgenic mice. *Nucleic Acids Research* 16, 1027–1041.

Lee, K.-F., Atiee, S.H. and Rosen, J.M. (1989) Differential regulation of rat β-casein-chloramphenicol acetyltransferase fusion gene expression in transgenic mice. *Molecular and Cellular Biology* 9, 560–565.

L'Huillier, P.J., Soulier, S., Stinnakre, M.-H., Lepourry, L., Davis, S.R., Mercier, J.-C. and Vilotte, J.-L. (1996) Efficient and specific ribozyme-mediated reduction of bovine α-lactalbumin expression in double transgenic mice. *Proceedings of the National Academy of Sciences USA* 93, 6698–6703.

Maga, E.A. and Murray, J.D. (1995) Mammary gland expression of transgenes and the potential for altering the properties of milk. *Bio/Technology* 13, 1452–1457.

Maga, E.A., Anderson, G.B., Huang, M.C. and Murray, J.D. (1994) Expression of human lysozyme mRNA in the mammary gland of transgenic mice. *Transgenic Research* 3, 36–42.

Maga, E.A., Anderson, G.B. and Murray, J.D. (1995) The effect of mammary gland expression of human lysozyme on the properties of milk in transgenic mice. *Journal of Dairy Science* 78, 2645–2652.

Maga, E.A., Anderson, G.B., Cullor, J.S., Smith, W. and Murray, J.D. (1998) Antimicrobial properties of human lysozyme transgenic mouse milk. *Journal of Food Protection* 61, 52–56.

Marziali, A.S. and Ng-Kwan-Hang, K.F. (1986) Effect of milk composition and genetic polymorphism on coagulating properties of milk. *Journal of Dairy Science* 69, 1793–1798.

McLean, D.M., Graham, E.R.B. and Ponzini, R.W. (1984) Effects of milk protein genetic variants on milk yield and composition. *Journal of Dairy Science* 51, 531–546.

McLean, D.M., Graham, E.R.B., Ponzini, R.W. and McKenzie, H.A. (1987) Effects of milk protein genetic variants and composition on heat stability of milk. *Journal of Dairy Research* 54, 219–235.

Meade, H., Gates, L., Lacy, E. and Lonberg, N. (1990) Bovine α_{s1}-casein gene sequences direct high level expression of active human urokinase in mouse milk. *Bio/Technology* 8, 443–446.

Mercier, J.C. (1981) Phosphorylation of caseins: evidence for an amino acid triplet code posttranslationally recognized by specific kinases. *Biochimie* 63, 1–17.

Mercier, J.C. and Vilotte, J.L. (1993) Structure and function of milk protein genes. *Journal of Dairy Science* 76, 3079–3098.

Murray, J.D., Nancarrow, C.D., Marshall, J.T., Hazelton, I.G. and Ward, K.A. (1989) Production of transgenic merino sheep by microinjection of ovine metallothionein–ovine growth hormone fusion genes. *Reproduction, Fertility and Development* 1, 147–155.

Muysson, D.J. and Verrinder-Gibbins, A.M. (1989) The alteration of milk content by genetic engineering and recombinant DNA-mediated selection techniques. *Canadian Journal of Animal Science* 69, 517–527.

Oh, S. and Richardson, T. (1991) Genetic engineering of bovine κ-casein to improve its nutritional quality. *Journal of Agriculture and Food Chemistry* 39, 422–427.

Paleyanda, R.K., Velander, W.H., Lee, T.K., Scandella, D.H., Gwazdauskas, F.C., Knight, J.M., Hoyer, L.W., Drohan, W.N. and Lubon, H. (1997) Transgenic pigs produce functional human factor VIII in milk. *Natural Biotechnology* 15, 971–975.

Palmiter, R.D., Brinster, R.L., Hammer, R.E., Trumbauer, M.E., Rosenfeld, M.G., Birnberg, N.C. and Evans, R.M. (1982) Dramatic growth of mice that develop from eggs microinjected with metallothionein-growth hormone fusion genes. *Nature* 300, 611–615.

Persuy, M.-A., Stinnakre, M.-G., Printz, C., Mahe, M.-F. and Mercier, J.C. (1992) High expression of the caprine β-casein gene in transgenic mice. *European Journal of Biochemistry* 205, 887–893.

Phillips, D.C. (1966) The 3-dimensional structure of an enzyme molecule. *Scientific American* 215, 78–90.

Platenburg, G.J., Kootwijk, E.P.A., Kooiman, P.M., Woloshuk, S.L., Nuijens, J.H., Krimpenfort, P.J.A., Pieper, F.R., de Boer, H.A. and Strijker, R. (1994) Expression of human lactoferrin in milk of transgenic mice. *Transgenic Research* 3, 99–108.

Richardson, T. (1984) Chemical modifications and genetic engineering of food proteins. *Journal of Dairy Science* 68, 2753–2762.

Rijnkels, M., Kooiman, P.M., Platenburg, G.J., van Dixhoorn, M., Nuijens, J.H., de Boer, H.A. and Pieper, F.R. (1998) High-level expression of bovine α_{s1}-casein in milk of transgenic mice. *Transgenic Research* 7, 5–14.

Roberts, B., DiTullio, P., Vitale, J., Hehir, K. and Gorden, K. (1992) Cloning of the goat β-casein encoding gene and expression in transgenic mice. *Gene* 121, 255–262.

Scharr, J. (1984) Effects of κ-casein genetic variants and lactation number on the renneting properties of individual milks. *Journal of Dairy Science* 51, 397–406.

Schmidt, D.G. (1982) Association of caseins and casein micelle structure. In: Fox, P.F. (ed.) *Developments in Dairy Chemistry*, Vol. 1. Applied Science Publishers, New York, pp. 61–82.

Schnieke, A.E., Kind, A.J., Ritchie, W.A., Mycock, K., Scott, A.R., Ritchie, M., Wilmut, I., Coleman, A. and Campbell, K.H. (1997) Human factor IX transgenic sheep produced by transfer of nuclei from transfected fetal fibroblasts. *Science* 278, 2130–2133.

Shamay, A., Pursel, V.G., Wilkinson, E., Wall, R.J. and Hennighausen, L. (1992) Expression of the whey acidic protein in transgenic pigs impairs mammary gland development. *Transgenic Research* 1, 124–132.

Shani, M., Barash, I., Nathan, M., Ricca, G., Searfoss, G.H., Dekel, I., Faerman, A., Givol, D. and Hurwitz, D.R. (1992) Expression of human serum albumin in the milk of transgenic mice. *Transgenic Research* 1, 195–208.

Simons, J.P., McClenaghan, M. and Clark, A.J. (1987) Alteration of the quality of milk by expression of sheep β-lactoglobulin in transgenic mice. *Nature* 328, 530–532.

Simons, J.P., Wilmut, I., Clark, A.J., Archibald, A.L., Bishop, J.O. and Lathe, R. (1988) Gene transfer into sheep. *Bio/Technology* 6, 179–183.

Sokol, D.L. and Murray, J.D. (1996) Antisense and ribozyme constructs in transgenic animals. *Transgenic Research* 5, 363–371.

Sokol, D.L., Passey, R.J., Mackinlay, A.G. and Murray, J.D. (1998) Regulation of CAT protein by ribozyme and antisense mRNA in transgenic mice. *Transgenic Research* 7, 41–50.

Soulier, S., Vilotte, J.L., Stinnakre, M.G. and Mercier, J.C. (1992) Expression analysis of ruminant α-lactalbumin in transgenic mice: developmental regulation and general location of important *cis*-regulatory elements. *FEBS* 297, 13–18.

Stromqvist, M., Houdebine, M., Andersson, J.O., Edlund, A., Johansson, T., Viglietta, C., Puissant, C. and Hansson, L. (1997) Recombinant human extracellular superoxide dismutase produced in milk of transgenic rabbits. *Transgenic Research* 6, 271–278.

Uusi-Oukari, M., Hyttinen, J.M., Korhonen, V.P., Vasti, A., Alhonen, L., Janne, O.A. and Janne, J. (1997) Bovine alpha s1-casein gene sequences direct high level expression of human granulocyte-macrophage colony-stimulating factor in the milk of transgenic mice. *Transgenic Research* 6, 75–84.

Velander, W.H., Page, R.L., Morcol, T., Russell, C.G., Canseco, R., Young, J.M., Dorhan, W.N., Gwazdauskas, F.C., Wilkins, T.D. and Johnson, J.L. (1992a) Production of biologically active protein C in the milk of transgenic mice. *Annals of the New York Academy of Sciences* 665, 391–403.

Velander, W.H., Johnson, J.L., Page, R.L., Russell, C.G., Subramanian, A., Wilkins, T.D., Gwazdauskas, F.C., Pittius, C. and Drohan, W.N. (1992b) High-level expression of a heterologous protein in the milk of transgenic swine using the cDNA encoding human protein C. *Proceedings of the National Academy of Sciences USA* 89, 12003–12007.

Vilotte, J.L., Soulier, S., Stinnakre, M.G., Massoud, M. and Mercier, J.C. (1989) Efficient tissue-specific expression of bovine α-lactalbumin in transgenic mice *European Journal of Biochemistry* 186, 43–48.

Wall, R.J., Pursel, V.G., Shamay, A., McKnight, R.A., Pittius, C.W. and Hennighausen, L. (1991) High level synthesis of a heterologous milk protein in the mammary gland of transgenic swine. *Proceedings of the National Academy of Sciences USA* 88, 1696–1700.

Wall, R.J., Rexroad, C.E., Jr, Powell, A., Shamay, A., McKnight, R. and Hennighausen, L. (1996) Synthesis and secretion of the mouse whey acidic protein in transgenic sheep. *Transgenic Research* 5, 67–72.

Waugh, D.F. (1971) Formation and structure of casein micelles. In: McKenzie, H.A. (ed.) *Milk Proteins Chemistry and Molecular Biology*. Academic Press, New York, pp. 3–85.

Wells, D.N., Misica, P.M., Day, A.M. and Tervit, H.R. (1997) Production of cloned lambs from an established embryonic cell line: a comparison between *in vivo*- and *in vitro*-matured cytoplasts. *Biology of Reproduction* 57, 385–393.

Whitelaw, C.B.A., Archibald, A.L., Harris, S., McClenaghan, M., Simons, J.P. and Clark, A.J. (1991) Targeting expression to the mammary gland: intronic sequences can enhance the efficiency of gene expression in transgenic mice. *Transgenic Research* 1, 3–13.

Wilmut, I., Archibald, A.L., McClenaghan, M., Simons, J.P., Whitelaw, C.B.A. and Clark, A.J. (1991) Production of pharmaceutical proteins in milk. *Experientia* 47, 905–912.

Wilmut, I., Schnieke, A.E., McWhir, J., Kind, A.J. and Campbell, K.H. (1997) Viable offspring derived from fetal and adult mammalian cells. *Nature* 385, 810–813.

Wright, G., Carver, A., Cottom, D., Reeves, D., Scott, A., Simons, P., Wilmut, I., Garner, I. and Colman, A. (1991) High level expression of active human α_1-antitrypsin in the milk of transgenic sheep. *Bio/Technology* 9, 830–834.

Yarus, S., Rosen, J.M., Cole, A.M. and Diamond, G. (1996) Production of active bovine tracheal antimicrobial peptide in milk of transgenic mice. *Proceedings of the National Academy of Sciences USA* 93, 14118–14121.

Yom, H.C. and Richardson, T. (1993) Genetic engineering of milk composition: modification of milk components in lactating transgenic animals. *American Journal of Clinical Nutrition* 58 (Suppl.), 299S-306S.

15

Comparison of Traditional Breeding and Transgenesis in Farmed Fish with Implications for Growth Enhancement and Fitness

R.A. Dunham[1] and R.H. Devlin[2]

[1]*Department of Fisheries and Allied Aquacultures, Alabama Agricultural Experiment Station, Auburn University, Alabama, USA;* [2]*Fisheries and Oceans Canada, West Vancouver, Canada*

Improvements in the performance of fish species used in aquaculture are being accomplished using a variety of approaches, including both traditional and molecular genetic methodologies. Historical gains in productivity have been achieved by domestication, selection, interspecific and interstrain crossbreeding, polyploidy, and synthesis of monosex populations. More recently, transgenesis has been explored as a technique to enhance growth rate and other performance characteristics.

Domestication of species, without directed selection, can yield improvement in production characteristics. Domesticated strains of farmed fish usually grow faster than wild strains, and this effect can be achieved fairly rapidly: for example in channel catfish, *Ictalurus punctatus*, domestication can improve the growth rate by approximately 2–6% per generation. In contrast, directed selection (mass selection) for body weight in fish has resulted in an up to 55% increase in body weight after four to ten generations of selection. In channel catfish, correlated responses to selection include higher dressing percentage, but a decreased ability to tolerate low concentrations of dissolved oxygen.

Intraspecific crossbreeding can increase growth in channel catfish, common carp and salmonids, but crossbreeding does not always result in heterosis. Interspecific hybridization seldom results in overdominant performance in fish. However, one catfish hybrid, channel catfish female × blue catfish (*I. furcatus*) male, exhibits improved performance for several traits including growth, disease resistance, survival, tolerance of low dissolved oxygen, angling vulnerability, seinability, dressing and fillet %.

Ploidy manipulation and sex-control technologies have also played an important role in enhancing production performance. Induction of triploidy does not improve performance in catfish hybrids, but in salmonids triploidy can enhance flesh quality by preventing sexual maturation, although growth rate is somewhat reduced

© CAB INTERNATIONAL 1999. *Transgenic Animals in Agriculture* (eds J.D. Murray, G.B. Anderson, A.M. Oberbauer and M.M. McGloughlin)

relative to diploids. Monosex male populations can increase growth rate in some strains of channel catfish, and monosex female populations of salmon have a reduced incidence of precocious maturation and thus overall improved flesh quality.

In comparison with traditional selective breeding, transgenesis in channel catfish can increase the growth rate by 30–40% by the introduction of salmonid growth hormone (GH) genes. For several species of salmonids, insertion of GH transgenes can result in dramatic weight increases of up to 11-fold after 1 year of growth. A variety of effects on commercially important characteristics other than growth are also observed in GH transgenic fish. Feed conversion efficiency is enhanced in transgenic catfish, common carp (*Cyprinus carpio*) and salmonids, an effect also observed in catfish improved for growth by traditional breeding approaches. Transgenic catfish and carp have increased protein levels and decreased fat, however, alterations in ratios among amino acids and among fatty acids in the flesh are slight or non-existent. Transgenic catfish demonstrate improved flavour and sensory scores, and GH transgenic common carp display improvements in dressing %.

Due to the potential for farmed transgenic fish to escape into natural ecosystems and breed with wild conspecifics, research has also been conducted into examining the fitness (morphological, physiological and behavioural characteristics) of transgenic animals relative to wild-type. As has been observed in other transgenic systems, body shape and physiological performance is altered in transgenic fish. In common carp, transgenic animals have larger heads, and deeper and thicker bodies. In GH transgenic salmonids, morphological disruptions analogous to acromegaly can be observed in the cranium of individuals with extraordinary growth rates. While reproductive traits are not affected in transgenic channel catfish and common carp, GH transgenic common carp display enhanced disease resistance and tolerance of low oxygen levels which could affect their ability to survive in natural systems. Foraging ability of transgenic and control catfish is similar, and under conditions of competition and natural food source, as would be the case in nature, growth is not different between transgenic and control catfish. Predator avoidance was also slightly impaired for GH transgenic catfish compared with control individuals. Swimming ability is reduced in transgenic salmon, which has implications for foraging ability and predator avoidance, as well as ability to complete arduous river migrations for spawning. Although transgenic fish may be released to nature by accident, it appears that ecological effects of transgenic fish developed and evaluated to date will be unlikely because of these examples of reduced fitness. However, each new variety of transgenic fish should be evaluated for potential environmental risk prior to utilization in aquaculture.

Great potential exists to improve production characteristics by transgenic and other molecular genetic approaches. However, future genetic improvements will continue to be achieved from traditional approaches, and, by utilizing a combination of both approaches simultaneously, maximum genetic gain should be accomplished.

Introduction

With the global escalation of human populations, world requirement and demand for high-quality protein are rising dramatically and an increasing proportion is being derived from aquatic sources. Currently, the quantity of animal protein harvested from global aquatic sources via the capture of natural fish populations is maximal: many major fish stocks are showing precipitous declines in harvest yield due to overfishing, and further increases in ocean productivity are not anticipated under the current global climate regime. As a consequence, future demands for aquatic protein must be met through the development of aquaculture-based production systems. In 1993, approximately 16 million tonnes of aquacultured animal protein was produced, representing some 13% of the total aquatic animal protein harvested or produced (Tacon, 1996). The production of aquacultured animal protein has been increasing at a rate of over 10% annually since the mid-1980s, compared with the more modest growth of terrestrial meat production which ranges from 0.7% (beef) to 5.2% (poultry). Similar trends are also observed on a smaller scale for the aquaculture production of various finfish species (e.g. catfish, salmonids) in North America, however, it should be noted that the most rapid growth of aquacultured finfish production is occurring in developing (11.5% annually) rather than developed (4.0% annually) countries (Tacon, 1996).

With increased demand for aquacultured foods has come a need for developing more efficient production systems. Major improvements have been recently achieved through enhanced husbandry procedures, improved nutrition, enhanced disease diagnostics and therapies, and the application of genetics to improve production traits. More recently, biotechnology has begun to play a role with the isolation of piscine genes influencing a variety of physiological processes (Donaldson and Devlin, 1996) and marker DNA sequences linked to particular production characteristics (e.g. sex differentiation; Devlin *et al.*, 1994a). Gene transfer methodologies have been explored in fish since the mid-1980s, both in model systems to study basic biological processes and in fish species that are cultured for food production (see reviews by Fletcher and Davies, 1991; Hackett, 1993; Gong and Hew, 1995; Devlin, 1996). Whereas the goals of transgenic fish research do not differ from those of traditional genetic selection (i.e. improvement of production efficiency, and enhancement of product quality or character), the scope for altering these traits has been enhanced dramatically by transgenesis due to the rapidity with which responses can be achieved and by the use of genes for useful biological processes not found in the host fish species (e.g. unique disease resistance genes found only in other phyla).

Several major areas of investigation are now under way in fish genetic research, based on the needs of aquaculture production systems. We briefly review some of these approaches using examples derived from work with channel catfish, common carp and salmonids.

Domestication

When wild fish are moved from the natural environment to the aquaculture environment, a new set of selective pressures are exerted on the population which result in changes of gene frequencies and consequent performance of the population. This process, termed domestication, occurs even without directed selection by man. Domestication effects are dramatic and can be observed in fish in as little as one or two generations after removal from the natural environment. Domesticated strains of fish almost always exhibit better performance than wild strains when placed in the aquaculture environment (Dunham, 1996). Domestication results in an increased growth rate of 3–6% per generation in captivity for channel catfish, *Ictalurus punctatus* and, not surprisingly, the oldest (86 years) domesticated strain of channel catfish (Kansas strain) has the fastest growth rate of all strains of channel catfish. Utilization of domestic strains instead of wild strains, and use of established, high-performance strains, are the first steps in applying genetic principles for improved fish culture management. Domestication will continue to play a key role in the establishment and expansion of aquaculture production as new species and strains are developed.

Strain Evaluation

Large strain differences exist for several traits of cultured fish. Channel catfish strains differ in growth, disease resistance, body conformation, dressing percentage, seinability, vulnerability to angling, age of maturity, time of spawning, fecundity and egg size (Dunham and Smitherman, 1984; Smitherman and Dunham, 1985). Similarly, rainbow trout (*Oncorhynchus mykiss*) strains also vary for numerous traits such as growth, feed conversion efficiency, survival, disease resistance, time of spawning, fecundity, hatching success and angling vulnerability (Kincaid, 1981).

Selection

Genetic selection has been a powerful tool for the improvement of production characteristics of agricultural species (crops, terrestrial animals and, more recently, aquatic species). Selection has allowed the development of several lines of fast-growing channel catfish, with the fastest growing select lines being derived from the fastest growing strains (Smitherman and Dunham, 1985) and growing twice as fast as average (Burch, 1986). For example, body weight of channel catfish has been improved by 12–20% in only one or two generations of mass selection (Bondari, 1983; Dunham and Smitherman, 1983a; Smitherman and Dunham, 1985). Similarly, after three generations of selection, the growth rate was improved by 20–30% when grown in ponds (Rezk, 1993), and four generations of selection in the Kansas strain resulted in a 55% improvement in growth rate (Padi, 1995).

In rainbow trout, six generations of selection increased body weight by 30% (Kincaid, 1983), whereas ten generations of selection of coho salmon resulted in an increased growth rate of 50% (Hershberger *et al.*, 1990). A single generation of selection of Atlantic salmon increased body weight by 7% (Gjedrem, 1979). Crandall and Gall (1993) have shown that heritabilities for body weight in rainbow trout can be high, ranging from 0.89 in the first generation to 0.27 in the second.

Different strains of fish may possess varying amounts of additive genetic variation which can affect responses to selection. Smisek (1979) estimated heritabilities for body weight of 0.15–0.49 in a Czechoslovakian strain of common carp, and Vietnamese common carp have demonstrated significant heritability (0.3) for growth rate (Tran and Nguyen, 1993). Responses can also differ depending on the direction of selection: the body weight of common carp in Israel was not improved over five generations, but could be decreased in the same strain selected for small body size (Moav and Wohlfarth, 1974a).

Body conformation of fish can be dramatically altered by selection. Rainbow trout selected for high growth rate have a higher weight to length ratio (condition factor) relative to wild strains. For common carp, heritability for body depth was high (0.40–0.80) and, as seen in rainbow trout, deep-bodied lines have been developed (Ankorion, 1966). Selection programmes for carcass quality and quantity have also been initiated for salmonids and catfish (Dunham, 1996). Heritabilities for fat percentage in catfish and trout are about 0.50, indicating that selection should work to decrease fat content. However, heritability estimates for dressout percentage are near zero for these two species, indicating that selection for this trait would probably be unsuccessful.

Intraspecific Crossbreeding

Interstrain crossbreeding plays an important role in many animal breeding programmes, including those for fish. Heterotic growth in excess of both parent strains occurred in 55% and 22% of channel catfish and rainbow trout crossbreeds evaluated, respectively (Dunham and Smitherman, 1983a; Dunham, 1996), whereas chum salmon crossbreeds have not shown increased growth rates (Dunham, 1996). Crossbred channel catfish can grow 10–30% faster than the largest parental strain, and some common carp crossbreeds also expressed heterosis in a low percentage of the crosses examined (Moav *et al.*, 1964; Moav and Wohlfarth, 1974a; Nagy *et al.*, 1984). When crossbred strains do display heterosis, they can play an important role in commercial strain development: some heterotic crossbred strains exhibiting heterosis are the basis for carp industries in both Israel and Vietnam.

Domestication is also an important factor in crossbreeding. Domestic × domestic channel catfish crosses were more likely to exhibit heterotic rates of growth than domestic × wild crosses (Dunham and Smitherman, 1983b).

Domestic crosses of rainbow trout were also more likely to express heterosis than domestic × wild crosses (Gall, 1969; Gall and Gross, 1978; Kincaid, 1981; Ayles and Baker, 1983).

Interspecific Hybridization

Many fish species possess abilities to produce viable offspring in crosses with other species within or close to their own genera and, while most hybrids do not perform well, performance traits can be affected favourably in some cases. For example, the channel catfish female × blue catfish male hybrid is the only one of 28 catfish hybrids evaluated that exhibited overdominance for economic traits with potential applications for aquaculture (Smitherman and Dunham, 1985). This channel–blue hybrid has increased growth, growth uniformity, disease resistance, tolerance of low oxygen, dressing percentage and harvestability. However, mating blocks between the two species have thus far prevented their commercial utilization. Characteristics of reciprocal catfish interspecific hybrids can also differ. Paternal predominance (where the hybrid possesses traits more like the male parent than its reciprocal) was observed in channel–blue hybrids (Dunham *et al.*, 1982).

Several salmonid hybrids have also been evaluated. As expected, salmonid hybrids were more viable when made within genera than between genera (Chevassus, 1979), but salmonid hybrids have not been found to express heterosis for growth rate. Some diploid salmonid hybrids are potentially valuable because of disease resistance inherited from the parent species that is usually not cultured, but these hybrids unfortunately have low viability. The synthesis of triploids containing one genome from the paternal species and two from the maternal species can increase the hatchability of these potentially important hybrids (Parsons *et al.*, 1986).

Correlated Responses

In many cases, selection for one trait will affect other traits in either positive or negative ways because the genetic and physiological processes affecting them are linked. For example, in rainbow trout, genetic correlations of spawning date and spawning size, egg size and egg volume have been observed (Su *et al.*, 1996). In channel catfish, correlated responses to selection have generally been positive: fecundity, fry survival and disease resistance all correlated positively to selection for increased body weight in channel catfish after one generation (Dunham and Smitherman, 1983a; Smitherman and Dunham, 1985), whereas dressout percentage, visceral percentage, head percentage, seinability, spawning date, spawning rate, hatchability of eggs or survival of sac fry were not affected (Dunham, 1981). After three and four generations of selection for increased body weight, increased dressout percentage, decreased tolerance of low oxygen, but no change in body composition and seinability were observed (Rezk, 1993; Padi, 1995). Progeny from select brood fish also had greater feed consumption and feeding vigour, more efficient feed conversion, and greater disease resistance than the random controls (Dunham, 1981; Al-Ahmad, 1983). Feed consumption had a greater effect on body weight than feed conversion (Al-Ahmad, 1983).

A positive genetic correlation between body weights at different ages in channel catfish is evident from selection experiments. Select progeny grew faster than their control population during fingerling production in the first season in all strains examined (Dunham and Smitherman, 1983a). Two of the three select groups grew more rapidly during winter, and all select lines grew slightly faster than controls during the second season of growth.

Polyploidy

Fish provide a remarkable system for ploidy manipulation which can yield very useful effects on reproductive characteristics for commercial aquaculture. In many fish species, triploid females are unable to produce viable gametes in large numbers and are therefore functionally sterile, and other traits also can be affected. Triploidy can be induced by either a temperature or pressure shock soon after fertilization to block extrusion of the second polar body. Tetraploid individuals can be produced by disrupting the first mitotic cleavage using similar simple physical treatments, and these animals produce diploid sperm and thus triploid offspring in crosses to regular diploid animals.

Channel catfish triploids become larger than diploids at about 9 months of age (at a weight of approximately 90 g) when grown in tanks (Wolters *et al.*, 1982), slightly later than the time that sexual dimorphism in body weight is first detected in channel catfish. However, genotype–environment interactions occur for growth rate in triploid channel catfish: when grown in earthen ponds, triploids are smaller than diploids after 18 months at a weight of approximately 454 g (Dunham and Smitherman, 1987).

Triploid channel catfish convert feed more efficiently than diploids in a tank environment (Wolters *et al.*, 1982), and have 6% more carcass yield than diploids when 3 years of age (Chrisman *et al.*, 1983) due to the lack of gonadal development in triploids. Other traits of triploids are affected, including a darkening of the natural pigmentation. Combining triploidy and hybridization in catfish has not provided strains that grow as rapidly as diploids in commercial settings, and they have no dressout advantage and have decreased tolerance of low dissolved oxygen relative to diploids. Thus, little or no advantage exists for using polyploid catfish.

The flesh quality of triploid rainbow trout females is improved relative to diploid females because postmaturational changes are prevented (Bye and Lincoln, 1986). Sex-reversal and breeding for production of monosex female and all-female triploid populations of trout is being evaluated with the goal of producing rainbow trout with both superior growth rate and flesh quality. Coho salmon triploids have reduced growth rates and survival relative to diploids (Withler *et al.*, 1995), but growth and survival of triploid Atlantic salmon was the same as diploids in both communal and separate evaluation (Dunham, 1996). Triploid salmonid hybrids have shown similar (Quillet *et al.*, 1987) or slower (Parsons *et al.*, 1986) growth than diploid hybrids but, similar to intraspecific triploids, interspecific salmonid triploids could grow faster than controls once the maturation period was reached (Quillet *et al.*, 1987). The rainbow trout × coho salmon triploid had decreased growth, but had increased resistance to IHN virus (Dunham, 1996).

Sex Reversal and Breeding

In many commercial agricultural species, one sex possesses more favourable production characteristics than the other, and it is often necessary to cull the population early to improve production efficiency. In contrast, single-sex populations of many fish species have been developed due to their labile sex determination process which can be influenced by exogenous sex steroids. For example, treatment of mixed-sex populations of salmonids with α-methyltestosterone produces regular XY males and sex-reversed XX males, the latter can be simply identified using sex-specific DNA markers and then used in crosses with regular XX females to produce all-female production populations (Devlin *et al.*, 1994a). All-female populations of salmonids are desirable because males display early maturity at a small size, and poorer flesh quality. A combination of sex-reversal and breeding to produce all-female XX populations is the basis for more than half the rainbow trout industry in the UK (Bye and Lincoln, 1986), as well as the chinook salmon industry in Canada. Monosex chinook salmon and coho–chinook hybrid salmon have also been produced (Hunter *et al.*, 1983).

All-male progeny would be beneficial for catfish culture since males grow 10–30% faster than females, depending upon strain (Benchakan, 1979; Dunham and Smitherman, 1984, 1987; Smitherman and Dunham, 1985). Sex reversal and breeding has allowed production of YY channel catfish males that can be mated to normal XX females to produce all-male XY progeny. Channel catfish were feminized with β-oestradiol (Goudie *et al.*, 1983), and XY phenotypic females (identified through progeny testing) were found to be fertile (Goudie *et al.*, 1985). The sex ratio of progeny from matings of XY female and XY male channel catfish was 2.8 males:1 female, indicating that most, if not all, of the YY individuals are viable. YY males are also viable in salmon and trout, Nile tilapia, goldfish and channel catfish (Donaldson and Hunter, 1982).

Genotype–Environment Interactions

Genotype–environment interactions are prevalent in aquaculture. Strains and select lines of channel catfish performed similarly in aquaria, cages and ponds (Smitherman and Dunham, 1985), and strains selected for increased body weight at one stocking density in ponds also grew faster than their control populations at other stocking densities (Brummett, 1986).

Genotype–environment interactions are large and significant when comparing growth of different species, intraspecific crossbreeds, interspecific hybrids or polyploids of catfish (Dunham, 1996). The best genotype for ponds, the channel catfish female × blue catfish male, has mediocre growth in aquaria, tanks and cages. The behaviours, nervousness and aggressiveness, are the factors causing genotype–environment interactions for the channel–blue hybrid and triploid channel catfish, respectively. Additionally, genotype–environment interactions can be related to low oxygen levels when comparing channel–blue hybrids and their parents.

Genetic Engineering

Recombinant DNA technology and genetic engineering are biotechnologies that began to be applied to aquacultural species in the 1980s, and are now complementing traditional breeding programmes for improvement of culture traits (Houdebine and Chourrout, 1991). A variety of techniques have been explored to introduce new DNA sequences into fish, including microinjection, electroporation, retroviruses, and liposome-mediated and biolistic transfer methods (Lin *et al.*, 1994; Gong and Hew, 1995). Since fish embryonic stem cells are still in the early stages of development (Wakamatsu *et al.*, 1994; Sun *et al.*, 1995), techniques have focused on direct transfer of DNA into gametes or fertilized eggs to produce transgenic embryos.

A number of genes have been transferred into fish. Early reports of transgenesis in fish were complicated by a lack of distinction between extrachromosomal persistence of DNA early after injection, and DNA integration into the host genome. Subsequent analyses in several fish systems have demonstrated that foreign DNA can be inserted into the genome, and Southern blot analysis has indicated that the DNA can be found at one or more loci (Dunham *et al.*, 1987, 1992; Guyomard, 1989a,b; Penman *et al.*, 1991). Copy numbers can range from one to several thousand at a single locus, and the DNA can also be found organized in all possible concatemeric forms (Tewari *et al.*, 1992) suggesting random end-to-end ligation of the injected DNA prior to integration.

Probably due to the cytoplasmic nature of DNA injection, virtually all founder transgenic fish are mosaic and the integrated DNA is found only in a subset of developmental cell lineages. Mosaicism has been demonstrated in somatic tissues based on molecular tests, and also can be inferred for the germline based on the observed frequencies of transgene transmission to F1 progeny being less than at Mendelian ratios. For salmonids, the frequency of transgene transmission from founder animals averages about 15%, suggesting that integration of the foreign DNA occurs on average at the two to four cell stage of development (see Devlin, 1996). Transmission of transgenes to F2 or later progeny occurs at Mendelian frequencies (Shears *et al.*, 1991), indicating that the DNA is stably integrated into the host genome and passes normally through the germline.

A variety of promoters have been shown to be active in fish cells, and most investigations have used promoters derived from non-piscine vertebrates and their viruses. Among others, reporter gene activity has been detected in fish or fish tissue culture cells from the Rous sarcoma virus long terminal repeat (RSV-LTR), simian virus SV40, cytomegaloviruses CMV-tk and CMV-IE, MMTV, polyoma viral promoters, human and mouse metallothionein, and human heat shock 70 promoters (Hackett, 1993).

Piscine genes and their promoters also have been utilized in fish transgenesis and in cell transfection studies. Promoters from flounder antifreeze, carp β-actin, and salmonid metallothionein-B and histone H3 have been found to be active (Liu *et al.*, 1990; Gong *et al.*, 1991; Chan and Devlin, 1993). In general, it appears that many eukaryotic promoters are able to function in fish cells, although if derived from non-homologous sources, the level of expression may be somewhat reduced.

Expression of uninterrupted coding regions from prokaryotic and eukaryotic sources has been successful in fish cells (e.g. Du *et al.*, 1992). However, it should be noted that expression in fish cells of gene constructs containing mammalian introns might be inefficient due to difficulties in RNA processing to yield functional mRNA (Bearzotti *et al.*, 1992; Bétancourt *et al.*, 1993). At present it is not possible to generalize about the activities of various gene promoters and gene constructs in different fish species. Such information needs to be empirically derived for the fish system of interest.

Biological Effects of Transgenesis

Positive biological effects have been exhibited by transgenic fish. Due to the lack of available piscine gene sequences, transgenic fish research in the mid-1980s utilized existing mammalian GH gene constructs, and evidence for growth enhancement was reported for some fish species examined (Zhu *et al.*, 1986; Enikolopov *et al.*, 1989; Gross *et al.*, 1992; Lu *et al.*, 1992; Zhu, 1992; Wu *et al.*, 1994). Peculiarly, mammalian gene constructs (e.g. mMT/rGH) failed to have any effect on growth of salmonids (Guyomard *et al.*, 1989a,b; Penman *et al.*, 1991), despite a comprehensive literature showing that salmonids are very responsive to growth stimulation by exogenously administered GH protein (McLean and Donaldson, 1993). This lack of response by salmonids to mammalian GH gene constructs may be due to difficulties associated with RNA processing as mentioned above, or due to the strains of trout utilized (see below).

Some investigators have developed gene constructs containing fish GH sequences driven by non-piscine promoters, and have observed growth enhancement in transgenic carp, catfish (Fig. 15.1), zebrafish and tilapia (Zhang *et al.*, 1990; Dunham *et al.*, 1992; Chen *et al.*, 1993; Zhao *et al.*, 1993; Martinez *et al.*, 1996). Growth stimulatory effects observed with the above constructs have ranged from no effect to approximately twofold increases in weight relative to controls, and provided the first convincing data demonstrating that growth enhancement in fish can be achieved by transgenesis.

More recently, GH gene constructs have been developed that are comprised entirely of piscine gene sequences. This was achieved using either an ocean pout antifreeze promoter driving a chinook salmon GH cDNA, or a sockeye salmon metallothionein promoter driving the full-length sockeye *GH1* gene. When introduced into salmonids, such gene constructs elevate circulating GH levels by 40-fold in some cases (Devlin *et al.*, 1994b; Devlin, 1996), and result in approximately a five- to 11-fold increase in weight (Fig. 15.1) after 1 year of growth (Du *et al.*, 1992; Devlin *et al.*, 1994b, 1995a). These GH gene constructs also induce precocious development of smoltification, a physiological transformation necessary for marine survival of salmonids.

Pleiotropic Effects

When a gene is inserted with the objective of improving a specific trait, that gene may affect more than one phenotypic character. Since pleiotropic effects could be positive or negative, it is important to evaluate commercially important traits in transgenic fish in addition to the trait intended for alteration.

The insertion of a rainbow trout GH transgene for growth enhancement may alter the survival of common carp. The number of F2 progeny inheriting this transgene is much less than expected, and differential mortality or loss of the transgene during meiosis are likely explanations. From fingerling size onwards, survival of the remaining transgenic individuals was higher than that of controls when subjected to a series of stressors and pathogens such as low oxygen, anchor worms and dropsy (Chatakondi, 1995). Where examined, reproductive traits such as fecundity or precocious sexual development in transgenic common carp, have not been affected in transgenic fish.

Increased growth rate of GH transgenic fish could arise from increased food consumption, feed conversion efficiency or both. Fast-growing transgenic common carp containing rainbow trout GH gene had lower feed conversion efficiency than controls (Chatakondi *et al.*, 1995) whereas other

Fig. 15.1. Growth-enhanced transgenic and control catfish (top panel) and coho salmon (bottom panel). For each species, transgenic individuals are shown on top relative to control animals of the same age (approximately 1 year of age for both species). Fish sizes: transgenic catfish 900 g, control catfish 350 g; transgenic salmon 730 g, control salmon 15 g. The gene construct used in catfish is RSV LTR-rtGH1 (Dunham *et al.*, 1992), and for salmon is OnMTGH1 (Devlin *et al.*, 1994b).

families had increased, decreased or no change in food consumption, and had improved feed conversion. Salmonids injected with somatotropins also display improved feed conversion (Devlin *et al.*, 1994c), and this effect is also anticipated in GH transgenic salmonids.

It is well known that GH has important physiological effects on energy absorption and utilization in vertebrates. Thus it is anticipated that GH transgenic fish may have altered body compositions relative to controls. Indeed, transgenic common carp containing a rainbow trout GH had more protein, less fat and less moisture than non-transgenic full-siblings (about a 10% change). GH promotes the synthesis of protein over fat, and the elevated levels of GH in transgenic fish thus increase the protein/lipid ratio, with fat level decreased by as much as 50%. The increased relative level of protein in transgenic common carp muscle also results in increased levels of amino acids, but the ratios among different amino acids and different fatty acids are virtually identical in control and transgenic common carp (Chatakondi *et al.*, 1995).

In GH transgenic salmon, the endocrine stimulation can be elevated to pathological levels in some cases, and excessive and deleterious deposition of cartilage analogous to the mammalian acromegaly syndrome has been observed (Devlin *et al.*, 1995b). This effect can be sufficiently severe such that impaired feeding and respiration may result in reduced growth and poor viability. Consequently, animals that ultimately display the greatest growth enhancement as adults are those that have been only moderately growth stimulated (Devlin *et al.*, 1995a).

In common carp, body shape changes as a result of expression of rainbow trout GH. The transgenic individuals have relatively larger heads, deeper and wider bodies and caudal areas when compared with controls. As growth differences increase, the body shape differences also increase to a point, and then plateau. The morphological change does not affect condition factor, but does improve the dressing percentage (Chatakondi *et al.*, 1994).

Environmental Risk and Fitness of Transgenic Fish

The use of transgenic fish in aquaculture has raised concerns regarding their potential interaction with other species as well as with wild members of the same species in natural ecosystems (Kapuscinski and Hallerman, 1991). Such potential impacts would depend on the degree of phenotypic change displayed by transgenic animals, as well as their fitness relative to wild-type and the number of individuals escaping from production facilities.

Transgenic channel catfish containing salmonid GH genes grow 33% faster than normal channel catfish in aquaculture conditions with supplemental feeding, but no significant difference was observed in ponds with only natural feed, indicating equal foraging abilities between transgenic and control animals. Thus, transgenic catfish require supplemental rations to exhibit their growth potential (Chitmanat, 1996), a condition which may not exist in natural environments. Relative to transgenic individuals, non-transgenic catfish fry and fingerlings had better predator avoidance of largemouth bass *Micropterus salmoides* and green sunfish *Lepomis cyanellus*. Spawning ability of transgenic and control channel catfish was equal, but GH transgenic Nile tilapia had decreased sperm production.

The swimming ability of some transgenic salmon is reduced compared with non-transgenic salmon (Farrell *et al.*, 1997) which could lead to greater vulnerability to predators, decreased ability to capture prey, and decreased ability to successfully complete spawning migrations. The increased vulnerability to predators, impaired swimming, lack of increased growth when foraging, unchanged spawning percentage and potential decreased sperm production indicate that some transgenic fish may not compete well under natural conditions which cause major ecological or environmental damage. However, due to the difficulty in predicting potential environmental impacts from laboratory fitness estimates, it would be desirable to implement physical and biological (sterilization) containment methods to reduce potential interactions between transgenic and wild fish populations (Devlin and Donaldson, 1992).

Dramatic Growth of Transgenic Fish: Explanations and Limitations

It is interesting to speculate why growth enhancement varies greatly among different transgenic fish systems (Devlin, 1996), and why salmonids in particular have shown the greatest response to stimulation to date. Several potential explanations exist which indicate that it may be difficult to duplicate these results in other fish species.

It is possible that completely homologous gene constructs (i.e. derived only from the same species, or from piscine sources) such as those used successfully in salmonids are expressed in fish more efficiently than gene constructs derived from other vertebrates. While this probably plays a role in efficient expression, it is also very likely that the differences in growth response observed in different transgenic systems are due to the vastly different physiologies and life-history characteristics that exist among the fish species examined.

The biology of salmon and their unique physiological adaptations undoubtedly play an important role in the dramatic growth enhancement observed in transgenic animals. Growth in salmonids is normally relatively slow throughout the year, and is seasonally extremely low when water temperatures are low and food resources in nature are scarce. This low growth rate appears to be controlled at least in part by the level of circulating GH and can be dramatically stimulated with exogenous GH protein (see above) and sufficient food. Thus, the dramatic growth stimulation observed in transgenic salmonids may arise, at least in part, by the seasonal deregulation of GH expression (Devlin *et al.*, 1994b, 1996) to allow high growth rates during winter months when control animals have very slow growth rates. This winter growth may also give them a large advantage that can later be magnified (Moav and Wohlfarth, 1974b). Additionally, salmonids are anadromous, and accelerated growth in transgenics allows them to reach a size where they smolt earlier than controls. Growth in the smolt stage is naturally increased, providing transgenic individuals with a further advantage to distance themselves in growth from controls.

In contrast, other non-salmonid fish species generally possess more rapid growth, and consequently may be much more difficult to stimulate by expression of GH in transgenic organisms (Devlin, 1996). These high growth rates occur naturally in some species (i.e. tilapia), whereas in others they have been enhanced through genetic selection and many years of domestication. Strains or species that have been selected to near maximal growth rates presumably have had many of their metabolic and physiological processes optimized, and would be expected to be more difficult to stimulate by a single factor such as GH.

Domestication is also important in transgenic growth responses. In this regard, we have observed that salmonid GH gene constructs that have a dramatic effect on growth in wild rainbow trout strains (with naturally low growth rates) have little or no effect in strains where growth rate has been enhanced by selection over many years (unpublished observation). Consistent with these observations, the dramatically growth-responsive salmon previously observed (Du *et al.*, 1992; Devlin *et al.*, 1994b, 1995a) were also derived from wild strains. Apparently, slow-growing wild strains can benefit much more from GH insertion than fish that already have growth enhancement from selective breeding.

By comparison, GH transgenic catfish derived from domesticated and selectively bred strains exhibit only a moderate growth enhancement (41%). However, if we extrapolate from a series of experiments starting with slow-growing wild strains of channel catfish and then improve their growth through domestication (Dunham, 1996), followed by further improvement from selective breeding (Padi, 1995), then further increases from interspecific hybridization (Jeppsen, 1995) or gene transfer, the overall growth enhancement is approximately tenfold, comparable to that observed with transgenic wild salmon. Thus the growth of wild fish apparently can be improved in one or two generations with the insertion of GH genes to the extent that would take many generations of selective breeding to achieve.

Similarly, it can be noted that dramatic growth stimulation in the mammalian system using GH transgenes has been observed in mice, but not in domestic livestock that have had many centuries of genetic selection (Pursel *et al.*, 1989; Palmiter *et al.*, 1992). In these domesticated and selected strains, the capacity for further growth improvement by GH may now be restricted by limitations in other physiological pathways, and other methods, including traditional breeding methods, may yield the greatest gains. For aquacultural species that have a much shorter history of domestication and selection, future genetic improvement will likely be accomplished by utilizing a combination of both approaches simultaneously.

References

Al-Ahmad, T.A. (1983) Relative effects of feed consumption and feed efficiency on growth of catfish from different genetic backgrounds. Ph.D. dissertation, Auburn University, Alabama.

Ankorion, Y. (1966) Investigations on the heredity of some morphological traits in the common carp, *Cyprinus carpio* L. M.S. thesis, the Hebrew University, Jerusalem (in Hebrew).

Ayles, G.B. and Baker, R.F. (1983) Genetic differences in growth and survival between strains and hybrids of rainbow trout (*Salmo gairdneri*) stocked in aquaculture lakes in the Canadian prairies. *Aquaculture* 33, 269–280.

Bearzotti, M., Perrot, E., Michard-Vanhée, C., Jolivet, G., Attal, J., Théron, M.C., Puissant, C., Dreano, M., Kopchick, J.J., Powell, R., Gannon, F., Houdebine, L.M. and Chourrout, D. (1992) Gene expression following transfection of fish cells. *Journal of Biotechnology* 26, 315–325.

Benchakan, M. (1979) Morphometric and meristic characteristics of blue, channel, white, and blue–channel hybrid catfishes. M.S. thesis, Auburn University, Alabama.

Bétancourt, O.H., Attal, J., Théron, M.C., Puissant, C., Houdebine, L.M. and Bearzotti, M. (1993) Efficiency of introns from various origins in fish cells. *Molecular Marine Biology and Biotechnology* 2, 181–188.

Bondari, K. (1983) Response to bi-directional selection for body weight in channel catfish. *Aquaculture* 33, 73–81.

Brummett, R.E. (1986) Effects of genotype × environment interactions on growth, variability and survival of improved catfish. Ph.D. dissertation, Auburn University, Alabama.

Burch, E.P. (1986) Heritabilities for body weight, feed consumption and feed conversion and the correlations among these traits in channel catfish, *Ictalurus punctatus*. M.Sc. thesis, Auburn University, Alabama.

Bye, V.J. and Lincoln, R.F. (1986) Commercial methods for the control of sexual maturation in rainbow trout (*Salmo gairdneri* R.). *Aquaculture* 57, 299–309.

Chan, W.K. and Devlin, R.H. (1993) Polymerase chain reaction amplification and functional characterization of sockeye salmon histone H3, metallothionein-B, and protamine promoters. *Molecular Marine Biology and Biotechnology* 2, 308–318.

Chatakondi, N.G. (1995) Evaluation of transgenic common carp, *Cyprinus carpio*, containing rainbow trout growth hormone in ponds. Ph.D. dissertation, Auburn University, Alabama.

Chatakondi, N., Ramboux, A.C., Nichols, A., Hayat, M., Duncan, P.L., Chen, T.T., Powers, D.A. and Dunham, R.A. (1994) The effect of rainbow trout growth hormone gene on the morphology, dressing percentage and condition factor in the common carp, *Cyprinus carpio*. *Proceedings of the V Congress of Genetics and Applied Livestock Production* 17, 481–484.

Chatakondi, N., Lovell, R., Duncan, P., Hayat, M., Chen, T., Powers, D., Weete, T., Cummins, K. and Dunham, R.A. (1995) Body composition of transgenic common carp, *Cyprinus carpio,* containing rainbow trout growth hormone gene. *Aquaculture* 138, 99–109.

Chen, T.T., Kight, K., Lin, C.M., Powers, D.A., Hayat, M., Chatakondi, N., Ramboux, A.C., Duncan, P.L. and Dunham, R.A. (1993) Expression and inheritance of RSVLTR-rtGH1 complementary DNA in the transgenic common carp (*Cyprinus carpio*). *Molecular Marine Biology and Biotechnology* 2, 88–95.

Chevassus, B. (1979) Hybridization in salmonids: results and perspectives. *Aquaculture* 17, 113–128.

Chitminat, C. (1996) Predator avoidance of transgenic channel catfish containing salmonid growth hormone genes. M.S. thesis, Auburn University, Alabama.

Chrisman, C.L., Wolters, W.R. and Libey, G.S. (1983) Triploidy in channel catfish. *Journal of the World Mariculture Society* 14, 279–293.

Crandall, P.A. and Gall, G.A.E. (1993) The genetics of age and weight at sexual maturity based on individually tagged rainbow trout (*Oncorhynchus mykiss*). *Aquaculture* 117, 95–105.

Devlin, R.H. (1996) Transgenic salmonids. In: Houdebine, L.M. (ed.) *Transgenic Animals: Generation and Use.* Harwood Academic Publishers, Amsterdam.

Devlin, R.H. and Donaldson, E.M. (1992) Containment of genetically altered fish with emphasis on salmonids. In: Hew, C.L. and Fletcher, G.L. (eds) *Transgenic Fish.* World Scientific, Singapore, pp. 229–265.

Devlin, R.H., McNeil, B.K., Solar, I.I. and Donaldson, E.M. (1994a) A rapid PCR-based test for Y-chromosomal DNA allows simple production of all-female strains of chinook salmon. *Aquaculture* 128, 211–220.

Devlin, R.H., Yesaki, T.Y., Biagi, C.A., Donaldson, E.M., Swanson, P. and Chan, W.-K. (1994b) Extraordinary salmon growth. *Nature* 371, 209–210.

Devlin, R.H., Byatt, J.C., McLean, E., Yesaki, T.Y., Krivi, G.G., Jaworski, E.G., Clarke, W.C. and Donaldson, E.M. (1994c) Bovine placental lactogen is a potent stimulator of growth and displays strong binding to hepatic liver receptor sites of coho salmon. *Genetic Comparative Endocrinology* 95, 31–41.

Devlin, R.H., Yesaki, T.Y., Donaldson, E.M., Du, S.-J. and Hew, C.L. (1995a) Production of germline transgenic Pacific salmonids with dramatically increased growth performance. *Canadian Journal of Fish and Aquatic Science* 52, 1376–1384.

Devlin, R.H., Yesaki, T.Y., Donaldson, E.M. and Hew, C.L. (1995b) Transmission and phenotypic effects of an antifreeze/GH gene construct in coho salmon (*Oncorhynchus kisutch*). *Aquaculture* 137, 161–169.

Donaldson, E.M. and Devlin, R.H. (1996) Uses of biotechnology to enhance production. In: Pennell, W. and Barton, B. (eds) *Principles of Salmonid Culture. Developments in Aquaculture and Fisheries Science,* Vol. 29. Elsevier Publishers, Amsterdam, pp. 969–1020.

Donaldson, E.M. and Hunter, G.A. (1982) Sex control in fish with particular reference to salmonids. *Canadian Journal of Fish and Aquatic Science* 39, 99–110.

Du, S.J., Gong, Z., Fletcher, G.L., Schears, M.A., King, M.J., Idler, D.R. and Hew, C.L. (1992) Growth enhancement in transgenic Atlantic salmon by the use of an 'all-fish' chimeric growth hormone gene construct. *Bio/Technology* 10, 176–181.

Dunham, R.A. (1981) Response to selection and realized heritability for body weight in three strains of channel catfish grown in earthen ponds. Ph.D. dissertation, Auburn University, Alabama.

Dunham, R.A. (1996) *Contribution of Genetically Improved Aquatic Organisms to Global Food Security.* International Conference on Sustainable Contribution of Fisheries to Food Security. Government of Japan and FAO, Rome, 150 pp.

Dunham, R.A. and Smitherman, R.O. (1983a) Response to selection and realized heritability for body weight in three strains of channel catfish, *Ictalurus punctatus*, grown in earthen ponds. *Aquaculture* 33, 88–96.

Dunham, R.A. and Smitherman, R.O. (1983b) Crossbreeding channel catfish for improvement of body weight in earthen ponds. *Growth* 47, 97–103.

Dunham, R.A. and Smitherman, R.O. (1984) *Ancestry and Breeding of Catfish in the United States.* Circular 273, Alabama Agricultural Experimental Station, Auburn University, Alabama.

Dunham, R.A. and Smitherman, R.O. (1987) *Genetics and Breeding of Catfish.* Regional Research Bulletin 325, Southern Cooperative Series, Alabama Agricultural Experimental Station, Auburn University, Alabama.

Dunham, R.A., Smitherman, R.O., Brooks, M.J., Benchakan, M. and Chappell, J.A. (1982) Paternal predominance in reciprocal channel–blue hybrid catfish. *Aquaculture* 29, 389–396.

Dunham, R.A., Eash, J., Askins, J. and Townes, T.M. (1987) Transfer of the metallothionein-human growth hormone fusion gene into channel catfish. *Transactions of the American Fish Society* 116, 87–91.

Dunham, R.A., Ramboux, A.C., Duncan, P.L., Hayat, M., Chen, T.T., Lin, C.M., Kight, K., Gonzalez-Villasenor, I. and Powers, D.A. (1992) Transfer, expression and inheritance of salmonid growth hormone in channel catfish, *Ictalurus punctatus*, and effects on performance traits. *Molecular Marine Biology and Biotechnology* 1, 380–389.

Enikolopov, G.N., Benyumov, A.O., Barmintsev, A., Zelenina, L.A., Sleptsova, L.A., Doronin, Y.K., Golichenkov, V.A., Grashchuk, M.A., Georgiev, G.P., Rubtsov, P.M., Skryabin, K.G. and Baev, A.A. (1989) Advanced growth of transgenic fish containing human somatotropin gene. *Doklady Akademii Nauk SSSR* 301, 724–727.

Farrell, A.P., Bennett, W. and Devlin, R.H. (1997) Growth-enhanced transgenic salmon can be inferior swimmers. *Canadian Journal of Zoology* 75, 335–337.

Fletcher, G. and Davies, P.L. (1991) Transgenic fish for aquaculture. *Genetic Engineering* 13, 331–369.

Gall, G.A.E. (1969) Quantitative inheritance and environmental response of rainbow trout. In: Neuhaus, O.W. and Halver, J.E. (eds) *Fish Research.* Academic Press, New York.

Gall, G.A.E. and Gross, S.J. (1978) Genetic studies of growth in domesticated rainbow trout. *Aquaculture* 13, 225–234.

Gjedrem, T. (1979) Selection for growth rate and domestication in Atlantic salmon. *Z. Tierz. Zuchtungsbiol.* 96, 56–59.

Gong, Z. and Hew, C.L. (1995) Transgenic fish in aquaculture and developmental biology. In: Pederson, R.A. and Schatten, G.P. (eds) *Current Topics in Developmental Biology*, Vol. 30. Academic Press, San Diego, pp. 175–214.

Gong, Z., Hew, C.L. and Vielkind, J.R. (1991) Functional analysis and temporal expression of promoter regions from fish antifreeze protein genes in transgenic Japanese medaka embryos. *Molecular Marine Biology and Biotechnology* 1, 64–72.

Goudie, C.A., Redner, B.D., Simco, B.A. and Davis, K.B. (1983) Feminization of channel catfish by oral administration of steroid sex hormones. *Transactions of the American Fish Society* 112, 670–672.

Goudie, C.A., Khan, G. and Parker, N. (1985) Gynogenesis and sex manipulation with evidence for female homogameity in channel catfish (*Ictalurus punctatus*). Annual Progress Report 1 Oct.–30 Sept. USFWS, Southeast. Fish Culture Laboratory, Marion, Alabama.

Gross, M.L., Schneider, J.F., Moav, N., Moav, B., Alvarez, C., Myster, S.H., Liu, Z., Hallerman, E.M., Hackett, P.B., Guise, K.S., Faras, A.J. and Kapuscinski, A.R. (1992) Molecular analysis and growth evaluation of northern pike (*Esox lucius*) microinjected with growth hormone genes. *Aquaculture* 103, 253–273.

Guyomard, R., Chourrout, D. and Houdebine, L. (1989a) Production of stable transgenic fish by cytoplasmic injection of purified genes. *Gene Transfer and Gene Therapy*, pp. 9–18.

Guyomard, R., Chourrout, D., Leroux, C., Houdebine, L.M. and Pourrain, F. (1989b) Integration and germline transmission of foreign genes microinjected into fertilized trout eggs. *Biochimie* 71, 857–863.

Hackett, P.B. (1993) The molecular biology of transgenic fish. In: Hochachka, P.W. and Mommsen, T.P. (eds) *Biochemistry and Molecular Biology of Fishes, Molecular Biology Frontiers*, Vol. 2. Elsevier, Amsterdam, pp. 207–240.

Hershberger, W.K., Myers, J.M., Iwamoto, R.N., Mcauley, W.C. and Saxton, A.M. (1990) Genetic changes in the growth of coho salmon (*Oncorhynchus kisutch*) in marine net-pens, produced by ten years of selection. *Aquaculture* 85, 187–197.

Houdebine, L.M. and Chourrout, D. (1991) Transgenesis in fish. *Experientia* 47, 891–897.

Hunter, G.A., Donaldson, E.M., Stoss, J. and Baker, I. (1983) Production of monosex female groups of chinook salmon (*Oncorhynchus tshawytscha*) by the fertilization of normal ova with sperm from sex-reversed females. *Aquaculture* 33, 355–364.

Jeppsen, T.S. (1995) Comparison of performance of channel catfish, *Ictalurus punctatus,* × blue catfish, *I. furcatus*, hybrids from Kansas select and Kansas random dams. M.S. thesis, Auburn University, Alabama.

Kapuscinski, A.R. and Hallerman, E.N. (1991) Implications of introduction of transgenic fish into natural ecosystems. *Canadian Journal of Fisheries and Aquatic Sciences* 48, 99–107.

Kincaid, H.L. (1981) Trout salmon registry. FWS/NFC-L/81–1, US Fish and Wildlife Services, Kearneysville, Wyoming.

Kincaid, H.L. (1983) Results from six generations of selection for accelerated growth rate in a rainbow trout population. Abst. *The Future of Aquaculture in North America.* Fish Culture Section of the American Fisheries Society 26–27.

Lin, S., Gaiano, N., Culp, P., Burns, J.C., Friedmann, T., Yee, J.-K. and Hopkins, N. (1994) Integration and germ-line transmission of a pseudotyped retroviral vector in zebrafish. *Science* 265, 666–669.

Liu, Z., Moav, B., Faras, A.J., Guise, K.S., Kapuscinski, A.R. and Hackett, P.B. (1990) Development of expression vectors for transgenic fish. *Biotechnology* 8, 1268–1272.

Lu, J.K., Chen, T.T., Chrisman, C.L., Andrisani, O.M. and Dixon, J.E. (1992) Integration, expression, and germ-line transmission of foreign growth hormone genes in medaka (*Oryzias latipes*). *Molecular Marine Biology and Biotechnology* 1, 366–375.

Martinez, R., Estrada, M.P., Berlanga, J., Guillin, I., Hernandez, O., Cabrera, E., Pimentel, R., Morales, R., Herrera, F., Morales, A., Pina, J., Abad, Z., Sanchez, V., Melamed, P., Lleonart, R. and de la Fuente, J. (1996) Growth enhancement of transgenic tilapia by ectopic expression of tilapia growth hormone. *Molecular Marine Biology and Biotechnology* 5, 62–70.

McLean, E. and Donaldson, E.M. (1993) The role of somatotropin in growth in poikilotherms. In: Schreibman, M.P., Scanes, C.G. and Pang, P.K.T. (eds) *The Endocrinology of Growth, Development and Metabolism in Vertebrates.* Academic Press, New York, pp. 43–71.

Moav, R. and Wohlfarth, G. (1974a) Carp breeding in Israel. In: Moav, R. (ed.) *Agricultural Genetics.* John Wiley & Sons, New York.

Moav, R. and Wohlfarth, G. (1974b) Magnification through competition of genetic differences in yield capacity in carp. *Heredity* 33, 181–202.

Moav, R. and Wohlfarth, G. (1976) Two-way selection for growth rate in the common carp (*Cyprinus carpio* L.). *Genetics* 82, 83–101.

Moav, R., Wohlfarth, G. and Lahman, M. (1964) Genetic improvement of carp. VI. Growth rate of carp imported from Holland, relative to Israeli carp, and some crossbred progeny. *Bamidgeh* 16, 142–149.

Nagy, A., Csanyi, V., Bakos, J. and Bercsenyl, M. (1984) Utilization of gynogenesis and sex-reversal in commercial carp breeding: growth of the first gynogenetic hybrids. *Aquacultura Hungarica (Szarvas)* IV, 7–16.

Padi, J.N. (1995) Response and correlated responses to four generations of selection for increased body weight in the Kansas strain channel catfish, *Ictalurus punctatus*, grown in earthen ponds. M.S. thesis, Auburn University, Alabama.

Palmiter R.D., Brinster, R.L., Hammer, R.E., Trumbauer, M.E., Rosenfeld, M.G., Birnberg, N.C. and Evans, R.M. (1982) Dramatic growth of mice that develop from eggs microinjected with metallothionein–growth hormone fusion genes. *Nature* 300, 611–615.

Parsons, J., Busch, R., Thorgaard, G. and Scheerer, P. (1986) Resistance of diploid and triploid rainbow trout, coho salmon and reciprocal hybrids to infectious hematopoietic necrosis (IHN). *Aquaculture* 57, 337–343.

Penman, D.J., Beeching, A.J., Penn, S., Rhaman, A., Sulaiman, Z. and Maclean, N. (1991) Patterns of transgene inheritance in rainbow trout (*Oncorhynchus mykiss*). *Molecular Reproduction and Development* 30, 201–206.

Pursel, V.G., Pinkert, C.A., Miller, K.F., Bolt, D.J., Campbell, R.G., Palmiter, R.D., Brinster, R.L. and Hammer, R.E. (1989) Genetic engineering of livestock. *Science* 244, 281–288.

Quillet, E., Chevassus, B. and Krieg, F. (1987) Characterization of auto- and allo tetraploid salmonids for rearing in seawater cages. In: Tiews, K. (ed.) *Selection, Hybridization and Genetic Engineering in Aquaculture*, Vol. 2. Heeneman, Berlin, p. 239.

Rezk, M.S. (1993) Response and correlated responses to three generations of selection for increased body weight in channel catfish, *Ictalurus punctatus*. Ph.D. dissertation, Auburn University, Alabama.

Shears, M.A., Fletcher, G.L., Hew, C.L., Gauthier, S. and Davies, P.L. (1991) Transfer, expression and stable inheritance of antifreeze protein genes in Atlantic salmon (*Salmo salar*). *Molecular Marine Biology and Biotechnology* 1, 58–63.

Smisek, J. (1979) Considerations of body conformation, heritability and biochemical characters in genetic studies of carp in Czechoslovakia. *Bulletin VÚRH*, Vodnany, no. 15, pp. 3–6. (*Animal Breeding Abstracts* 1980, 48, 302.)

Smitherman, R.O. and Dunham, R.A. (1985) Genetics and breeding. In: Tucker, C.S. (ed.) *Channel Catfish Culture*. Elsevier Scientific Publishing, Amsterdam, pp. 283–316.

Su, G.S., Liljedahl, L.E., and Gall, G.A.E. (1997) Genetic and environmental variation of female reproductive traits in rainbow trout (*Oncorhynchus mykiss*). *Aquaculture* 154, 115–124.

Sun, L., Bradford, C.S., Ghosh, C., Collodi, P. and Barnes, D.W. (1995) ES-like cell cultures derived from early zebrafish embryos. *Molecular Marine Biology and Biotechnology* 4, 193–199.

Tacon, A.G.J. (1996) Global trends in aquaculture and aqua feed production. *The International Milling Directory and Buyer's Guide*. Turret Group.

Tewari, R., Michard-Vanhée, C., Perrot, E. and Chourrout, D. (1992) Mendelian transmission, structure and expression of transgenes following their injection into the cytoplasm of trout eggs. *Transgenic Research* 1, 250–260.

Tran, M.T. and Nguyen, C.T. (1993) Selection of common carp (*Cyprinus carpio* L.) in Vietnam. *Aquaculture* 111, 301–302.

Wakamatsu, Y., Ozato, K. and Sasado, T. (1994) Establishment of a pluripotent cell line derived from a medaka *(Oryzias latipes)* blastula embryo. *Molecular Marine Biology and Biotechnology* 3, 185–191.

Withler, R.E., Beacham, T.D., Solar, I.I. and Donaldson, E.M. (1995) Freshwater growth, smolting, and marine survival and growth of diploid and triploid coho salmon (*Oncorhynchus kisutch*). *Aquaculture* 136, 91–107.

Wolters, W.R., Chrisman, C.L. and Libey, G.S. (1982) Erythrocyte nuclear measurements of diploid and triploid channel catfish, *Ictalurus punctatus* (Rafinesque). *Journal of Fisheries Biology* 20, 253–258.

Wu, T., Yang, H., Dong, Z., Xia, D., Shi, Y., Ji, X., Shen, Y. and Sun, W. (1994) The integration and expression of human growth gene in blunt snout bream and common carp. *Journal of Fisheries, China, Shuichan Xuebao* 18, 284–289.

Zhang, P., Hayat, M., Joyce, C., Gonzalez-Villasenor, L.I., Lin, C.M., Dunham, R.A., Chen, T.T. and Powers, D.A. (1990) Gene transfer, expression and inheritance of pRSV-rainbow trout-GH cDNA in the common carp, *Cyprinus carpio* (Linneaus). *Molecular Reproduction and Development* 25, 3–13.

Zhao, X., Zhang, P.J. and Wong, T.K. (1993) Application of Baekonization: a new approach to produce transgenic fish. *Molecular Marine Biology and Biotechnology* 2, 63–69.

Zhu, Z. (1992) Generation of fast growing transgenic fish: methods and mechanisms. In: Hew, C.L. and Fletcher, G.L. (eds) *Transgenic Fish*. World Scientific Publishing, Singapore, pp. 92–119.

Zhu, Z., Xu, K., Li, G., Xie, Y. and He, L. (1986) Biological effects of human growth hormone gene microinjected into the fertilized eggs of loach, *Misgurnus anguillicaudatus. Kexue Tongbao Academia Sinica* 31, 988–990.

16

Direct and Correlated Responses to Short-term Selection for 8-week Body Weight in Lines of Transgenic (oMt1a-oGH) Mice

F. Siewerdt[1], E.J. Eisen[1] and J.D. Murray[2]

[1] *Department of Animal Science, North Carolina State University, Raleigh, North Carolina, USA;* [2] *Department of Animal Science and Department of Population Health and Reproduction, University of California, Davis, California, USA*

The objective of this experiment was to evaluate the results of selection for increased 8-week body weight in lines of mice with or without a sheep metallothionein 1a sheep growth hormone (oMt1a-oGH) transgene in two genetic backgrounds. The transgene was introgressed into two lines of mice which had previously either been selected for rapid growth or randomly selected. Selection was practiced within families of full-sibs for seven generations. Selection was effective in increasing 8-week body weight in all non-transgenic lines and in some of the transgenic lines. The initial transgene frequency of 0.5 increased to about 0.6 in the lines with random selection background, but decreased to less than 0.2 in those lines from the selected background. Correlated responses in other growth and fitness traits were observed in some lines, and when present were chiefly in the desired direction. It was concluded that selection in the transgenic lines was successful, although the response was dependent on the genetic background. Realized response and realized heritability for 8-week body weight were lower in the transgenic than in the non-transgenic lines, but no significant differences were found between the selected and unselected background, nor was there any significant interaction. The lower response in the transgenic lines may have been due to reduced prenatal survival of transgenic embryos.

Introduction

Large-scale use of transgenics in animal breeding plans is not yet in effect, although incorporation of foreign DNA into commercial livestock species has already been achieved (Rexroad, 1992). One problem has been that

© CAB INTERNATIONAL 1999. *Transgenic Animals in Agriculture* (eds J.D. Murray, G.B. Anderson, A.M. Oberbauer and M.M. McGloughlin)

most transgene constructs developed for livestock have been poorly regulated. Another major difficulty lies in the method of gene transfer most commonly used, namely, microinjection. The number of transgenic individuals produced by microinjection is limited, and the site of insertion of the transgene construct is usually distinct for each transgenic founder animal. In addition, each line formed is partially inbred because all individuals descend from the same founder animal. Aside from specific expected benefits to production traits or disease resistance, a transgene should not have undesirable effects on fitness and should be able to be regulated. The activity of a transgene is affected both by its insertion site (Al-Shawi *et al.*, 1990) and by the background of the line in which the transgene is to be inserted (Eisen *et al.*, 1995; Siewerdt *et al.*, 1998).

The introduction of the sheep metallothionein 1a sheep growth hormone transgene (oMt1a-oGH) into the murine genome and its regulation are well documented (Shanahan *et al.*, 1989; Oberbauer *et al.*, 1992). The oMt1a-oGH transgene can be activated by adding supplementary zinc to the drinking water. Levels of circulating growth hormone become highly elevated upon activation of the transgene, but return to basal levels within 24 h after withdrawal of the zinc sulphate (Shanahan *et al.*, 1989). The oMt1a-oGH transgene incorporated into mice has been shown to increase growth rate, reduce fat content, and apparently has few unfavourable fitness problems (Pomp *et al.*, 1992; Eisen *et al.*, 1995; Murray and Pomp, 1995; Clutter *et al.*, 1996; Siewerdt *et al.*, 1998). The type of gene action of the oMt1a-oGH transgene was determined by Siewerdt *et al.* (1998). Dominance was the predominant form of action of the transgene on body weights and organ weights, although some differences were found according to the selection background of the lines into which the transgene was introgressed.

Sabour *et al.* (1991) reported the sole selection experiment with transgenic lines found in the literature, with lines carrying the rat growth hormone transgene. To our knowledge the present report is the first selection experiment on lines of mice carrying the oMt1a-oGH transgene insert. This paper reports the results of seven generations of selection for increased 8-week body weight in transgenic and non-transgenic lines of mice with different selection backgrounds. Direct and correlated responses to selection and patterns of change in frequency of the transgene are described.

Materials and Methods

Ten male mice from the MG101 line (Shanahan *et al.*, 1989) were tested for homozygosity for the oMt1a-oGH transgene insert. The structure considered an 'allele' of the oMt1a-oGH in the MG101 line consists of five copies of the insert (Shanahan *et al.*, 1989). Line MG101 originated due to an unequal crossover event in the original line, which carried 43 copies of the insert.

The transgenic males were mated to virgin females of a high growth line (M16), which has a history of 27 generations of selection for increased postweaning weight gain from 3 to 6 weeks (Eisen, 1975), and were also mated to females of a randomly selected control line (ICR) from which M16 originated. Hemizygous mice in the F1 were assumed to be transgenic and were reciprocally backcrossed, respectively, to the M16 and ICR lines. Tail-clips were obtained from backcross mice at 6 weeks of age as a source of DNA for a PCR analysis (Pomp and Murray, 1991). Mice testing positive for the presence of the oMt1a-oGH transgene insert were designated as founders of the TM and TC lines, respectively, from the crosses with M16 and ICR lines. Non-transgenic mice formed generation 0 of the NM and NC lines, in that order. A control line (CC) was formed from the same pool that originated the NC line. One male and one female from each litter in the CC line were selected randomly as parents of the next generation. The entire procedure was repeated after 4 weeks to form a second replicate. Each replicated line of the selection treatments comprised approximately 16 pairs of sires and dams.

Selection was practiced within families for seven generations on lines TM, TC, NM and NC. The selection criterion was large 8-week body weight. The heaviest male and female from each full-sib family were selected as parents for the next generation. Mice were pair-mated randomly but sib matings were not allowed. Line names were labelled with their corresponding replication number (1 or 2). Mice were fed *ad libitum* Purina Mouse Chow 5015 (17.5% crude protein, 11.0% fat, 4.35 kcal g^{-1} gross energy, 102.2 ppm zinc) from PMI Feeds, Inc. (St Louis, Missouri), and received tap water from mating until weaning at 3 weeks of age. From 3 to 8 weeks of age mice received Purina Lab Chow 5001 (23.4% crude protein, 4.5% fat, 4.00 kcal g^{-1} gross energy, 70.0 ppm zinc) and 25 mM zinc sulphate in distilled drinking water. Temperature (22°C) and humidity (55%) were kept constant in the laboratory. A light regime consisting of 12 h of light and 12 h of darkness (0700–1900) was used.

Data on body weights were collected at 3, 6 and 8 weeks of age (BW3, BW6 and BW8, respectively). Body weight gains were calculated for the periods between 3 and 6 weeks (GAIN36) and 6 and 8 weeks (GAIN68). Matings were done at about 10 weeks of age and cohabitation lasted for 17 days. Litters were standardized to eight pups or less within 24 h of birth. If less than eight live pups were born in a litter, crossfostering was used, provided that pups from other litters of the same line, born on the same day, were available. Litter sizes and dead pup numbers were recorded.

Body weight and weight gain data were analysed using PROC MIXED of SAS (SAS Institute, 1992). A mixed model that included the fixed effects of line, generation and sex, their interactions and the random effects of litter, nested within interaction of line and generation was fitted to the data of each replication. Replications were assumed to be random. Least-squares line means for each trait were expressed as deviations from the CC line, and were compared in the form of three orthogonal contrasts: transgenic versus non-transgenic lines ([TM+TC−NM−NC]/2), selected versus control background ([TM−TC+NM−NC]/2), and interaction between effects of the transgene and selection background ([TM−TC−NM+NC]/2). Realized heritabilities for BW8 were estimated by regressing the generation means of BW8 on the cumulative selection differentials (Hill, 1972). The selection differentials were weighted by the number of progeny with a record for BW8 produced by each individual. Realized heritabilities for individual selection were obtained by multiplying the within-family selection heritabilities by the factor $(1-t) \times (1-r)^{-1}$, where t is the full-sib intraclass correlation for the specific line and replication and $r = 0.5$ (Falconer and Mackay, 1996). The analysis assumes that there is no line–environment interaction. In this situation the randomly selected control populations account accurately for any environmental trends present (Muir, 1986). Correlated responses were obtained on BW3, BW6, GAIN36 and GAIN68 by regressing their corresponding deviations from CC line means on generation number.

Fitness traits measured on dams were: cohabitation to littering interval (CLI), litter size (LS), and preweaning mortality (MORT), defined as 100 × (number of live pups at weaning) × (number of pups after standardization at day 1)$^{-1}$. The proportion of infertile matings was also obtained. Two fitness indexes were defined as FI_1 = (litter size) × (proportion of fertile matings) × (proportion of preweaning pup survival) and $FI_2 = 0.8 \times FI_1 - 0.2 \times CLI$. The inclusion of CLI in the second fitness index favours the females that successfully mated and produced a litter in a shorter period of time. Data on fitness traits were analysed with a linear model including the effects of generation and replication. Each trait had its generation means for lines TM, TC, NM and NC expressed as deviations from corresponding mean for the line CC. Regression of these deviations over generation number provided estimates of correlated responses in fitness traits.

The regression coefficients for direct and correlated responses to selection in growth and fitness data were compared with the same three orthogonal contrasts used for mean body weights: transgenic versus non-transgenic lines, selected versus control background, and their interaction. When the interaction was significant, a further decomposition compared line TM with NM (effect of the transgene in the selected background) and line TC with NC (effect of the transgene in the control background).

In generations one through seven, tail-clips were collected on mice of lines TM and TC at 6 weeks of age as a source of DNA for PCR analyses. DNA samples from generations 2–6 were probed with a semi-quantitative PCR analysis (Schrenzel and Ferrick, 1995), which allows distinction between hemizygous (T/−) and homozygous (T/T) transgenics. The frequency of the transgene insert was calculated by allele counting. A qualitative PCR was run on DNA samples from generations 1 and 7. This analysis only makes distinction between non-transgenics (−/−) and transgenic mice. Since homozygous and hemizygous transgenics could not be distinguished, the frequencies of the transgene were calculated assuming the empirical genotypic ratios found by Siewerdt *et al.* (1998), which differed significantly from the genotypic proportions assumed when Hardy–Weinberg equilibrium holds. No PCR results were available for mice from replication 1 in generation 2, because the DNA samples were degraded.

Results

In the backcross generation, an overall percentage of 45.4% of transgenics was obtained. This percentage differs from 50% (P<0.01). In replications 1 and 2 percentages of 48.3 and 40.4% (P<0.01) of transgenics were observed, respectively. There was an under-representation of transgenic males (40.7%, P<0.01), but not of transgenic females (48.0%). With both selection backgrounds the percentage of transgenics was different from the 50% expected from the theoretical 1:1 ratio, 45.4% (P<0.01) in line ICR, and 44.1% (P<0.01) in line M16.

In generation 0, no significant line differences were found for percentage of infertile matings and percentage preweaning pup mortality, the overall means being 9.0% and 2.9%, respectively. However, the analysis of variance showed significant line differences for CLI (P<0.05) and LS (P<0.01). Mating between hemizygous transgenic mice (TC, TM) had a longer CLI (P<0.01) and a smaller LS (P<0.01) than non-transgenic mice (Table 16.1). The history of selection for high postweaning gain in the NM and TM lines explains the larger litter sizes in these lines compared with the control background lines (NC and TC), because selection for high postweaning gain led to a positive correlated response in litter size (Eisen *et al.*, 1973). Selection background had no significant effect on CLI. No significant interaction between the effects of selection background and transgene was detected for CLI and for LS. The CC and NC lines had similar means for these traits, as expected since no selection had yet occurred in NC.

The estimated frequencies of the transgene insert are presented in Fig. 16.1. In both replications of line TM the frequency declined from the initial value of 0.5. In line TM2 the frequency was below 0.05 as of generation 7, and there were no homozygous transgenic individuals in generations 6 and 7. In the TC line, the frequency of the transgene rose to values around 0.6 beginning in generation 3, except for small fluctuations observed on TC1 in generation 7 and on TC2 in generation 5.

All main effects and interactions in the analysis of variance affected BW8 (P<0.05). Least-square means of BW8 for all lines as of generation 7 are presented in Fig. 16.2, and the least-squares means of BW8 for the selected lines, as deviations from the control line, are presented in Fig. 16.3. The interaction between the presence of the transgene and selection background had a significant effect on 8-week body weight means in most

Table 16.1. Means ± SE and linear contrasts for cohabitation to littering interval (CLI) and litter size (LS) in matings of non-transgenic mice (CC, NC, and NM) and oMt1a-oGH transgenic mice (TC, TM) in generation zero, pooled over two replications.

Line	N[a]	CLI (days)	LS (pups)
CC	35	22.22±0.61	12.65±0.47
NC	38	22.00±0.58	12.68±0.45
NM	33	21.53±0.63	14.22±0.49
TC	34	23.59±0.62	9.64±0.48
TM	33	23.64±0.63	11.62±0.49
Contrast (L)		L ± SE	L ± SE
T vs. N[b]		1.85±0.62**	−2.82±0.48**
S vs. C[c]		−0.21±0.62	1.76±0.48**
Interaction[d]		0.26±0.62	0.23±0.48

* $P<0.05$, ** $P<0.01$.

[a] Sample sizes.

[b] Transgenic vs. non-transgenic. Contrast value is (TM+TC−NM−NC)/2.

[c] Selected vs. control background. Contrast value is (TM−TC+NM−NC)/2.

[d] Contrast value is (TM−TC−NM+NC)/2.

generations (Table 16.2). In the control background, line TC usually had higher means for BW8 than line NC. The opposite occurred in the selected background, where line NM had higher means for BW8 than line TM in most generations.

Response to selection for increased BW8 was different from zero ($P<0.05$) in both replications of lines NM and NC. Among the transgenic lines, genetic progress was obtained in TC1, TM2 ($P<0.01$) and in TM1

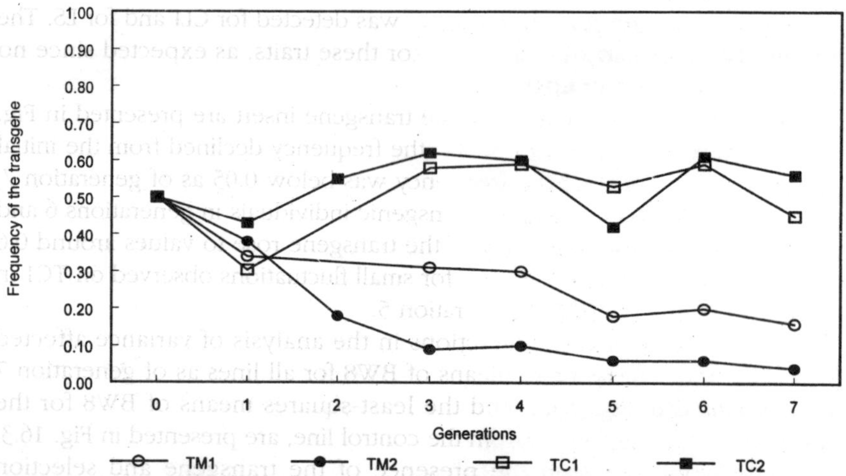

Fig. 16.1. Observed frequencies of the oMt1a-oGH transgene.

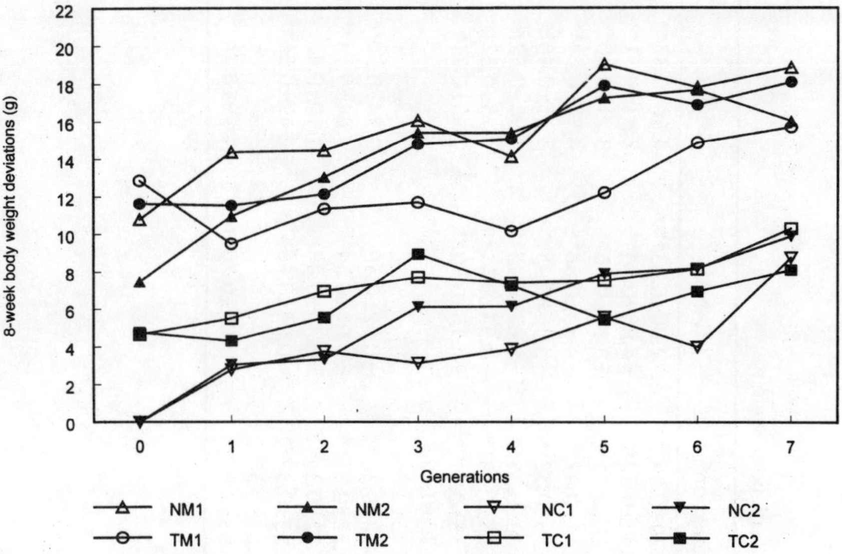

Fig. 16.2. Least-squares means ± SE for 8-week body weight as of generation 7.

(P<0.10). Estimates of genetic progress are presented in Table 16.3; realized heritabilities and the cumulative selection differentials (CSD) for each line are shown in Table 16.4. The CSD values were all around 24 g, except for line NC where the CSD was around 20 g. There was great variation in the

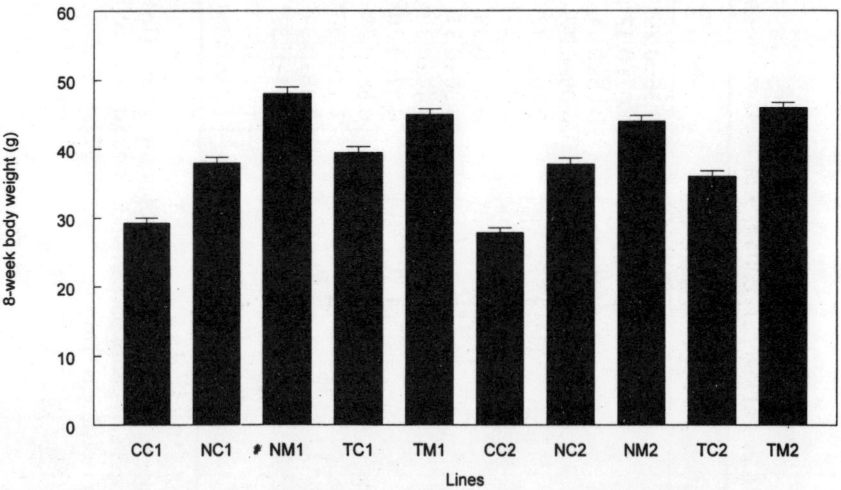

Fig. 16.3. Least-squares means for 8-week body weight, as deviations from the control lines.

Table 16.2. Orthogonal contrasts ± SE for comparisons among means of lines for 8-week body weight.

Replication	Generation	T–N[a]	M–C[b]	Interaction[c]	TM–NM	TC–NC
1	0	3.35±0.29**	9.50±0.29**	−1.29±0.29**	2.06±0.43**	4.56±0.37**
	1	−1.07±0.37**	7.79±0.37**	−3.80±0.37**	−4.88±0.54**	2.73±0.51**
	2	0.05±0.41	7.55±0.41**	−3.16±0.41**	−3.10±0.55**	3.21±0.60**
	3	0.09±0.47	8.46±0.47**	−4.47±0.47**	−4.38±0.68**	4.57±0.64**
	4	−0.20±0.32	6.51±0.32**	−3.76±0.32**	−3.96±0.45**	3.55±0.46**
	5	−2.43±0.35**	9.06±0.35**	−4.39±0.35**	−6.82±0.47**	1.96±0.52**
	6	0.61±0.39	10.26±0.39**	−3.55±0.39**	−2.94±0.53**	4.15±0.57**
	7	−0.80±0.51	7.76±0.51**	−2.35±0.51**	−3.15±0.73**	1.55±0.71*
2	0	4.47±0.29**	7.17±0.29**	−0.30±0.29	—	—
	1	0.92±0.36**	7.56±0.36**	−0.35±0.36	—	—
	2	0.66±0.38	8.14±0.38**	−1.58±0.38**	−0.92±0.56	2.24±0.50**
	3	1.12±0.38**	7.57±0.38**	−1.72±0.38**	−0.61±0.54	2.84±0.54**
	4	0.38±0.34	8.54±0.34**	−0.72±0.34*	−0.34±0.49	1.10±0.47*
	5	−0.91±0.37*	10.91±0.37**	1.55±0.37**	0.64±0.48	−2.47±0.56**
	6	−1.02±037**	9.71±0.37**	0.23±0.37	—	—
	7	−1.36±0.51**	8.06±0.57**	1.93±0.57**	2.07±0.70**	−1.79±0.74**

* $P<0.05$, ** $P<0.01$.

[a] Transgenic vs. non-transgenic. Contrast value is (TM+TC−NM−NC)/2.

[b] Selected vs. control background. Contrast value is (TM−TC+NM−NC)/2.

[c] Contrast value is (TM−TC−NM+NC)/2.

Table 16.3. Estimates of genetic progress[a] ± SE, in g per generation, in the selected lines, and linear contrasts for 8-week body weight.

Line	Replication 1	Replication 2	Pooled
NC	0.88±0.21**	1.35±0.11**	1.12±0.24[b]
NM	1.02±0.22**	1.32±0.27**	1.17±0.15
TC	0.65±0.11**	0.47±0.24	0.56±0.09
TM	0.57±0.27*	1.12±0.16**	0.85±0.28
Contrast (L)			**L ± SE**
T vs. N[c]			−0.44±0.20
M vs. C[d]			0.17±0.20
Interaction[e]			0.12±0.20

* P<0.10; ** P<0.01.

[a] Regressions of 8-week body weight, as deviations from the control lines, on generation number.

[b] Empirical standard errors, calculated as the standard error of replicate coefficients.

[c] Transgenic vs. non-transgenic lines. Contrast value is (TM+TC−NM−NC)/2.

[d] Selected vs. control background. Contrast value is (TM−TC+NM−NC)/2.

[e] Contrast value is (TM−TC−NM+NC)/2.

realized heritabilities across lines. The two heritability estimates within the NC line also differed greatly. Higher heritability estimates were observed in the non-transgenic lines than in the transgenic lines. Selection background did not affect heritability estimates.

Correlated responses in BW6 were observed in all selected lines except TC2, and in all lines for GAIN36 (Table 16.5). A significant correlated response in BW3 was found only in line NC2 and significant correlated responses were obtained in GAIN68 in lines NC1 and NM1. Very few significant correlated responses were observed in the fitness traits, and these were not consistent across replicates (Table 16.6). Changes in litter size were not significant in any line, except for a positive slope for litter size observed in TM2. An increase in infertility and in preweaning pup mortality over generations was observed in line NC2. Fitness index slopes were significantly negative in line NC2, and significantly positive in TM1.

Linear contrasts of slopes for body weight and weight gain traits are presented in Table 16.7. Significant contrasts were observed for BW8 in replication 2, where transgenic lines had smaller response to selection than non-transgenic lines and where lines from the selected background had a higher response to selection than lines from the control background. Three interactions were found to be significant. For BW3 in replication 2, further decomposition showed that line TC had a significantly smaller correlated response to selection than line NC, while the slopes of lines TM and NM were not different. For BW6 in replication 2 and for GAIN68 in replication

Table 16.4. Realized heritability (h^2) estimates and cumulative selection differentials (CSD), and linear contrasts for 8-week body weight, as of generation 7 of selection.

Line	Replication	Realized $h^{2\,b}$	Individual $h^{2\,b}$	CSD (g)
NC	1	0.31	0.46	19.23
	2	0.43	0.74	21.09
	Average	0.37±0.06[a]	0.60±0.14	
NM	1	0.29	0.38	23.58
	2	0.34	0.49	24.93
	Average	0.31±0.02	0.43±0.05	
TC	1	0.19	0.29	23.04
	2	0.12	0.19	22.94
	Average	0.16±0.03	0.24±0.05	
TM	1	0.17	0.26	25.55
	2	0.30	0.44	24.64
	Average	0.23±0.06	0.35±0.09	
Contrast (L)		L ± SE	L ± SE	
T vs. N[b]		−0.14±0.05	−0.22±0.09	
M vs. C[c]		0.00±0.05	−0.03±0.09	
Interaction[d]		0.07±0.05	0.14±0.09	

[a] Empirical standard errors, calculated as the standard error of replicate coefficients.
[b] Transgenic vs. non-transgenic lines. Contrast value is (TM+TC−NM−NC)/2.
[c] Selected vs. control background. Contrast value is (TM−TC+NM−NC)/2.
[d] Contrast value is (TM−TC−NM+NC)/2.

1, lines TC and NC had equivalent slopes but the average slope of the NM lines was larger than the average slope of the TM lines.

The corresponding contrasts among slopes for fitness traits are presented in Table 16.8. A significant interaction between the presence of the transgene and the selection background was found for INF, FI_1, FI_2 (both replications), and CLI (replication 1). No interaction was significant for LS and MORT. In replication 2, larger correlated responses in LS, FI_1, and FI_2, and a smaller correlated response in MORT were observed in the selected background when contrasted with the control background. Correlated responses on both fitness indexes were larger in the transgenic than in the non-transgenic lines.

Discussion

The impact of the introduction of foreign DNA into commercial livestock populations will be felt when production or disease resistance can be raised to higher levels. Since selection in transgenic populations will be done on

Table 16.5. Regression coefficients ± SE of correlated responses in body weight and weight gain traits, deviated from control lines, in the selected lines.

Line	Replication	BW3[a] (g)	BW6 (g)	GAIN36 (g)	GAIN68 (g)
NC	1	0.09±0.12[b]	0.69±0.20***	0.60±0.11***	0.18±0.04***
	2	0.46±0.10***	1.13±0.12***	0.66±0.13***	0.16±0.08
NM	1	−0.14±0.12	0.50±0.22*	0.67±0.11***	0.46±0.06***
	2	0.13±0.09	0.96±0.24***	0.86±0.23**	0.27±0.14
TC	1	0.18±0.11	0.54±0.13***	0.38±0.19*	0.09±0.11
	2	0.06±0.14	0.41±0.25	0.48±0.14**	0.01±0.14
TM	1	0.15±0.21	0.74±0.19***	0.60±0.14***	−0.11±0.17
	2	0.17±0.11	1.03±0.14***	0.86±0.17***	0.03±0.19

**P*<0.10, ***P*<0.05, ****P*<0.01.

[a] BW3, 3-week body weight; BW6, 6-week body weight; GAIN36, body weight gain from 3 to 6 weeks; GAIN68, body weight gain from 6 to 8 weeks.

[b] Standard error calculated from least squares regression.

nucleus or elite herds, the integration of a transgene must be compatible with overall selection goals. The maintenance of transgenic populations will depend on the economic value of the transgene and also on whether further selection causes a reduction in transgene frequency because of relatively poor viability of embryos or reproduction in adults. Regulated expression and stable integration of the transgene insert and transmission to offspring are critical in the development of useful transgenic livestock (Sabour *et al.*, 1991).

Reduction in the frequency of the transgene in both replications of the TM line and the lower response to selection in TM compared with NM may be explained by the fact that past selection may have increased the frequencies of other alleles involved in the production of growth hormone or may have affected genes downstream in the growth hormone cascade. Evidence favouring this interpretation is the negligible additive effect of the transgene for 8-week body weight in TM males and the smaller additive effect in TM compared with TC females (Siewerdt *et al.*, 1998). This factor would leave less opportunity for contribution to growth enhancement due to the expression of the transgene. A diametrically opposite explanation is that past selection for rapid 3- to 6-week body weight gain may have actually caused the reduction of the average level of circulating growth hormone, as observed by Medrano *et al.* (1991). The increase in circulating growth hormone induced by the transgene could be upsetting the physiological balance in the mice of the selected background. Another factor that may be partly responsible for the lower response to selection in TM compared with NM is a reduced fitness of transgenic embryos. The reduction in the frequency of the transgene is partially consistent with the results that Sabour *et al.* (1991) obtained in lines of mice carrying the rat growth hormone transgene. However, these authors found that the

Table 16.6. Regression coefficients ± SE of correlated responses in fitness traits, deviated from control lines, in the selected lines.

Line	Replication	INF (%)[a]	CLI (d)	LS (pups)	MORT (%)	FI_1	FI_2
NC	1	−0.62±1.42[b]	0.08±0.19	0.01±0.16	−0.56±0.89	0.16±0.19	0.11±0.17
	2	4.56±1.42***	0.17±0.19	0.06±0.16	1.99±0.89**	−0.73±0.19***	−0.62±0.17***
NM	1	−0.20±1.42	0.44±0.19	−0.04±0.16	−0.12±0.89	−0.01±0.19	−0.10±0.17
	2	0.98±1.42	0.24±0.19	0.23±0.16	0.19±0.89	0.03±0.19	−0.02±0.17
TC	1	2.23±1.42	0.18±0.19	−0.12±0.16	−1.22±0.89	−0.20±0.19	−0.19±0.17
	2	0.62±1.42	−0.19±0.19	0.09±0.16	1.81±0.89	−0.15±0.19	−0.08±0.17
TM	1	−2.46±1.42*	−0.32±0.19*	0.20±0.16	−1.32±0.89	0.57±0.19***	0.52±0.17***
	2	1.93±1.42	0.08±0.19	0.52±0.16***	0.89±0.89	0.09±0.19	0.06±0.17

*$P<0.10$, **$P<0.05$, ***$P<0.01$.

[a] INF, infertile matings; CLI, cohabitation to littering interval; LS, litter size; MORT, preweaning pup mortality; FI_1, fitness index 1; FI_2, fitness index 2. See text for definition of fitness indexes.

[b] Standard error calculated from least squares regression.

Table 16.7. Linear contrasts of slopes for body weight and weight gain traits (generations 0–7).

Contrast	Replication	BW3[a] (g)	BW6 (g)	BW8 (g)	GAIN36 (g)	GAIN68 (g)
T vs. N[b]	1	0.19±0.15[e]	0.05±0.19	−0.41±0.26	−0.15±0.14	−0.33±0.11***
	2	−0.18±0.11	−0.33±0.20	−0.74±0.21***	−0.09±0.17	−0.20±0.14
M vs. C[c]	1	−0.13±0.15	0.01±0.19	0.11±0.26	0.15±0.14	0.04±0.11
	2	−0.11±0.11	0.23±0.20	0.54±0.21**	0.29±0.17	0.07±0.14
Interaction[d]	1	0.10±0.15	0.20±0.19	−0.31±0.26	0.08±0.14	−0.24±0.11**
	2	0.22±0.11*	0.40±0.20*	0.27±0.21	0.09±0.17	−0.05±0.14
TM–NM	1	—	—	—	—	−0.57±0.13***
	2	0.04±0.10	−0.72±0.20***	—	—	—
TC–NC	1	—	—	—	—	−0.09±0.08
	2	−0.40±0.12***	0.07±0.20	—		——

*P<0.10, **P<0.05, ***P<0.01.

[a] See Table 16.5, footnote [a], for definition of traits.

[b] Transgenic vs. non-transgenic lines. Contrast value is (TM+TC−NM−NC)/2.

[c] Selected vs. control background. Contrast value is (TM−TC+NM−NC)/2.

[d] Contrast value is (TM−TC−NM+NC)/2.

[e] Standard error calculated from least squares regression.

Table 16.8. Linear contrasts of slopes for reproductive traits (generations 0–7).

Contrast	Replication	INF (%)[a]	CLI (d)	LS (pups)	MORT (%)	FI_1	FI_2
T vs. N[b]	1	0.29±1.15[e]	−0.33±0.14**	0.05±0.10	−0.92±0.69	0.11±0.16	0.16±0.14
	2	−1.50±1.15	−0.25±0.14*	0.16±0.10	0.26±0.69	0.32±0.16*	0.31±0.14**
M vs. C[c]	1	−2.13±1.15*	−0.07±0.14	0.13±0.12	0.17±0.57	0.30±0.17*	0.26±0.14*
	2	−1.14±1.15	0.17±0.14	0.31±0.12***	−1.36±0.57**	0.50±0.17***	0.37±0.14**
Interaction[d]	1	−2.55±1.25*	−0.43±0.12***	0.18±0.13	−0.27±0.88	0.47±0.13***	0.46±0.10***
	2	−2.45±1.25*	0.10±0.12	0.13±0.13	0.44±0.88	−0.26±0.13**	−0.23±0.10**

*$P<0.10$, **$P<0.05$, ***$P<0.01$.

[a] See Table 16.6, footnote [a] for definitions of traits.

[b] Transgenic vs. non-transgenic lines. Contrast value is (TM+TC−NM−NC)/2.

[c] Selected vs. control background. Contrast value is (TM−TC+NM−NC)/2.

[d] Contrast value is (TM−TC−NM+NC)/2.

[e] Standard error calculated from least squares regression.

reduction in the frequency of the rat growth hormone (rGH) transgene was independent of selection background, which disagrees with the results reported here. The different results can be explained by noting that females transgenic for the rGH have this growth hormone transgene chronically expressed, and they have a severe reduction in reproductive performance leading to a rapid loss of the transgene from the population.

Because of the large additive genetic effect of the oMT1a-oGH transgene that contributes to the total phenotypic value of 8-week body weight in both males and females in the unselected background (Siewerdt *et al.*, 1998), upward selection for this trait in the presence of the transgene at a starting frequency of 0.5 was expected to result in rapid genetic progress and an increase in the transgene frequency approaching one after seven generations. However, the selection response was actually higher in the unselected background lines without the transgene than in those with the transgene, and the frequency of the transgene only increased marginally in the TC replicates. A possible explanation for this finding is that embryos carrying the transgene construct have reduced viability (Eisen *et al.*, 1995; Clutter *et al.*, 1996; Siewerdt *et al.*, 1998).

In generation 0, there were no significant differences in pup mortality during the preweaning period between transgenic and non-transgenic lines (data not shown). The absence of line differences in preweaning pup mortality between non-transgenic and hemizygous transgenic crosses suggests that higher zygotic loss and prenatal and perinatal mortality in homozygous transgenic and/or hemizygous transgenic progeny genotypes may explain the lower litter size in hemizygous transgenic crosses. Perinatal mortality differences cannot be excluded entirely because litter size was recorded at 1 day of age, so any pups that may have died on the day of birth and were eaten by the mother before data were recorded would be included in the prenatal group. The perinatal period probably contributes minimally, however, since dams were checked daily for litter births and dead pups were routinely counted and removed.

An explanation for the reduced litter size in the hemizygous transgenic crosses is the reduced reproductive performance of transgenic females which could occur as a consequence of overexpression of the transgene (Bartke *et al.*, 1994). However, the transgenic female parents of generation 0 did not receive $ZnSO_4$ in the drinking water at any time so the transgene should not have been chronically expressed. Nevertheless, some leakage of the oMt1a-oGH transgene is known to occur (J.D. Murray, personal communication), which could have a negative effect on litter size. The 2-week withdrawal period could offset part of the negative effects of leakage. This is unlikely, however, because in cases where transgenic females are mated 2 weeks after the $ZnSO_4$ is withdrawn, litter size is actually enhanced due to an increase in ovulation rate (Murray and Pomp, 1995; Eisen *et al.*, 1995).

Fitness models affecting prenatal survival that may explain the difference in litter size between hemizygous transgenic matings and non-transgenic

matings within the same genetic background are given in Table 16.9. The
absolute ... hemizygous
transgen
backgrou
Althougl
fitness fc
more in

The
fashion
W_3. The
against
$W_1=W_2$
against
selectio
expecta
the tran
on earli
0.45 he
yielding
coeffici
differer
initial
frequer
models
compa
backgr
0.485 f
freque
embryo
than pr

A
reducti
line TC
1998),
indivic
transg
popul
other
increa
geneti
chanc
future
in the
every
long r

Table 16.9. Prenatal viability in fitness models that could explain the difference in litter size between mating of hemizygous transgenic mice versus matings of non-transgenic mice in generation 0.

	Unselected background							High-growth selected background						
	Relative fitness			Pup distribution				Relative fitness			Pup distribution			
Model[a]	T/T	T/–	–/–	T/T	T/–	–/–	Litter size (pups)	T/T	T/–	–/–	T/T	T/–	–/–	Litter size (pups)
1. $W_1=W_2=W_3$	1	1	1	3.17	6.34	3.17	12.68[b]	1	1	1	3.555	7.11	3.555	14.22[d]
2. $W_1<W_2=W_3$	0.04	1	1	0.13	6.34	3.17	9.64[c]	0.27	1	1	0.95	7.11	3.56	11.62[e]
3. $W_1=W_2<W_3$	0.68	0.68	1	2.16	4.31	3.17	9.64[c]	0.76	0.76	1	2.69	5.37	3.56	11.62[e]
4. $W_1<W_2<W_3$	0.40	0.82	1	1.27	5.20	3.17	9.64[c]	0.63	0.82	1	2.23	5.83	3.56	11.62[e]
5. $W_1>W_2<W_3$	0.90	0.57	1	2.85	3.61	3.17	9.64[c]	0.90	0.68	1	3.20	4.86	3.56	11.62[e]

[a] W_1, W_2, W_3 are relative fitness values for homozygous transgenics (T/T), hemizygous transgenics (T/–), and non-transgenics (–/–).
[b] Mean litter size for the NM line in generation 0.
[c] Mean litter size for the TM line in generation 0.
[d] Mean litter size for the NC line in generation 0.
[e] Mean litter size for the TC line in generation 0.

the contribution of all other loci involved in a specific phenotypic expression, then its frequency should only decay because of reduced fitness of the carriers. Whether it is a fitness handicap or a situation where the contribution of the transgene was diminished because of previous selection, the correct explanation for the reduction in the frequency of the transgene on lines TM1 and TM2 is a question left to be answered by future research.

No slopes for the fitness indices were significantly negative in the transgenic lines, and in line TM1 both slopes were even positive (Table 16.6). Also, when contrasted with non-transgenic lines, the slopes of the transgenic lines were significantly larger in replication 2 (Table 16.8). These fitness indexes, however, did not consider prenatal embryonic or fetal survival.

For BW8, the interaction between the effects of the transgene and the selection background was significant in most of the generations, as shown in Table 16.2. An identified pattern is that TC mice are larger than NC, but TM mice are smaller than NM. Results suggest that the oMt1a-oGH transgene has a greater effect when incorporated into populations with no past selection for increased body weight. Favourable correlated responses observed on BW6 and GAIN36 suggest pleiotropic effects associated with the transgene insert that can be advantageously exploited. Polygenic frequencies that affect the correlated traits may have also changed through indirect selection, but apparently the experimental design used does not allow isolation of the effects of polygenic inheritance from pleiotropic effects associated with the transgene.

The presence of an activated oMt1a-oGH transgene in lines TM and TC led to smaller realized heritability estimates when compared with the lines NM and NC, despite creating a larger phenotypic variance which resulted in higher selection differentials. In agreement with these results, Clutter *et al.* (1996) also found heritability estimates for BW8 to be larger in lines without the oMt1a-oGH transgene than in transgenic lines. The introduction of a transgene with large effect on a trait is expected to increase the heritability of the trait unless the transgene shows overdominance and is at or near the equilibrium gene frequency or the transgene introduced a large epistatic component. Although the oMt1a-oGH transgene exhibits overdominance effects (Siewerdt *et al.*, 1998), the equilibrium frequency of the transgene is about 0.9, which is much higher than the frequency in the present lines. Therefore the likely explanation for the reduced heritability is epistasis. In addition, environmental variation may have been increased due to a possible reduced buffering capacity of the transgenic mice. In line TC, genetic progress was obtained both in the form of a slight increase in the frequency of the transgene and in the accumulation of favourable alleles at other loci. In line TM the genetic progress was chiefly obtained at other loci that affect BW8, since the frequency of the transgene was reduced from the initial value of 0.5. This finding showed that there was still opportunity for genetic improvement in the crosses involving a line heavily selected in the past for growth. From the results reported here it is suggested that in choosing a transgene to be incorporated into commercial stocks one should aim at a construct which would affect traits not previously subjected to intense selection, since the effect of the transgene may not have as much impact as in traits which had received little selection pressure previously or may interact unfavourably with the artificially selected genotypes.

Acknowledgements

The research reported herein was funded by USDA National Research Initiative Competitive Grant No. 9403975 and the North Carolina Research Service, Raleigh, NC 27695–7643, USA. F.S. was supported by Universidade Federal de Pelotas, Pelotas, Brazil, and by a Fellowship (200387/94–4) from CNPq, Brasília, Brazil. We thank Janice S. Conrad-Brink, Tammy J. Schmitt and Ivy J. Parker for running the PCR analyses. Dr Daniel Pomp and two anonymous referees provided suggestions which greatly improved the manuscript.

References

Al-Shawi, R., Kinnaird, J., Burke, J. and Bishop, J.O. (1990) Expression of a foreign gene in a line of transgenic mice is modulated by a chromosomal position effect. *Molecular and Cellular Biology* 10, 1192–1198.

Bartke, A., Cecim, M., Tang, K., Steger, R.W., Chandreshekar, V. and Turyn, D. (1994) Neuroendocrine and reproductive consequences of overexpression of growth hormone in transgenic mice. *Proceedings of the Society for Experimental Biology and Medicine* 206, 349–359.

Clutter, A.C., Pomp, D. and Murray, J.D. (1996) Quantitative genetics of transgenic mice: components of phenotypic variation in body weights and weight gains. *Genetics* 143, 1753–1760.

Eisen, E.J. (1975) Population size and selection intensity effects on long-term selection response in mice. *Genetics* 79, 305–323.

Eisen, E.J., Hanrahan, J.P. and Legates, J.E. (1973) Effects of population size and selection intensity on correlated responses to selection for postweaning gain in mice. *Genetics* 74, 157–170.

Eisen, E.J., Murray, J.D. and Schmitt, T.J. (1995) An ovine-growth-hormone transgene model suitable for selection experiments for growth in mice. *Journal of Animal Breeding and Genetics* 112, 401–413.

Falconer, D.S. and Mackay, T.F.C. (1996) *Introduction to Quantitative Genetics*, 4th edn. Longman, Harlow.

Hill, W.G. (1972) Estimation of realised heritabilities from selection experiments. II. Selection in one direction. *Biometrics* 28, 767–780.

Medrano, J.F., Pomp, D., Sharrow, L., Bradford, G.E., Downs, T.R. and Frohman, L.A. (1991) Growth hormone and insuline-like growth factor-I measurements in high growth (hg) mice. *Genetical Research, Cambridge* 58, 67–74.

Muir, W.M. (1986) Estimation of response to selection and utilization of control populations for additional information and accuracy. *Biometrics* 42, 381–391.

Murray, J.D. and Pomp, D. (1995) Interaction between reproductive performance and dietary levels fed to female oMt1a-oGH transgenic mice. *Transgenics* 1, 553–563.

Oberbauer, A.M., Currier, T.A., Nancarrow, C.D., Ward, K.A. and Murray, J.D. (1992) Linear bone growth of oMt1a-oGH transgenic male mice. *American Journal of Physiology* 262, E936–942.

Pomp, D. and Murray, J.D. (1991) Single day detection of transgenic mice by PCR of toe-clips. *Mouse Genome* 89, 279.

Pomp, D., Nancarrow, C.D., Ward, K.A. and Murray, J.D. (1992) Growth, feed efficiency and body composition of transgenic mice expressing a sheep metallothionein 1a-sheep growth hormone fusion gene. *Livestock Production Science* 31, 335–350.

Rexroad, C.E. Jr (1992) Transgenic technology in animal agriculture. *Animal Biotechnology* 3, 1–13.

Sabour, M.P., Ramsey, U. and Nagai, J. (1991) Decreased frequency of the rat growth hormone transgene in mouse populations with or without selection for increased adult body weight. *Theoretical and Applied Genetics* 81, 327–332.

SAS Institute, Inc. (1992) *SAS® Technical Report P-229, SAS/STAT® Software: Changes and Enhancements*, Release 6.07. SAS Institute, Cary, Illinois.

Schrenzel, M.D. and Ferrick, D.A. (1995) Horse (*Equus cabalus*) T-cell receptor alpha, gamma, and delta chain genes: nucleotide sequences and tissue-specific gene expression. *Immunogenetics* 42, 112–122.

Shanahan, C.M., Rigby, M.W., Murray, J.D., Marshall, J.T.A., Townrow, C.A., Nancarrow, C.D. and Ward, K.A. (1989) Regulation of expression of a sheep metallothionein 1a-sheep growth hormone fusion gene in transgenic mice. *Molecular and Cellular Biology* 9, 5473–5479.

Siewerdt, F., Eisen, E.J., Conrad-Brink, J.S. and Murray, J.D. (1998) Gene action of the oMt1a-oGH transgene in two lines of mice with distinct selection backgrounds. *Journal of Animal Breeding and Genetics* 115, 211–226.

17

Ethics, Animal Welfare and Transgenic Farm Animals

J.A. Mench

Department of Animal Science and Center for Animal Welfare, University of California, Davis, California, USA

Introduction

The application of biotechnology to farm animals has the potential to benefit both humans and animals in significant ways (Seidel, 1986; Robinson and McEvoy, 1993; Müller and Brem, 1994; Powell *et al.*, 1994; Romagnolo and DiAugustine, 1994; Wilmut, 1995; this volume). However, the ethical ramifications of the development and implementation of new biotechnologies have been the subject of recent, and often heated, debate. These ethical concerns have been summarized by many others (e.g. Thompson, 1997a; Donnelley *et al.*, 1994; Mepham *et al.*, 1995; Rollin, 1995; Tannenbaum, 1995; Mepham *et al.*, 1998). Genetic engineering of animals and plants for agricultural use is probably the most contentious application of biotechnology, with the exception of cloning and other genetic manipulations of humans. Sociologist Frederick Buttel observed that medical technology accounts for about 90% of the products from genetic engineering, while food biotechnology accounts for 90% of the controversy (quoted in Thompson, 1997b). In this chapter, I will refer briefly to the spectrum of general ethical issues associated with food animal biotechnology, and then focus specifically on animal welfare concerns that arise as a consequence of transgenic manipulation of agricultural animals.

© CAB INTERNATIONAL 1999. *Transgenic Animals in Agriculture*
(eds J.D. Murray, G.B. Anderson, A.M. Oberbauer and M.M. McGloughlin)

General Ethical Concerns

Distribution of benefits

Many genetic manipulations of farm animals are intended to increase productivity. Yield increasing technologies like this, unlike labour-saving technologies, result in product surpluses and lower prices, and hence the removal of less efficient producers and a reduction in the number of production units (Dorner, 1983). Questions arise regarding whether the benefits from the use of transgenic technologies will be fairly distributed, particularly if transgenic animals are patented. Will small farmers be able to afford to purchase transgenic animals, or will they be driven out of business? If so, is this fair considering that much transgenic animal research has been publicly funded? How will the loss of small farms affect the quality of rural life in developed countries? What effects will the concentration of food production in increasingly fewer hands have on consumers? And what will be the effects of patenting and the increasing use of genetic technology on farmers in developing countries?

Dietary or food safety concerns

Although there seems to be a general consensus among scientists that genetically engineered foods are no more or less likely to pose food safety hazards than are traditionally produced foods (Berkowitz, 1993; Thompson, 1997b), food safety of bioengineered products is nonetheless a significant public concern. Market research in Britain has shown that 81% of consumers are concerned about genetically modified foods and would avoid them if possible (United Press International, 1998). In response, one of Britain's largest supermarket chains, Iceland, recently announced that it would not market genetically engineered foods under its own brand name.

Concerns about eating bioengineered foods can also arise for religious reasons. Chaudry and Regenstein (1994) have discussed some of the potential controversies that consuming such foods would create for Jews and Muslims who observe dietary laws. They conclude that, under Jewish law, gene transfer of one or two genes would be acceptable, even if that material came from prohibited species like swine, because the amount of material transferred would be considered trivial. However, 'mixing' characteristics to create something new could pose significant concerns. For Muslims, problems would arise if bioengineered products contain biologically active agents in high enough amounts, particularly if those agents are derived from prohibited animals.

Environmental impacts

The potential environmental impact of the intentional or unintentional release of genetically engineered organisms is a significant concern to many people (Hoban and Kendall, 1993). The greatest risks as far as transgenic agricultural animals are concerned are likely to be transgenic maricultured fish, which cannot easily be contained (Kapuscinski and Hallerman, 1990; Bruggemann, 1993). The wisdom of engineering farm animals to adapt to previously inhospitable habitats, thus displacing native wildlife and causing further damage to ecosystems, has also been questioned (Fox, 1989; Rollin, 1995).

Respect for life and 'unnaturalness' of genetic engineering

Ethical concern has also been voiced about the 'unnaturalness' of genetic engineering and the ways in which it might devalue nature and commercialize life. These concerns are more likely to be expressed by those who consider themselves religious, but are by no means restricted to religious individuals (Hoban and Kendall, 1993; Rollin, 1995; Thompson, 1997b). At the deepest level, these represent concerns about the ways in which technology might corrupt basic human values (Donnelly *et al.*, 1994). Strachan Donnelley of the Hastings Center (1993, p. 98) is a particularly eloquent advocate of this view:

> Animal biotechnology, inspired by often genuine and legitimate desires to meet human and animal need and interests, must beware that it does not pre-empt 'nature natural' in the minds and hearts of us human beings and replace it with its own 'nature contrived.' ... This would be the end of us as seekers after 'living' natural norms and ways of being human, and given the press of our present technological powers, no doubt the end of nature's richness and goodness itself. This would decidedly be a double moral disaster and irresponsibility.

Some critics of animal biotechnology explicitly claim that animals have a natural genetic integrity that must be respected (Rifkin, 1985; Fox, 1986). Notions of animal integrity and the challenges they raise for transgenic animal production will be discussed in more detail later in this paper.

Public Attitudes towards Agricultural Biotechnology

Ethical concerns are reflected in public attitudes towards biotechnology. In a recent Eurobarometer survey (Biotechnology and the European Public Concerted Action Group, 1997), over 16,000 people in the European Union were asked for their opinions about the use of biotechnology for genetic testing, production of medicines and vaccines, increasing crop pest resistance, food production, developing genetically modified animals for research, and xenotransplantation. Although all of these applications of biotechnology were thought to be useful, the last three, which involve genetic manipulation of animals, were viewed negatively. Perception of risk appeared to play relatively little role in this judgment, except in the case of food production. What was most important was whether or not the application of the technology was felt to be *morally* acceptable. The committee that interpreted the survey concluded that the results indicated that perceived usefulness was a precondition for support and that people were prepared to accept some risks for those benefits, but that moral doubts acted as a veto irrespective of views on risks and benefits.

The Eurobarometer survey also showed that opposition to applications of biotechnology was not based on a lack of knowledge about science and technology. The belief that biotechnology and genetic engineering can make a positive contribution to people's lives had declined since a 1993 survey, even though the level of knowledge about basic biology had increased slightly. Furthermore, level of knowledge was poorly correlated with support for particular applications of biotechnology.

Similar views were aired at a consensus conference held in Copenhagen in 1992 (Sandoe *et al.*, 1996). The welfare of genetically engineered animals was a major concern of the participants, but they also thought it morally unacceptable to induce genetic changes in animals in order to adjust the animals to existing agricultural methods or to produce cheaper food. Likewise, a 1993 (Hoban and Kendall, 1993) survey of approximately 1300 adults in the USA followed by focus group discussions revealed that, while most believed that biotechnology would be personally beneficial to them, 53% also believed that it was morally wrong to use biotechnology to change animals, while only 24% believed that changing plants was wrong. The least acceptable applications of biotechnology were those that changed the composition of meat or milk, or increased animal growth rates. In the focus groups, women were particularly concerned about the humane treatment of animals and animal welfare issues arising from biotechnology.

These opinions reflect a fundamental shift in the way society views the use and treatment of animals, a phenomenon that Rollin (1995) refers to as 'the new social ethic for animals.' Throughout history, animals have largely been viewed as property, of value only because of their usefulness to humans. The treatment of animals was governed by obligations to other humans (i.e. the obligation not to damage another person's property) rather than directly to animals. However, because of a confluence of social forces in industrialized countries in the 19th century (Ritvo, 1987) this view of animals began to change. The passage of anti-cruelty statutes like the 1897 Cruelty to Animals Act in Great Britain, which stipulated that animals used in painful experiments be provided with pain relief, are evidence of the emergence of a new view – that animals themselves are entitled to certain kinds of treatment. There is still a broad-based acceptance in industrialized countries that using animals for human benefit is appropriate. However, it is also clear that this acceptance is no longer unreserved. Numerous public opinion polls have shown that approximately 80% of people in the USA believe that animals have rights (Hoban and Kendall, 1993; Craig and Swanson, 1994). This belief seems to encompass the following views: that animal pain and suffering should be minimized whenever possible, that animals should only be used for sufficiently important reasons (that is, that the costs and benefits of animal use should be fairly weighed), that animals cared for by humans deserve to have some quality of life that goes beyond the minimization of pain and suffering, and that animals should have legal protection. These views have important implications for transgenic animal technology.

Animal Welfare Concerns and Transgenic Farm Animals

There are a number of potential animal welfare problems associated with the production of transgenic animals, both in terms of the animals that are used to produce transgenic offspring and the transgenic offspring themselves (Murphy, 1988; Fox, 1989; Van Reenen and Blokhuis, 1993; Moore and Mepham, 1995; Rollin, 1995; Tannenbaum, 1995, Mepham *et al.*, 1998). The nature of the transgene, the route by which the transgene is introduced, and the degree of control that is possible over the expression of the transgene are important determinants of the extent to which a transgenic animal's welfare might be compromised (Seamark, 1993).

Methods Used to Produce Transgenic Animals

Most transgenic farm animals are created using a technique called micro-injection, which involves the injection of purified DNA into the nucleus of a single-cell fertilized egg collected from a donor female. With microinjection, the site of insertion of the DNA into the genome and the number of copies of the DNA actually inserted cannot be predicted or controlled. Another method for producing transgenic animals is to use embryonic stem cells rather than fertilized eggs (Gordon, 1997). This method has the advantage of allowing the number of gene copies that are inserted and the site of insertion to be controlled, but at present this technology is developed only for mice. Although genes can be added to the genome using embryonic stem cell manipulation, this method is more commonly used to produce 'knockouts', that is animals that have had one of their own genes modified or deleted. Many different types of knockout mice are now used in biomedical research for the study of human diseases and disorders (Majzoub and Muglia, 1996). In livestock, potential uses for knockouts would include modifying the protein composition of milk, for example by removing proteins from bovine milk that are absent in human milk and that cause allergies (Eyestone, 1994).

Reproductive technologies

Reproductive manipulations including superovulation, semen collection, artificial insemination (AI), embryo collection and embryo transfer, are used in the production of transgenic farm animals. Whilst these manipulations are also used routinely by commercial breeders, they do raise a number of animal welfare concerns (Matthews, 1992; Seamark, 1993; Moore and Mepham, 1995), although their effects on animal welfare have thus far not been assessed systematically.

Handling and restraint, which are required for all of these manipulations, can be aversive to farm animals (Grandin, 1993). The administration of injections to induce ovulation or pseudopregnancy can cause transient distress, and the use of inserted controlled drug release devices or single injection regimes have been suggested in order to minimize this problem (Matthews, 1992). It has also been argued that electroejaculation can be stressful for some species, and that animals should be anaesthetized or tranquillized prior to this procedure being performed.

In cattle, AI and embryo collection and transfer can be accomplished using minimally invasive procedures, the latter under epidural anesthesia. However, these manipulations involve surgical or invasive procedures (laparotomy or laparoscopy) in sheep and pigs, and hence the potential for postoperative pain. Since livestock are valuable, they may be subjected to these procedures repeatedly during their lifetime. In particular, because of the problems involved in screening cattle embryos prior to implantation to ensure that they are actually carrying the transgene (Eyestone, 1994), cows found to be carrying non-transgenic offspring may be aborted and then reused as recipients. In poultry species, on the other hand, the hen is killed in order to obtain early-stage embryos.

Replacements for some of these manipulations are available (see Seamark, 1993; Moore and Mepham, 1995). A method has been devised for non-surgical embryo transfer in pigs. There has also been progress in developing *in vitro* oocyte maturation techniques and in obtaining ova from slaughterhouses, which would obviate the need for manipulation of live donor livestock females. However, lambs and calves produced using *in vitro* fertilization and embryo culture techniques tend to have higher birth weights and longer gestation lengths (Walker *et al.*, 1992; Van Reenen and Blokhuis, 1993), and difficult calvings (dystocia) can be a problem. In a recent study, Van Reenen and Blokhuis found that nearly 50% of cows carrying transgenic or non-transgenic offspring produced using *in vitro* techniques had calving difficulties. As a result, it is becoming more common to deliver offspring using Caesarean section (see Chapter 13). Again, the number of times that this procedure should be performed on any individual animal during her lifetime is a matter for scrutiny.

Efficiency of production and numbers of animals needed

Microinjection is an extremely inefficient method for producing transgenic offspring. Although the success of the method varies by species and gene construct, it has been estimated that fewer than 1% of injected livestock embryos result in transgenic offspring, and of those typically fewer than half actually express the transgene (Pursel *et al.*, 1989; Rexroad, 1994). About 80–90% of the mortality occurs very early during development, before the eggs are even mature enough to be transferred to the recipient female (Eyestone, 1994), but postnatal mortality also occurs (Pursel *et al.*, 1989). The effect of transgene insertion on mortality can be compounded by the use of certain reproductive technologies. Walker *et al.* (1992) found that more lambs produced by *in vitro* methods (20%) and/or carrying a transgene (22%) were dead at birth than were lambs produced *in vivo* (0–3.4%). Eyestone (this volume) reports that, in one study where cattle embryos were microinjected with a gene intended to cause expression of a human protein in the cow's milk, only 9% of 11,507 microinjected eggs developed to the stage where it was possible to transfer them to recipient cows. Only 19% of the cows produced calves, just nine of the 90 calves born were transgenic, and only one of those nine calves actually expressed the transgene. Even if an individual does express the transgene, it may not be transmitted to subsequent generations. Approximately 30% of transgenic animals are mosaics, which means that they carry the transgene in only some of their cells (Wilkie *et al.*, 1986). Mosaic animals may not pass the transgene to their offspring at all, or they may transmit it at a reduced rate.

In mice, the inefficiency associated with microinjection can be compensated for to a great extent by implanting recipient females with multiple embryos. In livestock, however, this can result in difficult births as well as masculinization of the female offspring if both a male and a female embryo are transferred to a cow. Most researchers therefore include an intermediate step in the production of transgenic cattle, which involves temporarily 'culturing' the embryos *in vitro* or in recipient cows or rabbit oviducts until the stage at which longer-term viability can be established (Eyestone, 1994). If cows are used, these developed embryos need to be recovered and then transferred to the recipient animals. Although this technique can therefore require the use of additional animals for the 'culturing' stage, it can reduce the number of recipient cows needed by up to 90%.

Mutations

Because microinjected DNA can insert itself in the middle of a functional gene, insertional mutations that alter or prevent the expression of that functional gene may inadvertently be created. Meisler (1992) estimates that

5–10% of established transgenic mice lines produced by microinjection have such mutations, and it is likely that similar rates would be found in microinjected livestock. Most (about 75%) of these are lethal prenatally, but those that are not are responsible for an array of defects in mice, including severe muscle weakness, missing kidneys, seizures, behavioural changes, sterility, disruptions of brain structure, neuronal degeneration, inner ear deformities and limb deformities. Individuals with such mutations can vary enormously with respect to the degree and type of impairment shown. Also, because many insertional mutations are recessive, their effects may not be obvious until subsequent generations. For example, even though mice engineered with a transgene for herpesvirus thymidine kinase were normal, 25% of their progeny had truncated hindlimbs, forelimbs lacking anterior structures and digits, brain defects, congenital facial malformations in the form of clefts, and a greatly shortened life expectancy (McNeish *et al.*, 1988).

Gene expression

Welfare problems can also arise because of poorly controlled expression of the introduced gene. The expression of genes is normally strictly regulated both with respect to the stage of the organism's development during which the gene is active and the cells or tissues in which the gene product is produced. In an attempt to control gene expression in transgenic animals, regulatory sequences including promoters, which allow genes to be switched on (and off) at specific times developmentally or expressed only in specific tissues, are attached to the gene to be inserted.

Nevertheless, transgenes may still be expressed inappropriately, since the efficiency of regulation can vary from one promoter to another, as well as among different species of animals even when the same promoter is used (Murray *et al.*, 1989; Rexroad, 1994). Furthermore, under certain circumstances the animal's own gene control sequences can influence the expression of the introduced gene. In addition, problems can arise because of the influence of the animal's own genes or gene products on the expression of the inserted gene (epigenetic effects), or because the inserted gene has multiple (and sometimes unexpected) effects on the animal (pleiotropy). It has been estimated that 80% of transgenic animals either do not express the gene or show variable or uncontrolled expression (Seamark, 1993), although the percentage of inappropriate expression is probably decreasing as genetic technologies are refined.

As Mepham *et al.* (1998) state, the potential welfare problems associated with any particular type of transgenic animal lie on a continuum from benign (e.g. the production of proteins that are biologically inactive for that particular species and that are secreted at a low level in specific tissues isolated from the bloodstream) to severe (biologically active proteins synthesized in large amounts in many tissues with abundant access to the bloodstream).

terms of its phenotype. Pigs transgenic for growth hormone, for example, vary enormously in the number of the DNA copies that they have per cell (from one to 490) and in the amount of growth hormone that they secrete (from 3 to 949 ng ml^{-1}). Only 50% of pigs transgenic for a gene (*c-ski*) intended to enhance muscle development experienced muscle weakness in their front legs, and in general the degree and site of muscle abnormality in these pigs varied considerably from one individual to another (Pursel *et al.*, 1992).

This makes the task of evaluating the welfare of transgenic animals particularly difficult, since adverse effects are almost impossible to predict in advance and each individual animal must be assessed for such effects. Van Reenen and Blokhuis (1993) describe the difficulties involved in such assessments. In most cases deleterious phenotypic changes in transgenic farm animals, particularly animals transgenic for growth hormone or other growth-promoting factors, have been easy to detect because they cause such gross pathologies. However, more subtle effects are also possible. Growth hormone, for example, has many systemic effects, including effects on the efficiency of nutrient absorption (Bird *et al.*, 1994). It has been reported that pigs injected with growth hormone have different nutrient requirements (Fox, 1989), and similar effects might be expected to occur in animals transgenic for growth hormone. Some types of knockout mice have also been found to have behavioural problems, like increased aggressiveness and impaired maternal and spatial behaviours (Nelson, 1997), that are not immediately apparent, but that could significantly affect housing and care requirements.

Sometimes adverse effects are seen only when animals are challenged in some way. The abnormal stress response of the Beltsville pigs when restrained is an obvious example. In addition, some problems may not become evident until later in development. Mice transgenic for an immune system regulatory factor, interleukin 4, develop osteoporosis, but not until about 2 months of age (Lewis *et al.*, 1993). This emphasizes the importance of monitoring the welfare of transgenic animals throughout their lifetime.

Welfare Benefits to Animals and Animal Integrity

Of course, genetic engineering also has the potential to improve the welfare of animals. Decreasing mortality and morbidity by increasing resistance to diseases or parasites is an obvious example of welfare benefits, and an area in which much transgenic research is focused (Müller and Brem, 1994). It has also been pointed out that transgenic animals may well receive a higher standard of care than non-transgenic animals because of their greater economic worth (Morton *et al.*, 1993).

Genetic engineering could also be used to deal with non-disease-related welfare problems. It might be possible, for example, to engineer hens or cows that produce only female offspring (Banner, 1995). This would eliminate the problems associated with surplus male chicks or calves, the former of which are killed at the hatchery and the latter of which are reared for veal, both practices that have been the target of a great deal of criticism. The need for the so-called standard agricultural practices like castration and dehorning could also be reduced or eliminated by genetic engineering. Pigs are castrated to prevent boar taint in the meat, but this trait is strongly genetically linked and thus amenable to genetic manipulation. Similarly, horns on cattle, which are removed because they cause injuries to humans and other cattle, are the result of a single gene that could be knocked out using genetic manipulation without affecting other desirable performance traits.

Applications of biotechnology to further adapt farm animals to intensive production systems might also be possible, but are likely to generate the most controversy. Should animals be engineered so that they are less responsive, tolerate crowding, or show fewer behavioural abnormalities, or to eliminate behaviours that are difficult to accommodate in intensive confinement while maintaining economic efficiencies? In an address to the Royal Society of Agriculture, Heap (1995) stated, 'Programmes which threaten an animal's characteristics and form by restricting its ability to reproduce normally, or which may in the future diminish its behaviour or cognition to improve productivity would raise serious intrinsic objections because of their assault on an animal's essential nature.'

But what is the animal's 'essential nature'? Some critics of genetic engineering argue that genetic engineering is inherently wrong because animals have a natural genetic integrity that should not be disturbed. Fox (1986), for example, states that biotechnology makes it possible for the first time for the boundaries that separate species to be breached, which will lead to the unique genetic make-up of particular species being drastically modified to serve human ends. Biologists (and others) find this concept particularly problematical, since it suggests that genetic traits constitutive of species are somehow 'fixed' through time, contrary to the theory of evolution. Species boundaries in nature are fluid. In addition, healthy, successful animal hybrids have also been produced using traditional breeding technologies (Pluhar, 1986; Russow, 1998; Singleton, 1998), and the genotypes and phenotypes of domesticated animals have already been changed significantly by selective breeding.

Pluhar (1985) argues that the difference between genetic engineering and traditional breeding is one of degree only. And it is clear that serious welfare problems can also arise because of traditional breeding techniques. Broiler chickens are a case in point. Breeding for increased growth has also led to serious physical disabilities, including skeletal and cardiovascular weakness. Ninety per cent of broilers have gait abnormalities (Kestin *et al.*, 1992), and these may be painful and make it difficult for the birds to walk to the feeders and waterers. In addition, the parents of these birds must be severely feed restricted to prevent obesity, and this feed restriction is associated with extreme hunger and a variety of behavioural problems (Mench, 1993). Banner (1995) notes that our special scrutiny of biotechnology in this regard creates anomalies – a transgenic modification that created the modern broiler chicken would have been closely regulated in Britain, while the broiler chicken produced using traditional breeding methods receives far less protection.

Since the genetic makeup of species is not fixed, then, is there any sense in which animal integrity raises genuine ethical issues? A perhaps more fruitful framing of the concerns raised by violations of animal integrity is that it is the *interests* of animals (Pluhar, 1986; Rollin, 1986) that need to be considered when making decisions not only about genetic engineering, but about any aspect of animal treatment. An animal's interests are a product both of its genetic background and its experience. Relevant interests as far as animal welfare is concerned might include avoiding pain, distress or suffering; experiencing pleasure; having social relationships; and pursuing (even short-term) goals (Mench, 1998). These are also the traits that are recognized as conferring moral relevance on animals, which may be the reason that there is particular controversy about interfering with them (Moore and Mepham, 1995).

Ideas about integrity are closely tied to ideas about the human responsibility to respect animals and treat them with dignity (Vorstenbosch, 1993). As Heap's statement above makes clear, changing fundamental animal interests to improve the 'fit' between animals and the human-created environment may well be construed as ethically problematical (see also Tannenbaum, 1995), regardless of whether that change occurs through transgenic technology or through more traditional means. Sapontzis (1991) reminds us that the traditional goal of ethics – a better world – refers to a world in which frustrations have been reduced by *fulfilling* interests rather than eliminating them.

Transgenic technology may consider and respect animal interests, in the same way that traditional selection techniques may fail to consider those interests (Pluhar, 1985). Improving disease resistance to decrease pain and suffering is an application of transgenic technology that considers animal interests. But it should be stressed that animal welfare is multi-faceted. Important elements of animal welfare include freedom from disease, pain or distress, physiological normality, and the opportunity to perform normal behaviours (Broom, 1993). While reducing disease is clearly beneficial, if this also permits animals to be more closely confined and thus decreases the opportunity for them to perform their normal behaviours then the net effect on welfare may be negative.

Is there anything special, then, about genetic engineering *vis à vis* traditional breeding? The primary difference between traditional breeding and genetic engineering is the speed at which change typically occurs (although naturally occurring mutations and recombination events can also cause rapid and dramatic change) and the single-gene nature of the change. As Russow (1998) notes, traditional methods of selection are more likely to be subject to the checks and balances imposed by natural selection. Many related and apparently unrelated traits are genetically correlated, and selective breeding thus involves selecting for a whole phenotype rather than a single gene product. Because most production and behavioural traits in livestock are polygenic and our understanding of livestock genomes is poor, few traits can reliably and predictably be modified or introduced by manipulating only one gene (Moore and Mepham, 1995). In addition, because changes occur more gradually with selective breeding, there is more time to figure out how to correct problems. Speed is also a strength of genetic engineering, however, since it can permit 'quick fixes' for problems arising from other practices. Given the special concern about transgenic technology, it would be ironic if genetic engineering turned out to be the fastest and best solution for some of the welfare problems that we have created using traditional breeding methods like leg problems in broilers.

What Future for Animal Transgenesis?

Transgenic animal technology, like other areas of applied science, has both risks and benefits. Whether public concerns about animal biotechnology prove groundless or not, it seems clear that this technology will be subjected to increasing scrutiny, and that the public will favour some form of regulation, either self-regulation or government regulation (Hoban and Kendall, 1993; Biotechnology and the European Public Concerted Action Group, 1997). Mechanisms are already in place in several European countries for the ethical evaluation of proposed genetic manipulations of animals. In The Netherlands, for example, no manipulation of animals is permitted until an independent committee has reviewed the ethics. The intrinsic value of the animal is taken as a starting point for such assessment, and the effects of the manipulation on animal health and welfare and animal autonomy are carefully considered (Brom and Schroten, 1993).

In making such assessments, costs and benefits need to be weighed carefully. Ignoring ethical costs for a moment, the financial cost of producing transgenic livestock is substantial. It has been estimated that it costs US$60,000 to produce one transgenic sheep and US$300,000 to produce one transgenic cow (Chapter 3). In the long term, are these costs proportional to the benefits that will be gained in terms of increased productivity? When expression of growth hormone is appropriately regulated in transgenic pigs, the increases shown in growth and feed efficiency are modest, and similar to the increases that can be attained by simply injecting pigs with porcine growth hormone (Pursel *et al.*, 1989; Chapter 11, this volume). Pursel *et al.* (1989) suggest that centuries of selection for growth and body composition may limit the ability of the pig to respond to additional growth hormone. Indeed, it is possible that we have already pushed some farm animals to the limits of productivity that are possible using selective breeding, and that further increases will only exacerbate the welfare problems that have arisen during selection.

Financial costs, of course, are only one of the many factors that need to be weighed in deciding whether it is appropriate to pursue particular applications of transgenic animal technology (or for that matter any other agricultural innovations). Short- and long-term impacts on animals, farmers, consumers and the environment all need to be carefully evaluated. Mepham (1995) outlines impacts that need to be considered within the framework of respect for well-being, autonomy and justice. For animals, these include welfare, behavioural freedom and respect for integrity. For farmers, these include adequate working conditions, freedom to adopt or avoid technologies, and fair treatment in trade and law. For consumers, these include the availability of safe and affordable food and consumer choice. Lastly, environmental considerations encompass protection and sustainability of populations and maintenance of biodiversity.

Many schemes are being proposed for a more formal approach to ethical evaluation and oversight of proposed biotechnologies (Appleby, 1988; Hoban and Kendall, 1993; Mepham, 1993; Donnelley *et al.*, 1994; Sandøe and Holtung, 1996; Mepham *et al.*, 1998). In the past, scientists have tended to isolate themselves from these debates. This posture needs to change. Scientists need to become full and fully informed participants in the debate about the ethical effects of the technologies that their work is instrumental in developing. Otherwise, consumer confidence in science and scientists may well be lost.

References

Appleby, M.C. (1988) Genetic engineering, welfare and accountability. *Journal of Applied Animal Welfare Science* 1, 255–275.

Banner, M.C. (1995) *Report of the Committee to Consider the Ethical Implications of Emerging Technologies in the Breeding of Farm Animals.* HMSO, London.

Berkowitz, D.B. (1993) The food safety of transgenic animals: implications from traditional breeding. *Journal of Animal Science* 71 (Suppl. 3), 43–46.

Berlanga, J., Infante, J., Capo V., de la Fuente, J. and Castro, F.O. (1993) Characterization of transgenic mice linkages. I. Overexpression of hGH causes the formation of liver intranuclear pseudoinclusion bodies and renal and hepatic injury. *Acta Biotechnology* 13, 361–371.

Biotechnology and the European Public Concerted Action Group (1997) Europe ambivalent on biotechnology. *Nature* 387, 845–847.

Bird, A.R., Croom, W.J., Black, B.L., Fan, Y.K. and Daniel, L.R. (1994) Somatotropin transgenic mice have reduced jejunal transport rates. *Journal of Nutrition* 124, 2189–2196.

Brom, F.W.A. and Schroten, E. (1993) Ethical questions around animal biotechnology. The Dutch approach. *Livestock Production Science* 36, 99–107.

Broom, D.M. (1993) Assessing the welfare of modified or treated animals. *Livestock Production Science* 36, 39–54.

Bruggemann, E.P. (1993) Environmental safety issues for genetically modified animals. *Journal of Animal Science* 71 (Suppl. 3), 47–50.

Chaudry, M.M. and Regenstein, J.M. (1994) Implications of biotechnology and genetic engineering for kosher and hallal foods. *Trends in Food Science and Technology* 5, 165–168.

Craig, J.V. and Swanson, J.C. (1994) Review: welfare perspectives on hens kept for egg production. *Poultry Science* 73, 921–938.

Donnelly, S. (1993) The ethical challenges of animal biotechnology. *Livestock Production Science* 36, 91–98.

Donnelly, S., McCarthy, C.R. and Singleton, R. Jr (1994) The brave new world of animal biotechnology. *Hastings Center Report* 24 (Suppl.), S3–S31.

Dorner, P. (1983) Technology and U.S. agriculture. In: Summers, G.F. (ed.) *Technology and Social Change in Rural Areas: Causes, Consequences, and Alternatives.* Westview Press, Boulder, Colorado, pp. 73–86.

Ebert, K.M. and Schindler, J.E.S. (1993) Transgenic farm animals: progress report. *Theriogenology* 39, 121–135.

Eyestone, W.H. (1994) Challenges and progress in the production of transgenic cattle. *Reproduction, Fertility and Development* 6, 647–652.

Fox, M.W. (1986) On the genetic manipulation of animals: a response to Evelyn Pluhar. *Between the Species* 2, 51–52.

Fox, M.W. (1989) Genetic engineering and animal welfare. *Applied Animal Behaviour Science* 22, 105–113.

Gordon, J.W. (1997) Transgenic technology and laboratory animal science. *ILAR Journal* 38, 32–40.

Grandin, T. (1993) *Livestock Handling and Transport.* CAB International, Wallingford, UK.

Heap, R.B. (1995) Agriculture and bioethics – harmony or discord? *Journal of the Royal Agriculture Society of England* 156, 69–78.

Hoban, T.J. and Kendall, P.A. (1993) *Consumer Attitudes About Food Biotechnology.* North Carolina Cooperative Extension Service Project Report, Raleigh, North Carolina.

Kapuscinski, A.R. and Hallerman, E.M. (1990) Transgenic fish and public policy: anticipating environmental impacts of transgenic fish. *Fisheries*, 2–11.

Kestin, S.C., Knowles, T.G., Tinch, A.E. and Gregory, N.G. (1992) Prevalence of leg weakness in broiler chickens and its relationship with genotype. *Veterinary Record* 131, 190–194.

Lewis, D.B., Liggitt, H.D., Effmann, E.L., Motley, S.T., Teitelbaum, S.L., Jepsen, K.L., Goldstein, S.A., Bonadio, J., Carpenter, J. and Perlmutter, R.M. (1993) Osteoporosis induced in mice by overproduction of interleukin 4. *Proceedings of the National Academy of Sciences USA* 90, 11618–11622.

Majzoub, J.A. and Muglia, L.J. (1996) Knockout mice *New England Journal of Medicine* 334, 904–907.

Matthews, L.R. (1992) Ethical, moral and welfare implications of embryo manipulation technology. *ACCART News* 5, 6–7.

McNeish, J.D., Scott, W.J. Jr and Potter, S.S. (1988) *Legless*, a novel mutation found in PHT-1 transgenic mice. *Science* 241, 837–839.

Meisler, M.H. (1992) Insertional mutation of 'classical' and novel genes in transgenic mice. *Trends in Genetics* 8, 341–344.

Mench, J.A. (1993) Problems associated with broiler breeder management. In: Savory, C.J. and Hughes, B.O. (eds) *Fourth European Symposium on Poultry Welfare*. Universities Federation for Animal Welfare, Potters Bar, UK, pp. 195–207.

Mench, J.A. (1998) Thirty years after Brambell: whither animal welfare science? *Journal of Applied Animal Welfare Science* 1, 91–102.

Mepham, T.B. (1993) Approaches to the ethical evaluation of animal biotechnologies. *Animal Production* 57, 1993.

Mepham, T.B. (1995) Ethical aspects of animal biotechnology. *Journal of the University of Wales Agricultural Society* 75, 3–22.

Mepham, T.B., Tucker, G.A. and Wiseman, J. (eds) (1995) *Issues in Agricultural Bioethics*. Nottingham University Press, Nottingham, UK.

Mepham, T.B., Combes, R.D., Balls, M., Barbiere, O., Blokhuis, H.J., Costa, P., Crilly, R.E., de Cock Bunting, T., Delpire, V.C., O'Hare, M.J., Houdebine, L.-M., van Kreijl, C.F., van der Meer, M., Reinhardt, C.A., Wolf, E. and van Zeller, A.-M. (1998) The use of transgenic animals in the European Union. The report and recommendations of ECVAM Workshop 28. *Alternatives to Laboratory Animals* 26, 21–43.

Moore, C.J. and Mepham, T.B. (1995) Transgenesis and animal welfare. *Alternatives to Laboratory Animals* 23, 380–397.

Morton, D., James, R. and Roberts, J. (1993) Issues arising from recent advances in biotechnology. Report of the British Veterinary Association Foundation Study Group. *Veterinary Record*, 17 July, 53–56.

Müller, M. and Brem, G. (1994) Transgenic strategies to increase disease resistance in livestock. *Reproduction, Fertility and Development* 6, 605–613.

Murphy, C. (1988) The 'new genetics' and the welfare of animals. *New Scientist*, 10 December, 20–21.

Murray, J.D., Nancarrow, J.T., Marshall, J.L., Hazelton, I.G. and Ward, K.A. (1989) The production of transgenic Merino sheep by microinjection of ovine metallothionein–ovine growth hormone fusion genes. *Reproduction, Fertility and Development* 1, 147–155.

Nelson, R.J. (1997) The use of genetic 'knockout' mice in behavioural endocrinology research. *Hormones and Behaviour* 31, 188–196.

Pluhar, E. (1985) On the genetic manipulation of animals. *Between the Species* 1, 13–18.

Pluhar, E. (1986) The moral justifiability of genetic manipulation. *Between the Species* 2, 136–137.

Powell, B.C., Walker, S.K., Bawden, C.S., Sivaprasad, V. and Rogers, G.E. (1994) Transgenic sheep and wool growth: possibilities and current status. *Reproduction, Fertility and Development* 6, 615–623.

Pursel, V.G., Rexroad, C.E., Bolt, D.J., Miller, K.F., Wall, R.J., Hammer, R.E., Pinkert, K.A., Palmiter, R.D. and Brinster, R.L. (1987) Progress on gene transfer in farm animals. *Veterinary Immunology and Immunopathology* 17, 303–312.

Pursel, V.G., Pinkert, K.A., Miller, K.F., Bolt, D.J., Campbell, R.G., Palmiter, R.D., Brinster, R.L. and Hammer, R.E. (1989) Genetic engineering of livestock. *Science* 244, 1281–1288.

Pursel, V.G., Sutrave, P., Will, R.J., Kelly, A.M. and Hughes, S.H. (1992) Transfer of *c-ski* gene into swine to enhance muscle development. *Theriogenology* 37, 278 (abstract).

Rexroad, C.E. (1994) Transgenic farm animals. *ILAR Journal* 36, 5–9.

Rifkin, J. (1985) *Declaration of a Heretic*. Routledge and Kegan Paul, Boston, Massachusetts.

Ritvo, H. (1987) *The Animal Estate*. Harvard University Press, Boston, Massachusetts.

Robinson, J.J. and McEvoy, T.G. (1993) Biotechnology – the possibilities. *Animal Production* 57, 335–352.

Rollin, B.E. (1986) On telos and genetic manipulation. *Between the Species* 2, 88–89.

Rollin, B.E. (1995) *The Frankenstein Syndrome*. Cambridge University Press, Cambridge.

Romagnolo, D. and DiAugustine, R.P. (1994) Transgenic approaches for modifying the mammary gland to produce therapeutic proteins. *Environmental Health Perspectives* 102, 846–851.

Russow, L.-M. (1998) *Genetic Engineering and Animal Welfare: Why it's Not Nice to Fool Mother Nature*. Scientists Center for Animal Welfare, Greenbelt, Maryland.

Sandøe, P. and Holtung, N. (1996) Ethical limits to domestication. *Journal of Agricultural and Environmental Ethics* 9, 114–122.

Sandøe, P., Forsman, B. and Hansen, A.K. (1996) Transgenic animals: the need for ethical dialogue. *Scandinavian Journal of Laboratory Animal Science* 23, 279–285.

Sapontzis, S.F. (1991) Why we should not manipulate the genome of domestic hogs. *Journal of Agricultural and Environmental Ethics* 4, 177–185.

Seamark, R.F. (1993) Recent advances in animal biotechnology: welfare and ethical implications. *Livestock Production Science* 36, 5–15.

Seidel, G.E. (1986) Characteristics of future agricultural animals. In: Evans, J.W. and Hollaender, A. (eds) *Genetic Engineering of Animals*. Plenum Press, New York.

Shamay, A., Pursel, V.G., Wilkinson, E., Wall, R.J. and Hennighausen, L. (1992) Expression of the whey acidic protein in transgenic pigs impairs mammary development. *Transgenic Research* 1, 124–132.

Singleton, R. (1998) *Transgenic Mammals: Science and Ethics*. Scientists Center for Animal Welfare, Greenbelt, Maryland.

Tannenbaum, J. (1995) *Veterinary Ethics*, 2nd edn. Mosby Yearbook, St Louis, Missouri.

Thompson, P.B. (1997a) *Food Biotechnology in Ethical Perspective*. Blackie, London.

Thompson, P.B. (1997b) Food biotechnology's challenge to cultural integrity and individual consent. *The Hastings Center Report* 27, 34–38.

United Press International (1988) Press release, 19 March, 1998.

Van Reenen, C.G. and Blokhuis, H.J. (1993) Investigating welfare of dairy calves involved in genetic modification: problems and perspectives. *Livestock Production Science* 36, 81–90.

Vorstenbosch, J. (1993) The concept of integrity. Its significance for the ethical discussion on biotechnology and animals. *Livestock Production Science* 36, 109–112.

Walker, S.K., Heard, T.M. and Seamark, R.F. (1992) *In vitro* culture of sheep embryos without co-culture: successes and perspectives. *Theriogenology* 37, 111–126.

Wilkie, T.M., Brinster, R.L. and Palmiter, R.D. (1986) Germline and somatic mosaicism in transgenic mice. *Development Biology* 118, 9–18.

Wilmut, I. (1995) Modification of farm animals by genetic engineering and immunomodulation. In: Mepham, T.B., Tucker, G.A. and Wiseman, J. (eds) *Issues in Agricultural Bioethics*. Nottingham University Press, Nottingham, UK, pp. 229–246.

18

The Future of Transgenic Farm Animals

G.E. Seidel, Jr

Animal Reproduction and Biotechnology Laboratory, Colorado State University, Fort Collins, Colorado, USA

Animal scientists have dreamed of applying transgenic technology to improve production characteristics of farm animals for nearly two decades. Except for the special case of producing pharmaceutical products in milk, efforts have been disappointing. In retrospect, this is not surprising in view of our limited knowledge of gene regulation and function, particularly interactions and pleiotropic effects. Furthermore, insertion of constructs at random sites has much in common with random germline mutations that occur naturally; most such mutations have negative, if any, consequences for the organism (Crow, 1997). There also are serious non-molecular limitations to generating transgenic farm animals, including high costs of animals and their care, lack of inbred lines, long generation intervals, small litter size in some species, expense of adequate replication and failure to develop usable embryonic stem cells.

Over the next few years, transgenic techniques are more likely to be successful for obtaining basic information about farm-animal biology than for improving production characteristics such as growth or lactation rates. Ultimately the resulting information will lead to improved production traits, but the application phase often will not require transgenic procedures. Transgenic technology with farm animals is rapidly becoming more reliable and flexible at the same time as our knowledge of genes is increasing, in great part due to information from other species. This combination will lead to remarkably insightful findings over the next decade, and probably will result in several applications to production-animal agriculture. Finally, we must continue to share information and procedures, even when developed in the private sector, or with private-sector funding (usually with considerable public-sector input). Few organizations can afford to waste valuable resources on protracted litigation over intellectual property or circumventing inventions derived from obvious procedures that are either inappropriately patented, or appropriately patented but unavailable for licensing.

© CAB INTERNATIONAL 1999. *Transgenic Animals in Agriculture*
(eds J.D. Murray, G.B. Anderson, A.M. Oberbauer and M.M. McGloughlin)

Introduction

Hundreds of millions of dollars were invested in transgenic farm-animal research between 1983 and 1997, much of it by the private sector, primarily for producing pharmaceutical products. Questions that arise are: What have investors and taxpayers received for this investment? What will they receive in the future? and What returns are likely from additional investment? Future transgenic farm-animal research will be dependent on perceived answers to such questions.

Research in transgenic farm animals has a unique character. Thousands of person-years of effort, much of it from the private sector, have been expended without yielding any product. Huge emphasis has been placed on refining techniques, rather than on using techniques to answer biological questions or to develop potential practical applications. To some extent, this may be explained by the newness of the endeavour and by the daunting number of unknown quantities. Like embryo transfer, transgenic research is pushed along by the soaring imagination of people who would apply the technology. One other oddity of transgenic farm-animal research is the large number of review papers (Wall, 1996). There is nearly one review paper for every three data papers, a situation that probably arises because programme chairpersons are excited about new approaches and techniques, and transgenic approaches certainly are exciting to think about.

Transgenic research with farm animals generally is undertaken with one of three broad goals (Fig. 18.1). One goal is to create animals for special non-agricultural purposes such as producing pharmaceuticals in milk (Wright *et al.*, 1991) or xenografts for replacing human tissues. When only a very few animals are needed to have a saleable pharmaceutical product, and the value of the product is high enough to justify costs both for creating and caring for the animals, this goal will be achievable and easily justified commercially.

A second goal is to produce either improved farm animals, for example, those that grow more efficiently, or improved animal products, for example, milk that yields more cheese. While such objectives start with creation of one or two animals, a resulting line of animals must survive and reproduce successfully with little or no further technological interference. This can be a formidable task and, because the line must be characterized, will not immediately result in a product that will cover the costs of developing the technology. The commercial advantage to the phenotype of individual transgenically altered animals usually will be less than 10% over herd mates (possibly excluding a transient spike of profit from novelty or exclusivity). Moreover, to be acceptable in production agriculture, the transgenic animals would have to be certified as healthy and not require special care. In the short term, few transgenic lines will meet the requirements for agricultural application. However, some decades from now, such applications may be relatively common.

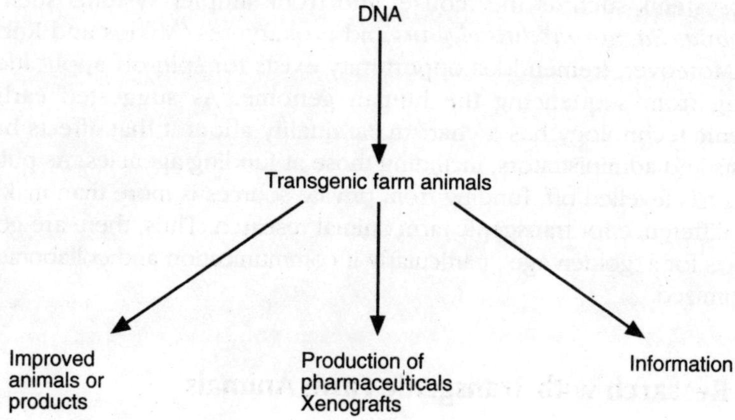

Fig. 18.1. Broad goals of transgenic research.

The third goal is to use transgenic procedures as a tool for basic research on the physiology of farm animals, for example, lactation, resistance to disease, or mechanisms of growth. Questions addressed in this paper include the following: Are transgenic farm animals a sensible approach to obtaining information, which then can be applied in a variety of ways, including making more appropriate transgenic farm animals? What sort of questions might be appropriate to ask using transgenic technology? Further: How useful might transgenic technology be for making improved farm animals? Are the problems insurmountable? In best-case situations, what traits might it be desirable to modify in farm animals using transgenic procedures?

A transgenic golden age?

The field of transgenic farm animals may be entering a 'golden age'. The power of transgenic technology has been proven in the mouse model; many hundreds of papers are published each year that effectively test hypotheses not easily tested by other methods (Wall, 1996). Also, we have accumulated an extensive foundation of technology for transgenic farm animals, both in making standard transgenic techniques more reliable (see other chapters in this volume), and in developing new techniques such as somatic-cell cloning (Campbell and Wilmut, 1997).

At this point in evolution, human society places great emphasis on *applications* of science and technology. Since use of transgenic technology in farm animals, almost by definition, is an application, this will be viewed favourably by both public and private funding sources. As a basis for developing applications, there currently is an explosion of information from model systems such as the mouse, and from simpler systems such as *Drosophila, Caenorhabditis elegans* and prokaryotes (Miklos and Rubin, 1996). Moreover, tremendous opportunity exists for spin-off applications resulting from sequencing the human genome. As suggested earlier, transgenic technology has a charismatic quality about it that affects both scientists and administrators, including those at funding agencies. As public funding has levelled off, funding from private sources is more than making up the difference for transgenic farm animal research. Thus, there are good prospects for a 'golden age', particularly if communication and collaboration are optimized.

Basic Research with Transgenic Farm Animals

Tools for basic research evolve

One great thing about scientific research is that new techniques and new approaches constantly come along to solve old problems or create entirely new possibilities. Some of these innovations have been anticipated by those working in a given area, for example sequencing DNA, cloning adult animals or developing mammalian artificial chromosomes. Of course, it is difficult to anticipate when individual techniques will become available (Gomory, 1983).

Some techniques or concepts that affect their application cannot be anticipated, or at least not by the majority of people working in the field. For example, few reproductive physiologists foresaw the polymerase chain reaction, gametic imprinting, embryonic stem cells or transgenic technology itself. I do not mean to imply that researchers were entirely ignorant of these possibilities. For example, many of us struggled with the fact that parthenogenetic embryos develop encouragingly for a while, but never to term, and some had vague notions of possible mechanisms approximating gametic imprinting (Markert, 1982). But it took considerable research to build a sufficient body of evidence to understand how imprinting would both constrain and explain experimental outcomes.

Specific characteristics of transgenic research with farm animals

Transgenic research with farm animals is limited by high costs, long generation intervals, lack of highly inbred lines, and lack of good culture systems and usable embryonic stem cells and other techniques commonly used in transgenic research with laboratory species. Wall and Seidel (1992) and Wall (1996), and numerous others, have reviewed this area thoroughly and suggested many improvements in transgenic methodology. Frequently there are simpler ways to get the desired information than by making a transgenic farm animal, for example, using radioisotopes to study metabolic pathways or laboratory-animal or cell-line models of farm animals. On the other hand, there are some distinct advantages to doing basic research with farm animals. For example, considerable amounts of tissue are available, and one can sample blood frequently, a process that can quickly be deleterious to small animals. There is also the ridiculous advantage that it is less expensive to care for sheep than rabbits in many research facilities.

A special advantage of ruminants and swine is that methods to clone embryos by nuclear transplantation have progressed more rapidly than in most other species, including rodents. This is due in part to their commercial value, but there also may be a biological basis, for example, later activation of the embryonic genome. Recent developments of cloning by somatic cell nuclear transplantation in sheep (Campbell and Wilmut, 1997) are particularly attractive for many transgenic experimental needs. The efficiency of this promising methodology, however, needs to be improved.

Another reason for using farm animals for some kinds of basic research is the lack of models for some tissues. For example, mice simply lack hooves, a rumen or a shell gland. Moreover, even when laboratory-species models are feasible, transgenic work needs to be confirmed in farm animals because differences in physiology of some systems dictate confirmation in the animal species of interest. As transgenic technology becomes more widely applied among species, we are likely to be in for some surprises.

There also are reasons for not using farm-animal models for some kinds of basic research. It would be inappropriate, in my opinion, to attempt to unravel how primordial follicles are selected to begin differentiation in the ovary with a transgenic farm-animal model because of the expense. Similarly, regulation of gametic imprinting, studies on sperm–oocyte receptor mechanisms, or regulation of certain aspects of meiotic maturation might best be undertaken first in the less expensive and better-characterized small animal models than in farm animals.

To summarize, transgenic farm animals are not a panacea, even for obtaining basic information. On the other hand, as outlined earlier, transgenic approaches are sensible for some basic research objectives with farm animals.

Examples of basic research with transgenic farm animals

An elegant example of the power of transgenic approaches is presented by Pursel *et al.* in Chapter 10. They arranged for IGF-1 to be produced in muscle, which enabled study in a paracrine rather than an endocrine mode; this avoided the confusion of exposing all tissues of the body to high concentrations of IGF-1. This is by no means an isolated example of how transgenic technology can be used for basic research.

In some ways, the most striking recent example of basic research with transgenic farm animals has been to transfect fibroblasts with the transgenic construct of interest, followed by fusing such a fibroblast with an oocyte to produce a transgenic animal with molecular properties already pre-characterized *in vitro* (Campbell and Wilmut, 1997; Schnieke *et al.*, 1997; also see various chapters in this volume). This, of course, circumvents one of the major disappointments in basic research in reproductive technology of farm animals, failure to develop usable embryonic stem cells (Chapter 4). Although much more needs to be done with this and parallel systems such as cultured primordial germ cells, these approaches likely can be used to make non-chimeric founder animals homozygous for the transgene, saving huge amounts of time in species with long generation intervals.

Applications of Transgenic Technology in Farm Animals

Considerations limiting application

As much as for basic research, the expense, long generation interval and long time lag from experimentation to observation of results in farm animals greatly limit commercial applications of transgenic technology (Wall, 1996). On top of these logistical constraints, there are special issues raised through commercialization such as: (i) safety to animals or consumers of the animals, especially regarding side effects; (ii) consumer acceptance, even if there are no known problems; (iii) possibilities of escape and contamination of wild-animal genomes; (iv) competition from simpler approaches and systems; and (v) problems with extreme phenotypes.

Transgenic procedures often produce extreme phenotypes, and nature tends to select against such extremes. There seems to be one best fit to the environment for most species, so with natural selection individuals within a species end up being similar in colour, size, shape and behavioural characteristics. Dramatic changes in physiology usually are incompatible with normal life cycles. For example, if cattle or horses were to superovulate naturally each reproductive cycle, the simultaneous development of multiple fetuses would lead to abortion. Very extreme phenotypes, such as sheep that produce an excess growth hormone, which leads to diabetes (Rexroad *et al.*, 1991), or excess growth in pigs, which may lead to arthritis (Pursel *et al.*, 1990), are not practicable or compensable by husbandry practices, and may be ethically inappropriate as well. Also, there is the serious limitation that animals with extreme phenotypes often fail to reproduce (Pursel *et al.*, 1990).

While there are costs (Box 18.1), there may be considerable benefits to extreme phenotypes, as long as they are not too extreme. Although animals with such phenotypes would not survive in nature, the farmers who use them in production agriculture may survive well economically.

One other constraint to development of new genetic variants, especially dramatic ones, is the need to study them in various genetic configurations.

Box 18.1. Examples of costs of extreme phenotypes in agriculture.

Dwarf wheat	Cannot compete with other plants
Large cows	Grazing insufficient for required nutrients
Multiplets in sheep	Lambs require extra feed
Docility	Protection from predators required
Twins in cattle	Major management changes required
Bovine somatotropin (BST) dairy cattle	For well managed herds *only*
Large beef carcasses	Do not fit standard transportation box

For example, all transgenic combinations of male, female, hemizygous, homozygous and controls should be studied for both beneficial and detrimental effects. This is as attractive an exercise as determining the factorial of a five-digit number without a calculator; unfortunately, homozygous transgenes frequently are lethal (Palmiter and Brinster, 1986). Also, there are cases in which a genetic change may be beneficial in one sex and detrimental in the other, or one sex may transmit the transgenic allele and not the other (Palmiter *et al.*, 1984). A further problem is introgressing a transgene from a single founder to the homozygous state while minimizing inbreeding (Smith *et al.*, 1987). Finally the transgene may be imprinted, causing further confusion.

Long-term prospects

The long-term prospects for application of transgenic technology are favourable. In a reversal of the law that everything that can go wrong will, things seem to be going unprecedentedly well. For example, as techniques for producing transgenic animals are becoming more efficient, the regulatory elements for genes are becoming easier to use (see Chapter 3). Indeed, relatively precise regulation of transgenes may be possible through feed additives or by injection (Pursel *et al.*, 1997). Breakthroughs in other species also will be integrated rapidly into the farm-animal technology.

One driving force for rapid development of this technology, apart from the fact that against the odds it *can* be done, is the pressing need for more efficient food production due to rapid population growth throughout the world. Moreover, it is entirely possible that there will be specific consumer demand for transgenic farm-animal products, despite the current recoiling. Witness the wide acceptance of vaccination and synthetic vitamin pills. In a way, these are far more radical products of biotechnology than animals transgenically modified to be resistant to disease. They have been accepted by consumers because of their history of success. Once the value and safety of, for example, nutraceuticals have been demonstrated, consumers will demand the products rather than hold them in suspicion. Moral scruples, as usual, will be sedated by convenience, economics and improved health.

Manipulation of genetic progress in non-production traits

One mistake that animal scientists are rightly accused of making is to emphasize production traits when low production is not a problem. More attention needs to be paid to non-production traits such as animal welfare, animal health, consumer acceptance and so on. A number of these non-production traits may be especially amenable to transgenic approaches. For example, some sheep are resistant to the spongiform encephalopathy, scrapie, because they have a particular allele (Westaway *et al.*, 1994). Homologous alleles might be transferred among breeds or even species, possibly making cattle resistant to bovine spongiform encephalopathy. Another example is the common practice of docking tails in lambs to minimize debris, faeces, etc. that accumulate, resulting in a haven for parasites. Genes that affect tail length have been identified in a number of species, and it is likely that appropriate alleles could be transferred to sheep transgenically to make tail docking unnecessary.

An intellectual exercise that illustrates important nuances in requirements for transgenic approaches to improving non-production traits is the elimination or diminution of odour from porcine faeces. Porcine faeces not only present a huge disposal problem, but the smell is also responsible for much ill will toward the swine industry. One could imagine a compound that could be fed to pigs that would end up in the faeces, neutralizing the odour. A more elegant solution would be to incorporate genes for such a compound in pigs that might, for example, be secreted into the bile or otherwise eliminated in the faeces, obviating the need to use feed additives. Such a compound should not decrease growth rates, carcass quality or other production traits, and conceivably might even enhance them. The manipulation also must have no detrimental effect on behaviour of the pig (e.g. reproductive behaviour or avoiding faeces); the flavour of the meat; the physiology of the pig; consumer safety (especially if eaten); or repulsion of faeces to pathogenic organisms or vectors such as insects. Clearly this is a rigorous set of requirements. Nevertheless, such chemicals may exist, and pigs with such transgenes conceivably could supplant pigs that did not have the odour-neutralizing chemical.

Manipulation of genetic progress in production traits

Use of transgenic procedures for production traits has been thoughtfully reviewed by Smith *et al.* (1987) and Hoeschele (1990) among others. Therefore, I will concentrate on some less conventional approaches here.

Transgenic technology might be used to decouple factors that inhibit animal production. For example, for continued egg production, hens eventually need to enter a starvation state that results in moulting of feathers; this somehow renews them reproductively, and they start egg production again. Currently, it is economically more practical to slaughter layers after peak egg production rather than invest in caring for them while they are unproductive. Reproduction in mammals similarly is affected negatively by a variety of situations. For example, lactation, weight loss, poor nutrition and season delay or inhibit reproduction in female mammals. We are starting to understand how these effects are regulated. In many cases, promoter regions of specific genes are inhibited or enhanced. Transgenic animals presumably could be made that did not have those inhibitory or enhancer elements in critical gene regulatory regions and, thus, would not be subject to such inhibitions. Of course, one would have to compensate with appropriate husbandry, for example, ensuring that an animal reaching puberty at a light weight eventually grows sufficiently to give birth normally. Such genetic changes could make animal agriculture much simpler and more profitable.

One special set of transgenic applications is to move genes from one species to another. Here are three of a wide spectrum of examples:

1. Tolerance to larkspur. Larkspur is a plant common to pastures in the foothills of the western USA. At certain times of the year its consumption is lethal to cattle, so the economic impact of this poisonous plant is significant. Sheep, on the other hand, are minimally affected by eating larkspur, possibly because they have an efficient enzyme for detoxifying the alkaloid in larkspur that is lethal to cattle. Simply replacing the bovine gene with the ovine gene for this enzyme might make cattle tolerant to larkspur.
2. Visual indicator of oestrus in farm animal species. When oestrogen concentrations in blood are high in baboons, they sport a bright red posterior, indicating that they are in oestrus. Likely this response is governed by only one or two genes. If pigs, for example, could be made to have bright posteriors when they were in oestrus, timing artificial insemination would be easier.
3. Omega-3 fatty acids in fish. Fish, although it has high concentrations of fat, appears to be a health-promoting food, and eating certain fish actually may decrease coronary disease in humans. It may be possible, for example, to modify pigs so that this healthful substance will be present in high concentrations in pork, although the pig may or may not smell and taste more like a fish than a pig.

Building on sex differences

An interesting aspect of animal husbandry is that what is good for the goose is not necessarily good for the gander. In most farm animals, the ideal female in a herd or flock has what are termed good maternal traits such as mothering ability, appropriate milk or egg production, small to moderate size, high fertility and early puberty. On the other hand, males and females to be slaughtered for meat should have a different set of traits. These are termed terminal-cross traits and include desirable carcass characteristics such as juiciness and tenderness, rapid growth rates, and moderate to large animal size. To some extent, maternal and terminal traits are antagonistic, and therefore inappropriate to have in the same animal or same breed. While these differences are, in fact, exploited by having different lines of animals for different purposes, gross inefficiencies still result. For example, in a maternal line, half of the offspring are of the 'wrong' sex. Because only a few males are needed for breeding, most males that have the maternal-cross characteristics have suboptimal carcasses and are, therefore, a by-product. To some extent this might eventually be circumvented with sexed semen, a product that is not available commercially at this time.

One could enhance these sex differences transgenically. To continue the above example, extra growth could be designed into the male. By adding androgen-response elements to the regulatory regions of growth genes and inseminating sexed semen, one could specify that the growth gene would be activated either by secretion of testosterone as the testis matures or, if the animal is castrated or female, by an implant with androgenic properties. Exploiting such naturally occurring methods of regulating genes would seem a high priority for transgenic research.

Sex differences also might be exploited by adding genes to the Y chromosome which, therefore, would only be expressed in males of the line (Wall and Seidel, 1992). The Y chromosome is nature's artificial chromosome. It is one of the smallest chromosomes in most species and has few genes; most of the chromosome has no known function. Also, most of the chromosome is hemizygous, which simplifies many aspects of application. The Y chromosome would seem to be a good place to add cassettes of genes. Another potentially exploitable fact is that XYY males are fertile and usually have normal XY sons. This allows for the possibility of transplanting an entire Y chromosome, custom modified in a cell line, to a one-cell embryo to serve as a vector.

Collegiality and Intellectual Exchange

Nature of the scientific enterprise

Science is a social enterprise, and most scientists love to discuss their findings. They genuinely do walk on the shoulders of giants. Scientists attend scientific meetings, participate in e-mail discussion groups, and even tolerate peer review of their work by granting agencies and journals. In many ways, the main currency of this social enterprise is communication, and the most rigorous form of communication in science is the refereed journal article, which involves increasing numbers of collaborators and co-authors compared with a decade or two ago. Scientists have special rules regarding ownership of ideas and the ethics of using the ideas of others. For example, while plagiarism is condemned, it is considered honourable and even flattering if someone uses another's findings, as long as proper attribution is made.

Science has a broadly international character since nationality is irrelevant to ideas. International collaboration takes a number of forms for a number of reasons, including circumventing or exploiting local constraints or sources of funding, costs of doing research, resources available, animal diseases, laws and even culture. International research often is encouraged by funding agencies and others, sometimes because there is a prestige element. I, and many others, believe that goodwill is *the* most important international commodity.

Decreasing scientific collegiality

In recent years, I have perceived a decreased collegiality amongst scientists, including those doing transgenic research. There has always been, in any human endeavour, a balance between competition and collaboration. This has particular significance in science because informal collaboration, including discussing new approaches with potential competitors, is part of the creative process, and also is a validating process. The balance between competition and informal collaboration may have changed to give more weight to competition, in part because sources of funding are less public and more private. Other factors influencing the degree of trust and communication among colleagues is the larger scientific community (more competitors), introduction into the process of the influence of uninformed public opinion through more rapid and efficient publicity given to new breakthroughs, and an increased potential for private gain.

Decreased collegiality sometimes results in gross inefficiencies. For example, scientists on the cutting edge frequently communicate their findings with each other well before they are published in formal journal articles. Those who rely on such articles generally are up to a year behind those who regularly communicate informally.

Inefficiency is perhaps the least of the harm as generation of new ideas is retarded by lessening of informal communication; for example, lack of informal constructive criticism increases inadvertent bad science. Another example of gross inefficiency is removing from general use the best, most efficient techniques by patenting them without then providing reasonable licensing terms, with the result that costs are driven up or fear of litigation becomes the basis of experimental design. Sequelae include slaughtering hundreds of animals for projects for which patented *in vitro* approaches are more sensible. To the extent that research funding is used for litigation rather than for the advancement of science, resources, especially time and intellectual energy are wasted.

Possible solutions

There are no simple solutions to these problems. Nor are the problems so different from those that have occurred with other human endeavours throughout history.

Communication could be encouraged by designing participatory, rather than passive scientific meetings. Another idea is the research consortium. Such consortia have been set up for companies working with high-speed computing and, in the USA, amongst the large automobile manufacturers to develop more efficient, safer automobiles. I suspect that private companies doing transgenic research would be much better off collectively if some types of research were done under the umbrellas of such consortia where findings were exchanged, thus making the whole field more efficient. Timing to commercialization, particularly of agricultural transgenic products, might be reduced by years, making certain transgenic endeavours profitable, instead of commercial failures.

Perhaps there should also be changes to the patent system. Patents currently are clumsy, time-consuming, and expensive to formulate and use. Inventors in the USA may wait 1 year from publication of their results before filing for a patent without losing potential patent rights. Such time delays are not permitted in all countries. Delays ideally might even be longer than 1 year in some circumstances. Perhaps it would be workable to have a two-phase time delay. For example, the right to patent for basic research use might expire after 1 year and for commercial applications after 2 years after publication. The principle would be to get information published expeditiously without having to give up the right to patent. Of course, most commercial entities would want to apply for patents sooner rather than later. It seems to me that misuse of 'submarine' patents and similar approaches aimed at undermining competitors (Petroski, 1998) are unethical when the health and nutrition of people and animals are at stake.

One other clearly unfair happenstance is that we give undue credit to timing of discoveries. For example, if one group files a patent application 1 day before another group, or publishes a paper 1 week before another group, priority is given to the first group in establishing ownership rights both legally and intellectually. Obviously, if ideas are to be used as commodities to gain wealth, there must be rules to define ownership of intellectual property as there must for real property, but the group that is 1 day ahead is not necessarily more deserving, and rushing to publication results in sloppy science. A rational example of how dating is dealt with on a routine basis is that when citing literature, we use the year, not the precise date that a publication appeared. Providing only this degree of precision is a way of saying that two groups publishing a particular concept within a year have substantially equivalent intellectual priority. One might broaden this in certain legal and intellectual situations by saying that any work reported within a 12-month period was, for practical purposes, simultaneous, and that a certain sharing of rights and credit would be appropriate.

Collegiality and sharing information contributes to making scientific work rewarding and, to the extent that these do not occur, less pleasant and less efficient. With all that science has to offer humanity, and with all of the problems that plague us, it seems to me that we have responsibilities to make scientific endeavours as efficient as possible. Therefore, we are obligated to invest a certain amount of energy in making it desirable to share information and collaborate as appropriate. Organizations that encourage cross collaboration and intra- and inter-organizational collegiality will thrive.

References

Campbell, K.H.S. and Wilmut, I. (1997) Totipotency and multipotentiality of cultured cells: applications and progress. *Theriogenology* 47, 63–72.

Crow, J.F. (1997) The high spontaneous mutation rate: is it a health risk? *Proceedings of the National Academy of Sciences USA* 94, 8380–8386.

Gomory, R.E. (1983) Technology development. *Science* 220, 576–580.

Hoeschle, I. (1990) Potential gain from insertion of major genes in dairy cattle. *Journal of Dairy Science* 73, 2601–2618.

Markert, C.L. (1982) Parthenogenesis, homozygosity and cloning in mammals. *Journal of Heredity* 73, 390–397.

Miklos, G.L.G. and Rubin, G.M. (1996) The role of the genome project in determining gene function: insights from model organisms. *Cell* 86, 521–529.

Palmiter, R.D. and Brinster, R.L. (1986) Germ-line transformation of mice. *Annual Review of Genetics* 20, 465–499.

Palmiter, R.D., Wilkie, T.M., Chen, H.Y. and Brinster, R.L. (1984) Transmission distortion and mosaicism in an unusual transgenic mouse pedigree. *Cell* 36, 869–877.

Petroski, H. (1998) An independent inventor. *American Scientist* 86, 222–225.

Pursel, V.G., Hammer, R.E., Boldt, D.J., Palmiter, R.D. and Brinster, R.L. (1990) Integration, expression, and germ-line transmission of growth-related genes in pigs. *Journal of Reproduction and Fertility* 41 (Suppl.), 77–87.

Pursel, V.G., Wall, R.J., Solomon, M.B., Bolt, D.J., Murray, J.D. and Ward, K.A. (1997) Transfer of an ovine metallothionein–ovine growth hormone fusion gene into swine. *Journal of Animal Science* 75, 2208–2214.

Rexroad, C.E., Jr, Mayo, K., Boldt, D.J., Elsasser, T.H., Miller, K.F., Behringer, R.R., Palmiter, R.D. and Brinster, R.L. (1991) Transferrin- and albumin-directed expression of growth-related peptides in transgenic sheep. *Journal of Animal Science* 69, 2995–3004.

Schnieke, A.E., Kind, A.J., Ritchie, W.A., Mycock, K., Scott, A.R., Ritchie, M., Wilmut, I., Coleman, A. and Campbell, K.H. (1997) Human factor IX transgenic sheep produced by transfer of nuclei from transfected fetal fibroblasts. *Science* 278, 2130–2133.

Smith, C., Meuwissen, T.H.E. and Gibson, J.P. (1987) On the use of transgenes in livestock. *Animal Breeding Abstracts* 55, 1–10.

Wall, R.J. (1996) Transgenic livestock: progress and prospects for the future. *Theriogenology* 45, 57–68.

Wall, R.J. and Seidel, G.E., Jr (1992) Transgenic farm animals – a critical analysis. *Theriogenology* 38, 337–357.

Westaway, D., Zuliani, V., Mirenda Cooper, C., Da Costa, M., Neuman, S., Jenny, A.L., Detwiler, L. and Prusiner, S.B. (1994) Homozygosity for prion protein alleles encoding glutamine-171 renders sheep susceptible to natural scrapie. *Genes and Development* 8, 959–969.

Wright, G., Carver, A., Cottom, D., Reeves, D., Scott, A., Simons, P., Wilmut, I., Garner, I. and Colman, A. (1991) High level expression of active human alpha-1-antitrypsin in the milk of transgenic sheep. *Bio/Technology* 9, 830–834.

Index

Abalone, Japanese 88
Acetyl-coenzyme A (acetyl-CoA) 160, 162, 172–174, 199
Acid glycylglycine acidification technique 135
Acid guanidinium thiocyanate–phenol–chloroform method 164
Allantocentesis 187
Altrenogest 133, 134
ALV *see* Avian leukosis virus
Amniocentesis 187
Ampicillin 29, 30
Animal welfare 13, 251–268
Antibodies, human 74–75
Atropine sulphate 134
Australia 159
Avian embryo, early development of 98–99
Avian leukosis virus (ALV) 118
Avian species, wild 106

BAC *see* Bacterial artificial chromosome
Bacillus thuringiensis 50
Backcrossing 80, 91, 233, 246
Bacterial artificial chromosome (BAC) 50
Bass, largemouth (*Micropterus salmoides*) 221
Beltsville Area Institutional Animal Care and Use Committee 133
Beltsville embryo culture medium 134
Beltsville Poultry Semen Extender 89
Benefits and costs 252, 263–264, 274–275
Binding region, *Hairy-wing* 48
Biochemistry, of domestic animals 157–176
Biomedicine 76
Bioreactors 5, 12–13
Biotechnology 73–74, 76, 193, 211, 251, 263
Blastocyst 71, 72, 84, 90, 181, 185–187
    injection 59, 60
    inner cell mass (ICM) 58, 59, 61, 62, 68, 81, 82, 84
    trophectoderm 58, 81–82
Blastomere 8, 39, 40
Blood vascular system, in poultry 119–120
Bluetongue 188
Body weight, selection for 231–250
Border elements 31, 32
Bovine spongiform encephalopathy 75, 276

Breeding of fish, traditional 209–224
Britain 252
Busulphan 92, 120
Butorphanol tartrate 134

Caenorhabditis elegans 25, 272
Carcass composition, in pigs 131, 132, 135–136, 139–141, 145, 152, 153
Carcinoma cells, embryonal 57–58
Carp, common (*Cyprinus carpio*) 20, 28, 209–211, 219–221
  Czechoslovakian strain 213
  Vietnamese strain 213
Casein protein 193, 194, 197, 198
Castration, of pigs 261
CAT *see* Chloramphenicol acetyltransferase
Catfish, blue (*Ictalurus furcatus*) 209, 214, 217
Catfish, channel (*Ictalurus punctatus*) 20, 209–212, 214–217, 220, 221–222, 223
  Kansas strain 212
Cattle 5–7, 12, 13, 37, 40, 59, 63, 64, 69, 72, 74, 79–86, 88–90, 132, 177–191, 193–195, 198, 256, 257, 261, 263, 274, 276
  calving and perinatal care 177, 180–185
  embryo production 183, 184
  freemartins 74
  Holstein 179, 186
  see also Dehorning; Embryo; Mammary gland; Oestrus; Pluripotential embryonic cells; Transgene; Transgene expression; Twinning; Zygote production
Cell cycle 70, 71, 81, 83
  in nuclear transfer 69–72
Ceratitis capita 25
Cheese 194, 270
Chicken *see* Fowl, domestic
Chimeras/chimerism 58, 60, 63, 64, 68, 69, 80–81, 84, 97, 98, 103, 106, 117, 118, 120, 122, 123
  embryonic stem cell 79–86
  germline 103–105, 119–125
Chloramphenicol acetyltransferase (CAT) gene 200
Cloning 10, 82, 86, 188, 199, 202, 251, 273
CMV *see* Promoters, human cytomegalovirus
Co-culture techniques 10
COC *see* Cumulus–oocyte complex
Codon bias 50
Collegiality 278–281
Commercialization 13–14, 79, 158, 213, 231, 240, 249, 253, 256, 274
Copenhagen conference, 1992 254
Costs and benefits 263–264, 274–275
Creutzfeldt–Jakob disease, in humans 75
Cross-breeding 46–47
  interspecific and interstrain 209
  intraspecific 209
  intraspecific, of fish 213–214
Cruelty to Animals Act, 1897 (Great Britain) 254–255
Cumulus–oocyte complex (COC) 179–180
Cysteine biosynthesis 157, 159, 160–171, 173
Cystic fibrosis, in humans 75

Death, adult, causes of 75–76
Dehorning, of cattle 261
Dietary concerns, in animals 252
Dietary laws, Jewish/Muslim 252
Diploidy *see* Polyploidy
Disease
  genetic basis of 4
  resistance to 4, 260
Disease models 75
DNA microinjection, pronuclear 2–6, 8–9, 37–39, 41, 51, 58, 67–68, 74, 79, 80, 83, 88, 117, 124, 131, 134, 146, 147, 148, 154, 158, 162, 177, 178, 180, 183, 184, 185, 186, 217, 232, 255, 257, 259–260
DNA replication 70
DNA transfection 117, 118
  ballistic 121
  of primordial germ cells 117–129
DNA uptake, by sperm 88–90

Domestic animals 5–8, 157–176
biochemistry of 158–159
as bioreactors 12–13
physiology of 158–159
productivity of 12
Domestication, of fish 209, 211–212
Doxycycline 45
Drosophila hydei 25, 31
Drosophila melanogaster 45, 48
Drosophila spp. 124, 272
Dual-energy X-ray absorptiometry (DXA) 131, 135, 136, 140
Dystocia 181, 186, 256

EC cell *see* Embryonal carcinoma cells
EG cell *see* Embryonic germ cells
Electroporation 9, 85–87, 89–90, 93, 105, 217
Embryo 38–42, 173
avian, early development of 98–99
culture, in cattle 180
development, in cattle 180–181
transfer, in cattle 181
Embryonal carcinoma (EC) cells 57, 58
Embryonic germ (EG) cells 62–64
Embryonic stem (ES) cells 9–10, 57–66, 68, 79–86, 255
discovery and use of 58–59
isolation of 60–62
murine 60–62
species specificity of 59–60
technology of 59–60
Environmental impact 14, 210, 221, 253, 264
Epigenetic effects 258
ES cells *see* Embryonic stem cells
Escherichia coli 28, 29, 41, 42, 45, 124, 157, 160, 161, 163, 164, 169
Ethical issues 13, 251–268
Expression vectors 31–32

Farm animals 1–18, 19, 67–86, 251–282
basic research 272–274
growth rates 254
limitations 7–8
production efficiency 257
production methods 255–260
uniqueness of 259–260
see also Oestrus
FCS *see* Fetal calf serum
Federal Drug Agency (USA) 50
Fertilization 71, 90
Fetal calf serum (FCS) 61, 177, 179, 180–181, 185, 187
Fibroblasts 85–86, 273–274
Fish 9, 14, 19–36, 47, 87, 88, 89, 124, 209–224
dramatic growth 222–224
environmental risk 221–222
fitness of 221–222
sterile 33
wild species 47, 210–214, 221–223
see also Breeding; Cross-breeding; Domestication; Gene transfer; Genetic engineering; Genotype–environment interactions; Hybridization; Monosex population; Omega-3 fatty acids; Polyploidy; Promoters; Salmonids; Selection; Sex reversal; Spawning; Transgenesis; Vaccine, genetic; *and* individual species
Fisheries, depletion of 19, 210–211
Foods, genetically engineered 252
Foot-and-mouth disease 188
Fowl, domestic (*Gallus gallus domesticus*) 88–90, 97, 98, 102, 104, 105, 108

Galactose analogue 43
Gene addition 200–201
Gene constructs 21–23, 147, 169, 174, 218–219, 232
Gene expression 4, 21, 76, 258–259
down-regulation 45, 200
in pigs 131–144
'Gene farming' *see* 'Genetic farming'
Gene mapping 76
Gene regulation 10–12
Gene removal 201
Gene Screen nylon membrane 135
Gene tagging *see* Insertional mutagenesis

Gene transfer 11
  biolistic methods 217
  in fish 211
  lipofection-mediated 9
  liposome-mediated 217
  in pigs 141–142
  sperm-mediated 9, 10, 87–95, 118, 124
  see also Transgenesis
Genes, bacterial 157–176
Genetic engineering 1–4, 21–23, 158, 193, 199, 251, 260–264
  in fish 217–218
  research funding 269–273
  'unnaturalness' of 253
'Genetic farming' 13, 178
Genetic tools, development of 19–36
Geneticin 83
Genotype–environment interactions, in fish 217
Germ cells, culture of 97–116
  see also Embryonic germ cells; Primordial germ cells
Germinal crescent 119, 120
Germline, in vertebrates 99
GFP *see* Green fluorescent protein
GH *see* Growth hormone
Glucose 172
Goats 6, 7, 13, 59, 178, 193–195, 198, 199, 259
Goldfish 216
Gonadotrophic hormones 132
Green fluorescent protein (GFP) 40, 51, 106, 187
GRF *see* Growth hormone-releasing factor
Growth hormone (GH) 3, 132, 145–148, 152, 210, 218–224, 260
  porcine 145–156
Growth hormone-releasing factor (GRF) 12, 132
Gypsy insulator 48

Hairy-wing binding region 48
Hamster 59
Hardy–Weinberg equilibrium 235
Herpes VP16 45
Heterochromatin formation 152
Histological markers, in poultry 103, 104
Homeostatic imbalance 158
Host factors 24–25
Humans 75, 193, 194
Hybridization, in catfish 215
Hybridization, interspecific, in fish 214
Hypoblast formation, in poultry 99, 100

Iceland supermarket chain, Britain 252
ICM *see* Blastocyst, inner cell mass
ICSI *see* Intracytoplasmic sperm injection
IGF-I *see* Insulin-like growth factor-I
IGF-II *see* Insulin-like growth factor-II
In vitro fertilization (IVF) 6, 178, 179, 186, 256, 257
In vitro maturation (IVM) 6, 177, 256, 257
Inbreeding 79–80
Incubation, in poultry 99, 101–103
India 33
Inhibitor, aurintricarboxylic acid 91
Insemination, intravaginal 90
Insertional mutagenesis 20
Insulator 47–49
  Gypsy 48
Insulin-like growth factor-I (IGF-I) 12, 131–145, 148, 149, 150, 153, 273
Insulin-like growth factor-II (IGF-II) 132
Integrase, murine retroviral 23
Integrase plus transgenes, co-delivery of 23–24
Integration site effects 47–50
Intellectual exchange 278–281
Intracytoplasmic sperm injection (ICSI) 90
Invertebrate species 87
IVF *see In vitro* fertilization
IVM *see In vitro* maturation

Kanamycin 29, 30
Ketamine HCl 134
KRAB transcription factor 42

α-Lactalbumin, human 177, 179, 182, 185, 193–200
Lactobacillus viscous 201
Lambs *see* Sheep
Large Animal Clinic, Virginia–Maryland Regional College of Veterinary Medicine 181
Larkspur 277
Leukaemia inhibitory factor (LIF) 107
LIF *see* Leukaemia inhibitory factor
Life, respect for 253
Lipid granules 81, 82
Lipofectin/lipofection 31, 83, 89, 90, 92, 105
Liposomes 88–89, 93
Loach 90, 91
Luciferase 40
Lysophosphatidylcholine 89
Lysozymes, in milk 201

Major histocompatibility complex (MHC) genes 75
Mammals, genetic modification of 67–69
Mammary gland 195, 196
  in cattle 178–179, 185, 193
  in ruminants 199
MAR *see* Matrix attachment region
Mast cell growth factor *see* Stem cell factor
Matrix attachment region (MAR) 40, 48, 49
Meat production 45, 79
Meiosis promoting factor (MPF) 69, 70
Metallothionein *see* Promoters
Methylation 38, 41–42, 152
MHC *see* Major histocompatibility complex
Mice 2–9, 12, 13, 30, 40, 48, 57, 59, 60, 63, 68, 69, 71, 72, 75, 80–82, 87, 92, 104, 107, 124, 132, 133, 139, 142, 146, 148, 152, 158, 162–167, 169, 171–174, 193, 195, 198, 199, 200–202, 223, 231–250, 255, 258
  body weight 233, 237–239, 240, 241, 243
  cumulative selection differentials (CSD) 237, 240
  defects in 258, 259
  fitness traits 234, 239, 242, 245–247
  hair growth 166, 168
  knockout 255, 260
  oMt1a-oGH 231–250
  reproductive traits 244
  see also Zygote production
Micelle 197, 200, 201
Microinjection *see* DNA microinjection
Milk 7, 12, 67, 74, 79, 177–191, 193–208, 254, 255, 269, 270, 278
  altering properties of 199
  gene products and functions 195–198
  infant formula applications 188
Mink 59
Minnesota Transgenic Fish Group (MTFG) 21
mMT *see* Promoters, mouse metallothionein
'Molecular farming' 4, 13
Monkey 59
Monosex populations, in fish 209, 210, 216
Mosaicism 22–23, 39, 91, 92, 139, 142, 146, 152, 154, 217–218, 257
MPF *see* Meiosis promoting factor
MTCEK1 160–163, 165–167, 169, 170, 171, 173
Multiple copy insertion 91–92
Muscle, skeletal 131–144
Mutation 73, 257–258

Native regulation 49–50
Nematode 124
Neomycin analogue G418 40
Neomycin resistance gene (*neo*) 40
NLS *see* Nuclear localization signal
North America 211
Northern blot analysis 131, 135, 137, 138, 161–164
Nuclear localization signal (NLS) 26, 30

Nuclear transfer 5, 37, 51, 61–64, 67–78, 85–86
applications 73–74
cell cycle in 69–72
limitations 72–73
method of 69
potential value of 68–69
response in different species 72
of somatic cells 10, 79–86, 179, 188
'Nutraceutical' applications 179

O-acetylserine sulphydrylase 157, 160, 162, 163, 165
Oestrus
in cattle 181
in farm animals 277
in pigs 133–134
Omega-3 fatty acids, in fish 277
Oncoviruses 24
Organ replacement, human *see* Xenotransplantation
Ovulation
in pigs 133–134
in poultry 98–99

Parasites, resistance to 260
PAS *see* Periodic acid–Schiff
PEC *see* Pluripotential embryonic cells
Peptide nucleic acid (PNA) 40
Periodic acid–Schiff (PAS) 100, 101, 119
PGC *see* Primordial germ cells
Pharmaceuticals 4, 6, 7, 14, 74, 193, 195, 259, 269, 270
Physiological studies 76
Pigs 5–8, 12, 13, 59, 61, 63, 69, 72, 74, 75, 87, 88, 131–156, 178, 195, 198, 256, 259, 260, 263–264, 273, 274, 276
Beltsville 259, 260
DK-43 hybrid 133
infertility 146, 148
lameness 146, 148
see also Carcass composition; Castration; Gene expression; Gene transfer; Oestrus; Ovulation; Transgene
Pike, northern 21, 22
Pleiotropy *see* Transgenesis
'Pluripotent', definition of 57
Pluripotential embryonic cells (PECs) 80
in transgenic cattle 81–85
PNA *see* Peptide nucleic acid
Polyploidy, in fish 209, 215–216
Poultry 97–129, 256, 260, 262
egg production 277, 278
see also Beltsville Poultry Semen Extender; Blood vascular system; Histological markers; Hypoblast formation; Incubation; Ovulation; Transgenesis
Primordial germ cells (PGCs) 4, 62, 63, 68, 84, 97, 98, 100–110, 117–129, 274
avian 97–116
blood PGCs 121–123
characteristics of 119
culture of 62–64, 106–110
embryonic blood PGCs 121–123
germinal crescent 105, 120–121
germline chimeras 103–105
gonadal PGCs 106, 123–125
origin of 99–103, 119
transfer of 119–120
Product efficacy testing 4
Prokaryotic vector sequences 22
Promoters 11, 42–43, 45, 49, 122, 258
in fish 218–219
human cytomegalovirus (CMV) 28, 31, 82–83
metallothionein-Ia (Mt-Ia) 160, 171
metallothionein IIA (MT-IIA) 12, 145–154
mouse metallothionein (mMT) 132
phosphoenolpyruvate carboxykinase 148
WAP 198
Protein
recombinant 177–191
whey 177, 193, 194, 197–198
Pseudomonas fragi 201

Public attitudes 14, 118, 253–255, 263–264
  Eurobarometer survey 253–254

QTL *see* Quantitative trait loci
Quail 119–121
  Japanese 97, 103–105, 108
Quantitative trait loci (QTL) 33
Quiescence 70, 71

Rabbit 59, 72, 195, 198
Rat 59, 60
Rat growth hormone (rGH) transgene 232, 241, 245
Receptor
  ecdysone 46
  retenoic acid X (RXR) 46
Recipient costs 38, 80, 187, 269
Recombinant DNA technology 217
Rennin 197
Reproductive technologies 256
Research, transgenic, goals of 270–271
Respect for life 253
Retroviral transduction 117, 118, 120, 123, 217
Reverse transcriptase (RT)-PCR analysis 109, 163, 164, 172
Royal Society of Agriculture 261
RT-PCR *see* Reverse transcriptase–PCR analysis
Ruminants 198–199, 202, 273
  see also Mammary gland
RXR *see* Receptor, retenoic acid X

Salmon, Atlantic 22, 213, 216
Salmon, chinook 91, 216
Salmon, coho (*Oncorhynchus kisutch*) 20, 212–213, 216, 220–221
  defects in 259
Salmonella typhimurium 160, 162, 169, 171
Salmonids 89, 209–211, 214, 216, 218, 219, 221–222
SAR *see* Scaffold attachment region
SB *see* Transposon, *Sleeping Beauty*
Scaffold attachment region (SAR) 40, 48
SCF *see* Stem cell factor
Scrapie 75, 276
Sea urchin 88, 124
Selection, of fish 209
Serine transacetylase 157, 162, 163, 165
Sex differences 277–278
Sex reversal, in fish 216
Sexual maturation 46
Sheep 5, 6, 10, 12, 13, 37, 59, 62, 69, 71, 72, 75, 80–81, 132, 157, 167–174, 178, 179, 188, 193–195, 198, 199, 256, 257, 263, 273, 276
  birth weight 73
  congenital abnormalities 72–73
  defects in 259
  gestation 73
  lung development 73
  Merino 159
Sheep metallothionein 1a sheep growth hormone transgene (oMt1a-oGH) 231–233, 236, 245, 248
Smoltification 219, 223
Somatic cells, nuclear transplantation of 79–86, 179
Somatostatin 132
Somatotropin 43, 275
Southern blot analysis 22, 32, 40, 83–85, 90, 91, 120, 124, 125, 135, 137, 162, 163, 169, 170, 217
Spawning, in rainbow trout 214
Species considerations 8–9
Sperm binding 9
Sperm, DNA uptake by 88–90
Sperm-mediated delivery methods 87–95
Staphylococcus aureus 201
Stearylamine 89
Steel factor *see* Stem cell factor
Stem cell factor (SCF)/c-kit 107–110
Stem cells *see* Embryonic stem cells
Strain considerations 8–9
Sulphur 157, 160, 165–167, 174
Sunfish, green (*Lepomis cyanellus*) 222
Susceptibility models 75–76
Swine *see* Pigs
'Switches' *see* Transgene 'switches'

TcE *see* Transposable element
Technology, transgenic 178, 274–278
Telozol® 134
Teratocarcinoma 9, 57
Tetracycline analogue 43, 45
Tetraploidy *see* Polyploidy
Thymidine kinase 146
Thyrotrophic hormone 132
Tilapia, Nile 216, 222, 223
Topoisomerase II 91
Transgene 23–24, 37–56, 124, 149
  detection of 37–56, 90–91
  integration, in cattle 182
  integration, in pigs 137
  'rescue' 49
  sheep metallothionein 1a sheep growth hormone (oMt1a-oGH) 231–233, 236, 245, 248
  'switches' **42–47**, 51
Transgene expression 37–56, 145, 146
  in cattle 182
  in mammary gland 7, 198
  in milk 185
  'position effects' 148
Transgenesis 9–12, 87–95
  avian 117–118
  biological effects 218–219
  in fish 209–224
  in poultry 97–116
  pleiotropic effects 219–221, 258, 269
  using primordial germ cells 105–106
'Transgenic', definition of 2
Transgenic Research (journal) 3
Transposable element (TcE) 24–25
  Minos 25
  structural and functional features 25–26
  vector system 26–32
Transposon 31–32
  salmonid 25–27
  Sleeping Beauty (SB) 26–32
Tricarboxylic acid (TCA) 172
Triploidy *see* Polyploidy
Trout, rainbow (*Oncorhynchus mykiss*) 21, 22, 212–216, 219, 220, 221, 223
Turkey (*Meleagris gallopavo*) 98, 102–104, 119, 123, 124
Twinning, in cattle 74

US Federal Drug Agency 50

Vaccine, genetic, for fish 33
Veal 261
Vertebrate species, transgenic 3, 26–32, 87
Viral diarrhoea, bovine 188
Virginia–Maryland Regional College of Veterinary Medicine 181
Virus, infectious haematopoietic necrosis 33

Walleye 21, 22
Western blot analysis 90, 182
Whey protein 177, 193, 194, 197–198
Wool production 7, 12, 159–160
  in Australia 159

Xenopus laevis 88
Xenopus spp. 2, 124
Xenotransplantation 4, 74–76, 254, 270
Xylazine 134

Yeast artificial chromosome (YAC) 50

Zebrafish 21, 27–32, 89
Zinc 145, 146, 148–150, 152–154, 162–165, 232, 233
Zygote production
  in cattle 179–180, 186
  in mice 245–247